Understanding and minimising fungicide resistance

It is widely recognised that agriculture is a significant contributor to global warming and climate change. Agriculture needs to reduce its environmental impact and adapt to current climate change whilst still feeding a growing population, i.e. become more 'climate-smart'. Burleigh Dodds Science Publishing is playing its part in achieving this by bringing together key research on making the production of the world's most important crops and livestock products more sustainable.

Based on extensive research, our publications specifically target the challenge of climate-smart agriculture. In this way we are using 'smart publishing' to help achieve climate-smart agriculture.

Burleigh Dodds Science Publishing is an independent and innovative publisher delivering high quality customer-focused agricultural science content in both print and online formats for the academic and research communities. Our aim is to build a foundation of knowledge on which researchers can build to meet the challenge of climate-smart agriculture.

For more information about Burleigh Dodds Science Publishing simply call us on +44 (0) 1223 839365, email info@bdspublishing.com or alternatively please visit our website at www.bdspublishing.com.

Related titles:

Achieving sustainable cultivation of maize Volume 2: Cultivation techniques, pest and disease control
Print (ISBN 978-1-78676-012-8); Online (ISBN 978-1-78676-014-2, 978-1-78676-015-9)

Achieving sustainable cultivation of wheat Volume 1: Breeding, quality traits, pests and diseases
Print (ISBN 978-1-78676-016-6); Online (ISBN 978-1-78676-018-0, 978-1-78676-019-7)

Integrated disease management of wheat and barley
Print (ISBN 978-1-78676-216-0); Online (ISBN 978-1-78676-219-1, 978-1-78676-218-4)

Integrated management of diseases and insect pests of tree fruit
Print (ISBN 978-1-78676-256-6); Online (ISBN 978-1-78676-258-0, 978-1-78676-259-7)

Achieving durable disease resistance in cereals
Print (ISBN 978-1-78676-601-4); Online (ISBN 978-1-78676-603-8, 978-1-78676-604-5)

Chapters are available individually from our online bookshop:
https://shop.bdspublishing.com

climate•SMART•publishing
IN AGRICULTURAL SCIENCE

BURLEIGH DODDS SERIES IN AGRICULTURAL SCIENCE
NUMBER 132

Understanding and minimising fungicide resistance

Edited by Dr Francisco J. Lopez-Ruiz, Curtin University, Australia

burleigh dodds
SCIENCE PUBLISHING

Published by Burleigh Dodds Science Publishing Limited
82 High Street, Sawston, Cambridge CB22 3HJ, UK
www.bdspublishing.com

Burleigh Dodds Science Publishing, 1518 Walnut Street, Suite 900, Philadelphia, PA 19102-3406, USA

First published 2023 by Burleigh Dodds Science Publishing Limited

Library of Congress Control Number: 2022949293

British Library Cataloguing in Publication Data
A catalogue record for this book is available from the British Library

ISBN 978-1-80146-198-6 (print)
ISBN 978-1-80146-201-3 (PDF)
ISBN 978-1-80146-200-6 (ePub)
ISSN 2059-6936 (print)
ISSN 2059-6944 (online)

DOI: 10.19103/AS.2022.0116

Typeset by Deanta Global Publishing Services, Dublin, Ireland

Contents

Series list

Title	Series number
Achieving sustainable cultivation of maize - Vol 1 From improved varieties to local applications Edited by: Dr Dave Watson, CGIAR Maize Research Program Manager, CIMMYT, Mexico	001
Achieving sustainable cultivation of maize - Vol 2 Cultivation techniques, pest and disease control Edited by: Dr Dave Watson, CGIAR Maize Research Program Manager, CIMMYT, Mexico	002
Achieving sustainable cultivation of rice - Vol 1 Breeding for higher yield and quality Edited by: Prof. Takuji Sasaki, Tokyo University of Agriculture, Japan	003
Achieving sustainable cultivation of rice - Vol 2 Cultivation, pest and disease management Edited by: Prof. Takuji Sasaki, Tokyo University of Agriculture, Japan	004
Achieving sustainable cultivation of wheat - Vol 1 Breeding, quality traits, pests and diseases Edited by: Prof. Peter Langridge, The University of Adelaide, Australia	005
Achieving sustainable cultivation of wheat - Vol 2 Cultivation techniques Edited by: Prof. Peter Langridge, The University of Adelaide, Australia	006
Achieving sustainable cultivation of tomatoes Edited by: Dr Autar Mattoo, USDA-ARS, USA and Prof. Avtar Handa, Purdue University, USA	007
Achieving sustainable production of milk - Vol 1 Milk composition, genetics and breeding Edited by: Dr Nico van Belzen, International Dairy Federation (IDF), Belgium	008
Achieving sustainable production of milk - Vol 2 Safety, quality and sustainability Edited by: Dr Nico van Belzen, International Dairy Federation (IDF), Belgium	009
Achieving sustainable production of milk - Vol 3 Dairy herd management and welfare Edited by: Prof. John Webster, University of Bristol, UK	010
Ensuring safety and quality in the production of beef - Vol 1 Safety Edited by: Prof. Gary Acuff, Texas A&M University, USA and Prof. James Dickson, Iowa State University, USA	011
Ensuring safety and quality in the production of beef - Vol 2 Quality Edited by: Prof. Michael Dikeman, Kansas State University, USA	012
Achieving sustainable production of poultry meat - Vol 1 Safety, quality and sustainability Edited by: Prof. Steven C. Ricke, University of Arkansas, USA	013
Achieving sustainable production of poultry meat - Vol 2 Breeding and nutrition Edited by: Prof. Todd Applegate, University of Georgia, USA	014

Introduction

The emergence of fungicide resistance is a major challenge facing agriculture. With increasing regulation and costs limiting the development of new fungicides, farmers remain reliant on a relatively small group of actives, many of which are decreasingly effective as major crop disease pathogens develop resistance to them. This volume provides an authoritative review on the wealth of research on understanding the development of fungicide resistance in agricultural crops and the establishment of preventative measures which can be implemented to limit its spread and the consequent impact of disease on yields.

The chapters are split into two parts: Part 1 chapters focus on understanding and managing resistance, discussing topics such as the factors driving resistance emergence, the molecular basis of the resistance, the tools developed for tracking resistant pathogens as well as the fine balance between disease control and resistance selection. Chapters also review fungicide resistance risk assessment, minimising fungicide resistance in crop cultivation, the evolutionary biology of fungicide resistance, as well as the role of extension in fungicide resistance management and key challenges in developing new fungicides. Chapters in Part 2 examine resistance in key groups of fungicides as case studies, specifically drawing attention to sterol biosynthesis inhibitor (SBI), quinone outside inhibitor (QoI), succinate dehydrogenase inhibitor (SDHI), Anilinopyrimidine (AP) and oxysterol binding protein inhibitor (OSBPI) fungicides.

Part 1 Understanding and managing resistance

Part 1 opens with a chapter that provides an overview of how pathogens develop resistance to fungicides. Chapter 1 begins by first focusing on how fungicide resistance can be detected and measured. It then moves on to analyse the different mechanisms of resistance, drawing specific attention to target site mutation, target site overexpression and multidrug resistance. This is followed by a section on the evolution of resistance, highlighting how this evolution has been divided into two stages: emergence and selection. The chapter concludes by emphasising the importance of improving fungicide effectiveness to in turn, decrease their usage.

The next chapter reviews molecular evolution and mechanisms of fungicide resistance in plant pathogenic fungi. Chapter 2 reviews mechanisms of resistance to the most widely-used, broad-spectrum groups of fungicide over the last 50 years, including methyl benzimidazole carbamate (MBC) fungicides, azoles, amines including morpholines and piperidines, succinate dehydrogenase inhibitors (SDHI) and quinone-outside inhibitors (QoI, often known as strobilurins). For each class of fungicide the chapter discusses the fungal target and explores how resistance operates at the gene and protein level. It also focuses on powdery mildews as an example because they are notorious for the speed at which they have evolved resistance to most of the major classes of fungicide.

The subject of Chapter 3 is tracking the development of fungicide resistance. It starts by first providing an overview of the key objectives and requirements in tracking fungicide resistance. The chapter then moves on to review phenotyping tracking of fungicide resistance, which is followed by a discussion on developments in genotyping techniques. A section on in-field detection and quantification techniques is also included. The chapter also highlights developments in sequencing techniques, such as the use of genome-wide association studies as well as nanopore and other long-read sequencing approaches.

Chapter 4 draws attention to crop disease control efficacy and selection for resistance. It first focuses on control versus selection, highlighting the theory behind fungicide application programmes and provides examples of predictions and a test using already published data. The chapter also reviews reducing selection while maintaining control, focusing specifically on fungicide mixtures. Building on this discussion as a foundation, the chapter moves on to focus on the possibility of using two at-risk fungicides and what needs to be considered if mixtures such as these were to be developed. The chapter also review integrated pest management measures and mutual protection, before concluding by emphasising the importance of developing long-term disease control methods.

The next chapter focuses on fungicide resistance risk assessment. Chapter 5 reviews the development and scientific rationale of the widely used risk matrix approach to fungicide resistance risk assessment. This type of approach relies on previous experience of resistance with a particular fungicide mode of action, pathogen and agronomic system to assess the level of risk. In contrast to this, the chapter also reviews a more recent trait-based approach that doesn't rely on prior knowledge. A section that considers future trends in fungicide resistance risk assessment is also provided.

The subject of Chapter 6 is good practice in minimising the development of fungicide resistance in crop pathogens. The chapter begins by first providing an overview of resistance management guidance and how this has evolved over time. It then moves on to consider the three key evolutionary phases of fungicide resistance: the emergence phase, selection phase and adjustment phase. The chapter also describes governing principles of resistance evolution as a foundation of guidance, then addresses future development possibilities for resistance management guidance. A section on the complementary roles of experiments and modelling to inform resistance management guidance is also included, followed by an overview of how resistance management guidance can be updated.

Chapter 7 examines the evolution of fungicide resistance and identifies the practical implications of this. The chapter begins by first identifying the evolutionary origins of fungicide resistance, focusing specifically on standing variation and de novo mutation and how this can impact resistance risk and management. The chapter moves on to review the adaptive potential and overall risk of fungicide resistance for pathogens, then describes trait complexity and how this can affect fungicide resistance in certain genotypes. A section on the fitness penalties within fungicide-resistant genotypes is also described, focusing specifically on the costs and trade-offs and the impact on risk levels and management. The predictability of resistance is also reviewed, before the chapter concludes with an overview on how fungicide resistance is still an ongoing challenge and emphasises the importance of improved evolutionary understanding and how this can lead to better decision making.

The next chapter draws attention to the role of extension in fungicide resistance management. Chapter 8 features three case studies that illustrate how the extension service has been involved in (i) identifying resistance-related problems in producer fields, (ii) formulating and implementing solutions, and (iii) continued education of producers to adjust to ever changing resistance landscapes. Extension agents and specialists have two goals associated with resistance management: (1) provision of non-biased, accurate and current information to producers relative to resistance management best practices, and (2) rapid identification of resistance when it occurs, to be quickly followed by new recommendations – the feedback loop. The strong working relationship between extension agents and producers educates both parties, continues to build trust, and generates the foundation for change.

The final chapter of Part 1 focuses on the key challenges in developing new fungicides for effective control of plant diseases. Chapter 9 first provides an overview of past and present fungicide development, focusing on how chemical disease control became a key concept within the nineteenth century and how these methods have developed up to this point. The chapter also highlights the current challenges facing fungicide development and registration, before it moves on to consider fungicide resistance as a driver for innovation. A section on how fungicide development can change in the future as the industry adapts to different fungicide requirements is also provided.

Part 2 Case studies: resistance in key groups of fungicides

Part 2 opens with a chapter that discusses understanding resistance to SBI fungicides. Chapter 10 starts by first highlighting the current market and trends for this type of fungicide, then goes on to provide a short history of de-methylation inhibitor (DMI) fungicides, one of the four subclasses SBI fungicides can be divided into. This is then followed by an overview of resistance risk and general resistance characteristics of SBI fungicides. A section on mode of action and mechanisms of resistance is also included, followed by a discussion that provides specific examples of resistance to DMIs. The chapter also includes a section that highlights recommended uses for resistance management.

The next chapter draws attention to QoI fungicide resistance, focusing specifically on selection patterns and the current situation. Chapter 11 begins by describing the history of resistance and resistance mechanisms, then goes on to provide three case studies to describe different QoI resistance evolutionary patterns and dynamics. Each case study draws attention to a specific crop and fungicide, the first draws attention to fungicide use in soybean in Brazil, the second focuses on grapes in France and the final discusses sugar beet in Europe. A section on current global sensitivity to QoI groups 11, 11A and 45 is also provided.

The subject of Chapter 12 is understanding resistance to SDHI fungicides. The chapter begins by first describing the mode of action of SDHI fungicides, focusing specifically on succinate dehydrogenase (succinate-ubiquinone oxidoreductase, Complex II) target and the succinate dehydrogenase inhibitor (carboxamide, Group 7) fungicides. It then moves on to describe resistance mechanisms such as target site and non-target site, which is followed by a section on fitness cost and resistance risk assessment. The chapter concludes with a case study on the control of net-form net blotch in barley in South Australia.

Chapter 13 focuses on understanding resistance to AP fungicides. The chapter first describes the mode of action of AP fungicides, drawing attention to enzyme secretion, methionine biosynthesis and mitochondrial function. It goes on to discuss mechanisms of resistance in the target sites and non-target sites of AP fungicides, which is followed by an overview of the fitness cost of AP fungicide resistance. The chapter also highlights the importance of monitoring AP fungicide resistance and general fungicide resistance management practices.

The final chapter of the book discusses understanding resistance to OSBPI fungicides. Chapter 14 first examines the current status of fungicide resistance in oomycetes. It then moves on to review the mode of action and biological activity of OSBPI fungicides. A case study on OSBPI resistance risk assessment is also included, followed by an overview of sensitivity monitoring. The chapter also highlights the characterisation of resistant mutants, discusses molecular monitoring and also provides use recommendations for these fungicides.

Acknowledgements

Cover image acknowledgement:

- Barley plants showing symptoms of spot form net blotch, net form net blotch and smut. 2021. © Francisco J. Lopez-Ruiz.

We wish to acknowledge the following for their help in reviewing a particular chapter:

- Chapter 3: Dr Nichola Hawkins, NIAB, UK

Part 1

Understanding and managing resistance

Chapter 1

How pathogens develop resistance to fungicides: an overview

Richard Oliver, University of Nottingham, UK

1 Introduction

Fungicides form one of the four major pillars used by growers/farmers to protect crops from diseases, the others being genetics, agronomy and biosecurity. It is difficult to reliably estimate the contributions of each of these methods to crop protection, but there is no doubting the sustained demand from farmers for fungicides that reliably and economically deliver a healthy crop. The global annual market for fungicides was close to US$20 billion in 2020 and is growing at a steady pace.

Scientific fungicide use goes back to the eighteenth century, pre-dating the controlled use of pure chemicals in clinical therapeutic settings by more than 100 years. Resistance to fungicides was not a significant issue until the 1960s when systemic fungicides with single modes of action (MOA) were introduced. Resistance to single-site MOA is now a dominant factor in the fungicide industry. It impacts all aspects of fungicide development and is a major issue in determining the useful life of the products. I refer here not just to fungi *sensu strictu* but to all eukaryotic pathogenic microorganisms targeted by crop protection products (CPP).

Most fungicides work by inhibiting a key biochemical process that the fungus needs to colonise the plant, induce symptoms and complete its life cycle. As with other cases of 'Natural Selection', evolution to resistance occurs

http://dx.doi.org/10.19103/AS.2022.0116.01

when fungicides exert an inhibitory effect on the pathogen population growth. If the pathogen population varies in its sensitivity to the fungicide, the more sensitive isolates will be relatively inhibited, whilst the more resistant ones will grow and reproduce at a higher rate. As a result of this selection pressure, evolution towards resistance will inevitably occur. The speed at which resistance develops, the degree of resistance and the pathogenic fitness of the resistant population in the absence of the fungicide (the fitness penalty) all depend on features of the pathogen, the disease and the fungicide. Resistance may occur and render the fungicide useless in a matter of weeks. Alternatively, the useful economic life of fungicides can realistically be extended indefinitely if its properties are suitable and its use is managed wisely.

2 Detecting and measuring resistance

The fungicide resistance era began with the discovery of resistance to the methyl-benzimidazole carbamate (MBC) fungicides in sugar beet fields in Greece in 1972. Benomyl was introduced to control *Cercospora beticola* and, at first, was a substantial improvement on the organo-tin products then in use. In just the third season of use, catastrophic disease levels were noted, which were substantially worse than in organo-tin-protected controls. At first, it was believed the benomyl was being washed off the leaves by the irrigation equipment. The response was to increase the dose but to no avail. It was noted that fields that had received benomyl in each of the previous seasons were the most and earliest affected. It was only when isolates were tested *in vitro* that it was fully accepted that this was a clear case of evolved resistance to a specific fungicide (Georgopoulos and Dovas, 1973). It was later shown that resistant isolates carried an E198A mutation (i.e. a glutamate (E) at position 198 in the protein was changed to an alanine (A)) in the *b-tubulin* gene and that this genotype fully accounted for the phenotype (Mair et al., 2016). The mutant strains were as pathogenic as the wild type on untreated plants.

This first case of fungicide resistance contains many elements which have become familiar over the ensuing decades; blaming product application failure, repeated use, higher doses, solo use, delays in resistance detection and inappropriate responses.

There are many occasions when fungicides produce disappointing results, and only a small fraction of these cases prove to be due to evolved fungicide resistance. In most cases, the product was inappropriately applied or unsuited to the task. Furthermore, if the first evidence of resistance was a field failure, options for mitigating the outbreak would be limited. It is much better to detect resistance prior to the evolution of populations that create epidemic outbreaks on crops.

The classic way to detect resistance is to make large-scale, random isolations of the pathogen/s of interest and measure the EC50 phenotype – the concentration of fungicide that reduces the growth rate by 50% in an *in vitro* assay. The process of population-level phenotypic assessment is slow and expensive but irreplaceable, at least in the initial phases of releasing a new active. The first requirement is to establish a baseline of naïve and presumably fully sensitive isolates of the pathogen. Such studies typically uncover a surprisingly large range of EC50s covering a factor of 10 or more (Fig. 1). The presence and molecular basis of the hypersensitive isolates is a poorly studied aspect of fungicide resistance research. The EC50 values of the naïve population can be plotted on a cumulative frequency graph and typically forms a sigmoid shape. The EC50 can be defined as the inflexion point of the population cumulative frequency graph. Similar plots of isolate EC50s in successive growing seasons can be readily examined for shifts in the mid-point EC50s.

The methods used to measure EC50s vary between pathogens and labs. Obligate pathogens are particularly difficult, and the results obtained have a low degree of precision. The easiest pathogens are ones that grow in cultures as yeasts such as *Zymoseptoria tritici*. Then, turbidimetric methods can be used reliably to measure growth rates in liquid cultures, typically in a microtiter well plate (Oliver and Beckerman, 2022). As growth proceeds via a sigmoid progression, first slow, then fast and later slow again, estimates of the *maximum growth rate* are preferred but are more time-consuming than growth accrued at a *particular time point*. Filamentous organisms are not well suited to liquid

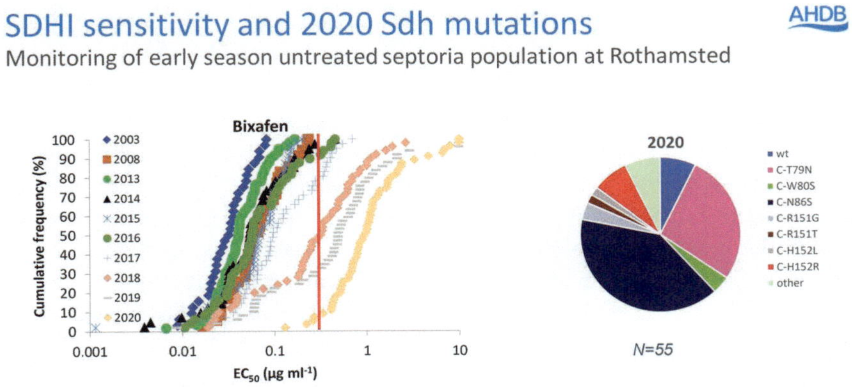

Figure 1 An example of EC50 analyses of a fungal population over time and during the evolution of fungicide resistance. The blue diamonds show the range of EC50 values of a naïve wild-type population prior to the use of a fungicide. The cumulative frequency plot shows a typical sigmoid shape, with the population having EC50 values varying from about 0.01 µg ml^{-1} to 0.1 µg ml^{-1}. Populations sampled in successive years have sigmoid curves shifted to higher EC50 values. The pie chart shows the distribution of genotypes of the SDH-C subunit in the 2020 samples. Figure courtesy of Bart Fraaije and funded by AHDB, with permission.

culture methods; the best results are radial growth rate tests on solid media but involve the use of formidable quantities of petri dishes, media and incubator space (Oliver and Beckerman, 2022). The ratio of the EC50 of the resistant isolate and a naïve one is called the resistance factor (RF). The choice of naïve strain obviously impacts the calculated value of the RF.

It is important to recognise that any measurable and heritable increase in EC50 is evidence of the evolution of resistance. There has been much debate about whether smaller RFs should be called 'tolerance' or 'reduced sensitivity' and that the use of the word 'resistance' should be restricted to cases of field failure. This is a false dichotomy. Even if isolates with small RFs are still well controlled by the label rate of the fungicide, the increase in EC50 indicates that the population is evolving under the selection pressure of the fungicide. This may be a stable situation or may be a prelude to the further increase in EC50s.

Determining EC50s and RFs is costly both in time and money. A cheaper method is to use a single concentration of fungicide and determine whether isolates can grow more than a defined threshold – so-called *discriminatory dose* methods (DD). The choice of dose would normally be equivalent to a full field dose. These are difficult to estimate, so more realistically, the concentration would be two to five times higher than the most resistant naïve isolate. DD tests give significantly less information than full EC50s.

It has been generally assumed that evolved resistance to fungicides detected as increased growth on an infected plant compared to a control isolate can be replicated when testing the isolates *in vitro* via liquid or solid culture methods. Recently there has been a significant case of fungicide failure *in planta* that could not be fully replicated *in vitro* (Van De Wouw et al., 2017). Seedlings grown from treated seeds could be infected by the resistant isolates but not by sensitive ones, but the resistance factor *in vitro* was insufficient to account for the breakdown. The mechanism has not yet been fully documented.

It is well recognised that the ideal scenario is to be able to detect resistance prior to field failure. This is important not just after the first introduction of a new active but also when resistance has been detected in one territory or region but not yet in a neighbouring region. If resistance were present in 1% of isolates, it would be necessary to test 300 isolates to have a 95% chance of detecting at least one. A cheaper way to detect rare cases of resistance in previously reliable fungicides is to set up a 'bait test' (Fig. 2). Bait trials reduce the number of isolates needed to detect resistance but only give a qualitative impression of the resistance frequency. The test is set up in a field treated with the standard fungicide regime. Within the field, a small plot is sprayed with fungicides suspected of resistance development, at concentrations ranging up to two times the max label dose. Any lesions on the higher doses is evidence that resistance is developing. Furthermore, the highest dose on which the

Field sown to susceptible variety and allowed to develop disease without inoculation

Fungicide A	0.5x	0.75x	1x	2x
Fungicide B	0.5x	0.75x	1x	2x

Figure 2 Schematic design of a bait trial. A field of the crop is sown with a susceptible variety. Small plots within the field are sprayed at a range of doses (it may be necessary to obtain special permission to spray above the maximum label rate). When the disease has established in the field, samples are taken from the small plots with the highest dose on which disease is apparent.

disease is seen indicates the resistance factor. Confirmation of resistance will need either phenotypic or genotypic tests.

3 Mechanisms of resistance

Once isolates have been acquired that have significantly increased RFs, the next step is to determine the mechanism (or mode) of resistance (MOR). Studies over the last 25 decades have identified three types of fungicide resistance - target site mutation (TSM), target site overexpression (TSO) and multidrug resistance (MDR). A fourth MOR, where the fungicide is converted to a less toxic form by the pathogen, has only sporadically reported and in just one pathogen *Sclerotinia homoeocarpa* (Sang et al., 2019; Green et al., 2018). This MOR, called detoxification or metabolic resistance, is commonplace in herbicide resistance (Yu and Powles, 2014) and is used by fungi to overcome plant defence compounds (Pedras and Ahiahonu, 2005). Given that many fungi are quite capable of detoxifying various types of xenobiotics, the more widespread detection of this MOR may well be expected.

3.1 Target site mutation

TSM was the first detected and remains the most important type of fungicide resistance. Mutant versions of the target site protein bind the fungicide less tightly so that a greater fungicide concentration is needed to exert an adequate inhibitory effect. Fitness penalties may be apparent if the mutation impacts the activity of the target site gene product. As expected, TSM affects all fungicides with the same biochemical MOA, although the degree of resistance normally varies between actives. In some fungicide classes, a single mutation gives a large RF. In others, multiple mutations are needed before the RF is large enough to impact field performance. Different mutations may alter the degree of resistance, alter the resistance to other fungicides with the same MOA, or ameliorate the fitness penalty.

Not all mutations in the coding regions of target site genes will have a significant impact on fungicide resistance. It is therefore necessary to determine whether the mutation directly impacts the resistance phenotype. There are many ways to do this. The most definitive strategy is to create and test fungicide sensitivity in isogenic strains; two fungal strains with identical genotypes except for the single difference in the fungicide target gene (see e.g. (Cools et al., 2010)). Other methods involve testing large numbers of non-isogenic strains with and without the mutation and determining the correlation with the resistance phenotype.

These methods will often link amino acid residues to resistance. It has often been observed that these amino acids are conserved between fungal species. Presumably they are essential for the structure and function of the protein. Furthermore, it has been frequently observed that the 'same' (i.e. evolutionarily orthologous and functionally identical) amino acids in different species are mutated in resistant isolates. If the proteins are well conserved, the affected amino acids may have the same number – e.g. G143A in *cytochrome b* genes, the target for QoI fungicides. In other cases, the target gene sequences are less well conserved, and the numbering varies between species. These differences in numbering can obscure the frequent observation that orthologous amino acids are being changed to the same or similar mutant amino acids as multiple pathogens evolve resistance to fungicide classes (Mair et al., 2016; Lucas et al., 2015). Some fungicide classes have a very consistent pattern of selection for mutations at one or a few amino acid positions. QoI and MBC fungicides are good examples. In other cases, many sites are affected, and different actives within the MOA group select different combinations of changes resulting in different levels of resistance. This is very clear in both SDHI and DMI fungicides.

In the case of DMIs, different actives select mutations along distinct pathways. Two or more mutations are needed to achieve resistance levels that have a noticeable impact in the field, and these are often different from the pathway needed for resistance to another active. This phenomenon leads to a 'rugged adaptive landscape' (Hawkins and Fraaije, 2016) and could potentially be exploited to achieve longer-term useful lifespans of the entire fungicide group. It also means that it is possible to predict from the genotype of the pathogen population which active will be the most useful in controlling the disease.

The RFs associated with TSM vary widely and depend on the properties of the fungicide and the target. RFs can be as high as 200; these are often difficult to measure precisely as the fungicide may not be soluble in such high concentrations. RFs may be less than 1 when the strain is hypersensitive to the fungicide.

It is quite common for pathogen isolates to carry TSM for more than one fungicide MOA. As many as seven have been detected in *Botrytis cinerea* (Samarakoon et al., 2017; Weber and Entrop, 2016; Fernández-Ortuño et al., 2015; Harper et al., 2022), but two or three are quite common. This would seem to be a response to the use of multiple fungicide groups, either in mixtures or sequentially. To distinguish this phenomenon from MDR, I suggest we call cases of two or more TSM in the same isolate 'oligo drug resistance' (ODR).

3.2 Target site overexpression (TSO)

The second method noted by which fungi have evolved resistance is to overexpress the target gene. If the target protein accumulates to a higher concentration in the pathogen's cells, more fungicide is needed to achieve the same level of pathway inhibition. This could be achieved by decreasing the turnover of the protein or by increasing the efficiency of mRNA translation. However, to date, most cases of TSO are linked to insertions in the promoters of target site genes leading to increased rates of transcription. The insertion sequences are typically fragments of transposable elements found elsewhere in the pathogen's genome. In a few cases, the TSO is induced by the presence of the fungicide that affects the target. This means the target site gene is produced at a suitable level at all fungicide concentrations. It also means that the EC50 measured *in vitro* can be increased if the strain is pre-treated with fungicide levels that activate TSO. Another form of TSO occurs when there is an increase in the copy number of the target site gene (Heckel, 2022).

It would be expected that all actives targeting the overexpressed gene would be affected similarly. TSO particularly affects DMIs. The MBCs, QoIs and SDHIs have not been impacted by TSO. This may be because the target site in these cases is a multimeric protein complex, so coordinated overexpression of all the component genes would be needed to increase the concentration of the complex.

RFs associated with TSO are comparable with the degree of overexpression. Hence, they are rarely more than 20. Fitness penalties might be associated with TSO, but in all cases so far studied, they have been undetectable in field situations.

3.3 Multidrug resistance

The third major method underlying resistance is the upregulation of efflux pumps that decrease the internal concentration of the fungicide. As with TSO, more fungicide is needed to achieve the same level of inhibition; RFs are typically 5–15. Unlike TSO and TSM, fungicides from more than one MOA class are affected. The EC50 for the drug tolnaftate is often used to confirm cases of MDR (Omrane et al., 2015; De Waard et al., 2006). There are two classes of efflux pump found in fungi; the ATP-binding cassette (ABC) class and the major facilitator superfamily (MFS) class (Hahn, 2015). Many major pathogens and especially *Botrytis cinerea*, are affected by MDR and all classes of fungicides are impacted. As with TSO, insertion elements in the promoter have often been found. Disappointingly, the fitness penalties associated with MDR are not particularly high.

4 The evolution of resistance

We have conventionally divided the process of fungicide resistance evolution into two stages – emergence and selection (Hobbelen et al., 2014; Hobbelen et al., 2011a; van den Bosch et al., 2014). In this scheme, the original naïve population is implicitly assumed to be uniformly sensitive. A mutant emerges from this population carrying a genotype conferring resistance. In the second phase, fungicide use *selects* the resistant strain(s), which can then take over the population in due course. There has been much debate about the role of dose, refugia (in time or space), mixtures and alternations in the rate of both emergence and selection. This division into emergence and selection applies to uniform populations found *in vitro* but, as we have seen, does not reflect the situation in agricultural settings. Naïve populations are typically very diverse in both genotype and phenotype, affecting many relevant properties including fungicide resistance but also virulence and survival – see Fig. 1.

It is also conventional to consider that the fungi are exposed to a single dose. This also is an oversimplification. Fungi in the field will be exposed to a range of doses depending on the dose applied, their position in the canopy, their state of development when the spraying takes place and the weather in the aftermath.

A more realistic view of the evolution of resistance under field conditions is to consider that a naïve population will typically comprise strains with a 10- or 100-fold difference in EC50, with the highest EC50s in the 1–10 ppm range. The first impact of fungicide use is to select against the most sensitive strains. The genotypic basis of fungicide hypersensitivity is understudied but initial studies indicate that many genes are involved (Pereira et al., 2020; Mohd-Assaad et al., 2016). This first step of fungicide resistance evolution has little, if any, visible effect on fungicide effectiveness. Almost all remaining strains have EC50s well below the dose typically applied in the field and the disease is still well controlled. However, given the large population size, we can predict that strains carrying the by-now well-known TSM, TSO and MDR mutations are present but at levels of 10^{-9} or even lower. If the same fungicide (or one from the same MOA group) is applied repeatedly, the resistant population will be preferentially selected and will eventually take over the overall population. Sooner or later, the fungicide will be ineffective, and the disease will be poorly controlled.

It is now well-established that the higher the dose, the greater the selection coefficient (van den Bosch et al., 2014). Furthermore, if doses are repeated, selection will be repeated. The factor that matters the most is the duration in a growing season, that the fungicide dose is high enough to control the sensitive population differentially. A single large dose is qualitatively as bad as the same dose spread over two applications. The evolution of resistance to single MOA can be likened to a firearm that is sold with a limited supply of ammunition. When all the bullets are used, the weapon is useless. Every use of the fungicide brings forward the time when resistance will be significant.

This analysis supposes that the fitness penalty associated with resistance is small or negligible. If the fitness penalty is high, the gaps between fungicide applications will select for the sensitive population. In these cases, refugia and fungicide withdrawal will prolong the effective life of a fungicide.

Faced with this apparently inevitable march of resistance as seen successively with MBC, QoI, DMI and SDHI fungicides, the industry promoted the use of mixtures and alternations (M&A) of fungicides (Gullino and Garibaldi, 1982; Hobbelen et al., 2013; Mavroeidi and Shaw, 2006). The rationale for using M&A is simple; if strains emerge with resistance to one fungicide, they will still be controlled by a different active. The chances that strains with resistance to both fungicides would emerge simultaneously seemed to be vanishingly unlikely. By 2010, most fungicides, especially new actives, were exclusively sold as mixtures to delay or prevent resistance. Mixtures were generally preferred to alternations because the latter would be subject to manipulation by growers, implying that the resistance-delaying effect would be diminished (Corkley et al., 2022; Oliver and Beckerman, 2022).

The rationale behind M&A made it clear that actives with different MOA must be used. Here, the mutations conferring resistance to one MOA would be expected to have no impact on the second MOA. However, in the case of DMI fungicides (and, to a lesser extent SDHIs), different actives select for different TSMs, and hence some of the benefits of M&A can be retained by DMI mixes.

Some fungicides have a low risk of resistance development and have been used in M&A with high-risk fungicides. Often the low-risk fungicide has a poorer activity than the high risk, but its use protects the latter. Furthermore, the low-risk fungicide can be used at the maximum permitted dose, allowing the use of the high-risk fungicide at a lower, less selecting dose whilst retaining adequate disease control (Hobbelen et al., 2011b). The need for low-risk (often multisite) fungicides as mixing partners has created a demand for such products. Chlorothalonil filled this role in Europe for some time but has recently been banned because of its poor environmental safety profile. It has been replaced in this role by folpet. Biological fungicides (biological control agents) with non-specific MOA also have potential in this role. Research to find a new low-risk multisite fungicide would be desirable but is stymied by the need to determine a specific MOA for any new product.

Mixtures, whether mixed products or tank mixes of solo products, represent one extreme case of product alternations (Fig. 3). The second product in an alternation strategy can be applied a few days later, a few months, in subsequent growing seasons or even after several years. In the meantime, any necessary sprays can use the first product. Similarly, we can consider the geographic scale of the alternation. We typically think of a single field, but the alternation regime could apply to an entire farm, a region (a county, state or province, small country) or even across a continent. The potential for longer and larger scale M&A regimes has not been intensively studied and is, of course, limited by issues of farmer cooperation.

The problem with this strategy is that the same crop will require different fungicides based on the spectrum of diseases that need to be controlled in a particular farm, the technical requirements associated with the delivery system

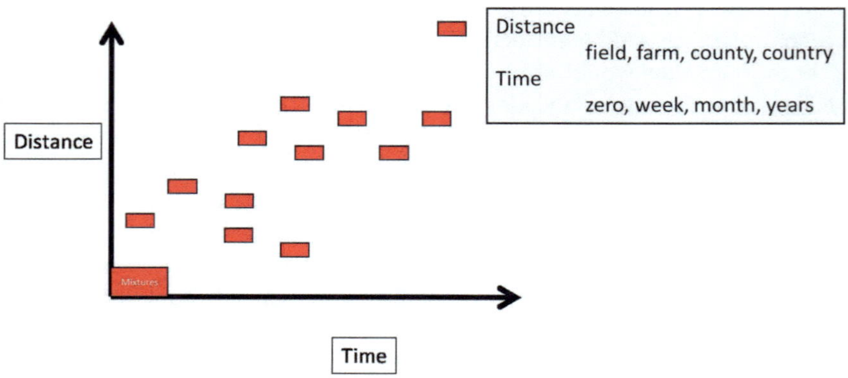

Figure 3 Schematic diagram of M&A options. Mixtures represent an alternation with zero time between fungicides and zero distance between different treatments. The time between application can be increased to a few days, months, years or more. The distance between application can increased from the same field (~ha), the entire farm (~10^3 ha), to the local area (~10^5 ha), to a count(r)y (~10^7 ha) and to entire continents (~10^9 ha).

(i.e. liquid in-furrow vs foliar) and the price range within which a fungicide provides a reasonable ROI in each case.

Larger M&As would guard against the 'cafeteria' effect. If neighbouring fields are using different fungicides, one field will support the growth of strains resistant to fungicide 'A', whereas the other field will support the growth of strains resistant to fungicide 'B'. If the fungicides are then alternated, a population is already present nearby that can grow despite the second fungicide. In effect, the fungi 'select' the field that best supports its growth. Larger M&As might work to suppress the population resistant to fungicide 'B', so that it was more effective when finally used.

The argument that is used to support the use of M&A strategies applies equally well when one of the 'fungicides' is genetic resistance. Pathogens that overcome a resistance gene should still be controlled by the fungicide and vice versa. This concept is very important and underpins the consistent advice that disease-resistant cultivars should always be preferred.

Three (linked) aspects of fungicide resistance M&A strategies have become abundantly clear:

1. M&A strategies rely on both the fungicides retaining full effectiveness. If one fungicide is subject to resistance, it cannot protect the second fungicide, which is effectively being used as a solo product. It is imperative that the development of resistance to both actives is actively monitored.
2. The MOR for both fungicides must be independent. If resistance to one fungicide also confers resistance to the second, the strategy will be nullified. This clearly applies to fungicides with the same MOA (with the proviso that some DMIs have distinct TSMs) but is undermined by MDR.
3. Whatever happens, the disease must be controlled. No fungicide resistance management strategy that allowed substantial disease would ever be tolerated by farmers or desirable from a theoretical perspective.

The widespread adoption of M&A as an anti-resistance strategy has probably been a success (it is difficult to get quantitative data), but it is now clear that the pathogens are evolving in response to the selection environment used to control them. If fungi face a range of different fungicides, they evolve resistance to the entirety. This is most obviously reflected in the increased detection of MDR, which threatens the usefulness of all fungicides but also applies to cases of ODR. The rise of MDR and ODR demands a rethink of the M&A resistance management strategy. One option is to withdraw all affected fungicides from the sale from a region for a period of time – a refugia strategy. Refugia have been used successfully to protect GM insecticide resistance conferred by Bt genes (Mallet and Porter, 1992). The refuge needs to support a large pathogen population that could dominate the resistant MDR/ODR population within one or, at most, a few seasons in the local area (Zhan and McDonald, 2013). The deliberate creation of a large pathogen population would run counter to a generalisation of resistance management: overall pathogen population size should be minimised. Agricultural areas using organic disease control methods possibly serve this role.

A second strategy is to exploit negative cross-resistance in which fungi resistant to one fungicide are simultaneously hypersensitive to another. Such strategies have been much discussed in relation to methyl-benzimidazole carbamates and N-phenylcarbamates (Leroux, 1992) and between DMIs but have never progressed to field implementation.

The traditional response to fungicide resistance was to develop a new fungicide, preferably with a new MOA. More than ever, this is the Holy Grail of the fungicide industry. Product efficacy and safety are paramount when developing a new fungicide, but it is also critical that the fungicide is at a low risk of resistance development. Hence, much thought goes into the question, 'can we predict the risk of resistance?'

All new fungicides are tested *in vitro* screening for mutants that are resistant. Such mutants are genetically analysed to determine the MOR. If the RFs are high and fitness penalties are low, the risk is probably going to be high. A method to predict the RF and fitness of resistant mutants would be highly desirable. The spectrum of genotypes linked to resistance is also important. It has often been observed that different mutants are found in the field compared to the lab. It is assumed that some resistant mutants have hidden fitness penalties that are only revealed in the field. A field-scale microcosm could be used to test the fungicides over multiple life cycles of the pathogen and would have a better chance of uncovering the relevant mutations. A national or regional facility that allowed for screening for mutants in a realistic cropping situation would be desirable. However, it is hard to imagine a universal facility that covers all crops, arable, horticultural and pasture in all regions that could serve as an effective field laboratory for testing new fungicides. In reality, the study of resistance occurs alongside the first commercial deployment, which is often a region of maximal disease pressure (Dooley et al., 2016).

5 Conclusion and future trends

Fungicide resistance is already a substantial problem but has often been exaggerated, particularly by those who wish to promote an alternative product or research agenda. The detection of resistance is only the beginning of the end of the use of a fungicide. MBC, DMI, QoI, SDHI, CAA and other fungicides have suffered from resistance for decades but remain in use somewhere in the world and are often market leaders. Resistance has been observed to follow a consistent pattern in terms of time to resistance, and mutants that are observed when products are introduced to new markets (Lucas et al., 2015; Tucker et al., 2020; McDonald et al., 2019). These studies emphasise that resistance evolves locally and follows inappropriate use patterns. Migration of resistant strains is a relatively local problem. This means that the resistance management strategy that is used in the primary market (use of genetically resistant cultivars, lowest useful dose, M&A) must be followed in secondary markets.

The key change that has impacted the fungicide resistance field in the last few years is the rapid development of different methods to detect, differentiate and quantify genotypic changes in pathogen populations conferring resistance (Miles et al., 2021; Huang et al., 2020; Liu et al., 2019; Hu et al., 2019; Dodhia et al., 2021; Zulak et al., 2018; Radhakrishnan et al., 2019; Bueno-Sancho et al., 2017; Saunders, 2015). These methods mean that it is realistic to screen infected plants for both pathogen presence and for known fungicide resistance mutations. These assays can be optimised for speed so that farmers can be given useful information to direct fungicide spray programmes in the coming days and weeks. Farmers can then make informed decisions about whether to omit a fungicide spray, alter the dose and change the actives to be used. Alternatively, the assays can be optimised for genotypic coverage so that the total content of pathogens can be ascertained – the microbiome of the sampled tissue. Entire genome sequences of all the microbes present can be sampled. These broader assays can be used to assess the status of the pathogen population on a season-by-season and a regional basis, allowing optimised choice of crop, cultivar and planting date and agronomy package for the following season.

Both types of assays are now cheap enough for sampling to be carried out on a comprehensive scale. If implemented, the data could be used to monitor the changes in pathogen species profiles and resistance genotypes. Such data could form the basis of a rational regional crop protection strategy whereby both fungicides and disease-resistance genes are monitored for breakdown.

The case of canola blackleg in South Australia illustrates the potential. Isolates gathered annually were used to infect canola cultivars carrying currently effective resistance genes in laboratory tests during the off-season. The 2010 batch had many isolates that could infect the cultivar Hyola50. Growers agreed to avoid sowing such a cultivar. Test plots sown in 2011 in the affected area suffered devastating disease. This simple procedure saved the canola growers an estimated $Au13m (or ca. 12% of yield) in each of the next few seasons (Howlett, 2018; Van De Wouw et al., 2014; Van De Wouw et al., 2018). This test used a relatively cumbersome phenotypic test but still had a cost-effectiveness of the order of 100:1. With genotypic tests now available for many virulence and fungicide resistance genes, and the potential is massive.

Two questions remain. The first concerns fungicide resistance and virulence gains not yet associated with a specific genotype. In these cases, there is no substitute for phenotypic tests. Annual fungicide performance assays are carried out by the AHDB in the UK, but I know of no other country carrying out publicly available fungicide performance assays. Many fungicide companies carry out screens with a view to protecting the performance of their actives and much of this data is made available. Cases of poor performance can be studied, which will, in most, identify the genes involved and permit a genotypic test.

The second question concerns who should pay for comprehensive pathogen prevalence monitoring. The ultimate beneficiaries are the public which would benefit from greater food security and minimised fungicide use. The prolongation of fungicide effectiveness might seem to benefit the company that invested in the discovery and development of the fungicide. However, the current length of patent protection – typically 20 years from the initial filing – means that much of the benefit of the prolonged life will fall to generic manufacturers. Farmers have the most to gain from prolonged active lives of fungicide products. Fungicide use will be minimised, and their costs will fall as patents expire. There is much to be gained for us all from global investment in pathogen genotypic monitoring and not just for fungicide resistance evolution.

6 Abbreviations

CAA	carboxylic acid amides
CPP	crop protection product
DD	discriminatory dose
DMI	demethylation inhibitor
EC50	effective concentration that inhibits growth by 50%
M&A	mixtures and alternations
MBC	methyl-benzimidazole carbamate
MOA	mode of action
MOR	mode of resistance
ODR	oligo drug resistance
QoI	quinone out inhibitor
RF	resistance factor
ROI	return on investment
SDHI	succinate dehydrogenase inhibitor
TSM	target site mutation
TSO	target site overexpression
MDR	multidrug resistance

7 Acknowledgements

I have benefited from research funding from GRDC, BBSRC, Syngenta, EU on fungicides over the last 25 years. I have also benefited from discussions and collaborations with many people. These include Fran Lopez-Ruiz, Frank van den Bosch, Lise Nistrup Jorgensen, Derek Hollomon, Nichola Hawkins, Bart Fraaije, Rosie Bryson, Andrew Milgate, Ken Young, Hans Cools, Mike Csukai, Gerd Stammler, the late Andy Leadbeater and many others.

8 Where to look for further information

Websites:

- FRAC: www.frac.info.
- https://ahdb.org.uk/knowledge-library/fungicide-performance.
- http://www.genomesonline.org.
- FRAG: http://www.pesticides.gov.uk/guidance/industries/pesticides/advisory-groups/Resistance-Action-Groups/frag.

9 References

Bueno-Sancho, V., Persoons, A., Hubbard, A., Cabrera-Quio, L. E., Lewis, C. M., Corredor-Moreno, P., Bunting, D. C. E., Ali, S., Chng, S., Hodson, D. P., Madariaga Burrows, R., Bryson, R., Thomas, J., Holdgate, S. and Saunders, D. G. O. 2017. Pathogenomic analysis of wheat yellow rust lineages detects seasonal variation and host specificity. *Genome Biology and Evolution* 9(12), 3282–96.

Cools, H. J., Parker, J. E., Kelly, D. E., Lucas, J. A., Fraaije, B. A. and Kelly, S. L. 2010. Heterologous expression of mutated eburicol 14α-demethylase (CYP51) proteins of Mycosphaerella graminicola to assess effects on azole fungicide sensitivity and intrinsic protein function. *Applied and Environmental Microbiology* 76(9), 2866–72.

Corkley, I., Fraaije, B. and Hawkins, N. 2022. Fungicide resistance management: maximizing the effective life of plant protection products. *Plant Pathology* 71(1), 150–69.

De Waard, M. A., Andrade, A. C., Hayashi, K., Schoonbeek, H. J., Stergiopoulos, I. and Zwiers, L. H. 2006. Impact of fungal drug transporters on fungicide sensitivity, multidrug resistance and virulence. *Pest Management Science* 62(3), 195–207.

Dodhia, K. N., Cox, B. A., Oliver, R. P. and Lopez-Ruiz, F. J. 2021. Rapid in situ quantification of the strobilurin resistance mutation G143A in the wheat pathogen Blumeria graminis f. sp. tritici. *Scientific Reports* 11(1), 4526.

Dooley, H., Shaw, M. W., Mehenni-Ciz, J., Spink, J. and Kildea, S. 2016. Detection of Zymoseptoria tritici SDHI-insensitive field isolates carrying the SdhC-H152R and SdhD-R47W substitutions. *Pest Management Science* 72(12), 2203–7.

Fernández-Ortuño, D., Grabke, A., Li, X. and Schnabel, G. 2015. Independent emergence of resistance to seven chemical classes of fungicides in Botrytis cinerea. *Phytopathology* 105(4), 424–32.

Georgopoulos, S. G. and Dovas, C. 1973. A serious outbreak of strains of Cercospora beticola resistant to benzimidazole fungicides in Northern Greece. *Plant Disease Reporter* 57, 321–4.

Green, R., Sang, H., Im, J. and Jung, G. 2018. Chlorothalonil biotransformation by cytochrome P450 monooxygenases in Sclerotinia homoeocarpa. *FEMS Microbiology Letters* 365(19).

Gullino, M. L. and Garibaldi, A. 1982. Use of mixture or alternation of fungicides with the aim of reducing the risk of appearance of strains of Botrytis cinerea resistant to Dicarboximides1. *EPPO Bulletin* 12(2), 151–6.

Hahn, M. and M. 2015. Multidrug efflux transporters. In: Ishii, H. H. D. W. (Ed.). *Fungicide Resistance in Plant Pathogenes*. Japan: Springer, 233–48.

Harper, L. A., Paton, S., Hall, B., McKay, S., Oliver, R. P. and Lopez-Ruiz, F. J. 2022. Fungicide resistance characterized across seven modes of action in Botrytis cinerea isolated from Australian vineyards. *Pest Management Science* 78(4), 1326–40.

Hawkins, N. J. and Fraaije, B. 2016. Adaptive landscapes in fungicide resistance; fitness, epistasis, constraints and predictability. In: 18th International Reinhardsbrunn Symposium; Modern Fungicides and antifungal compounds. Friedrichroda, Germany (vol. 91).

Heckel, D. G. 2022. Perspectives on gene copy number variation and pesticide resistance. *Pest Management Science* 78(1), 12–8.

Hobbelen, P. H. F., Paveley, N. D., Fraaije, B. A., Lucas, J. A. and van den Bosch, F. 2011a. Derivation and testing of a model to predict selection for fungicide resistance. *Plant Pathology* 60(2), 304–13.

Hobbelen, P. H. F., Paveley, N. D. and van den Bosch, F. 2011b. Delaying selection for fungicide insensitivity by mixing fungicides at a low and high risk of resistance development: A modeling analysis. *Phytopathology* 101(10), 1224–33.

Hobbelen, P. H. F., Paveley, N. D., Oliver, R. P. and van den Bosch, F. 2013. The usefulness of fungicide mixtures and alternation for delaying the selection for resistance in populations of Mycosphaerella graminicola on winter wheat: A modeling analysis. *Phytopathology* 103(7), 690–707.

Hobbelen, P. H. F., Paveley, N. D. and van den Bosch, F. 2014. The emergence of resistance to fungicides. *PLoS ONE* 9(3), e91910.

Howlett, B. J. 2018. 2017. Daniel McAlpine memorial lecture. A 'genome to paddock' approach to control plant disease. *Australasian Plant Pathology* 47(3), 239–43.

Hu, Y., Green, G. S., Milgate, A. W., Stone, E. A., Rathjen, J. P. and Schwessinger, B. 2019. Pathogen detection and microbiome analysis of infected wheat using a portable DNA sequencer. *Phytobiomes Journal* 3(2), 92–101.

Huang, X., Wang, X., Kong, F., Van Der Lee, T., Wang, Z. and Zhang, H. 2020. Detection and characterization of carboxylic acid amide-resistant Plasmopara viticola in China using a TaqMan-MGB real-time PCR. *Plant Disease* 104(9), 2338–45.

Leroux, P. 1992. Negative cross-resistance in fungicides: from the laboratory to the field. In: Denholm, I., Devonshire, A. and Hollomon, D. (ed.), *Resistance '91: Achievements and Developments in Combating Pesticide Resistance*. Dordrecht: Springer.

Liu, X., Zeng, R., Gao, S., Xu, L. and Dai, F. 2019. Rapid detection of SdhBP225F and SdhBH272R mutations in Boscalid resistant Botrytis cinerea Strains by ARMS-PCR. *The Plant Pathology Journal* 35(1), 71–6.

Lucas, J. A., Hawkins, N. J. and Fraaije, B. A. 2015. The evolution of fungicide resistance. In: *Advances in Applied Microbiology* 90, 29–92.

Mair, W., Lopez-Ruiz, F., Stammler, G., Clark, W., Burnett, F., Hollomon, D., Ishii, H., Thind, T. S., Brown, J. K., Fraaije, B., Cools, H., Shaw, M., Fillinger, S., Walker, A. S., Mellado, E., Schnabel, G., Mehl, A. and Oliver, R. P. 2016. Proposal for a unified nomenclature for target-site mutations associated with resistance to fungicides. *Pest Management Science* 72(8), 1449–59.

Mallet, J. and Porter, P. 1992. Preventing insect adaptation to insect-resistant crops: are seed mixtures or refugia the best strategy? *Proceedings of the Royal Society of London. Series B: Biological Sciences* 250(1328), 165–9.

Mavroeidi, V. I. and Shaw, M. W. 2006. Effects of fungicide dose and mixtures on selection for triazole resistance in Mycosphaerella graminicola under field conditions. *Plant Pathology* 55(6), 715–25.

McDonald, M. C., Renkin, M., Spackman, M., Orchard, B., Croll, D., Solomon, P. S. and Milgate, A. 2019. Rapid parallel evolution of azole fungicide resistance in Australian populations of the Wheat Pathogen Zymoseptoria tritici. *Applied and Environmental Microbiology* 85(4).

Miles, T. D., Neill, T. M., Colle, M., Warneke, B., Robinson, G., Stergiopoulos, I. and Mahaffee, W. F. 2021. Allele-specific detection methods for QoI fungicide-resistant Erysiphe necator in vineyards. *Plant Disease* 105(1), 175–82.

Mohd-Assaad, N., McDonald, B. A. and Croll, D. 2016. Multilocus resistance evolution to azole fungicides in fungal plant pathogen populations. *Molecular Ecology* 25(24), 6124–42.

Oliver, R. P. and Beckerman, J. L. 2022. *Fungicides in Practice*. CABI Publishing.

Oliver, R. P. and Hewitt, H. G. 2014. *Fungicides in Crop Protection* (2nd edn.).

Omrane, S., Sghyer, H., Audéon, C., Lanen, C., Duplaix, C., Walker, A. S. and Fillinger, S. 2015. Fungicide efflux and the MgMFS1 transporter contribute to the multidrug resistance phenotype in Zymoseptoria tritici field isolates. *Environmental Microbiology* 17(8), 2805–23.

Pedras, M. S. C. and Ahiahonu, P. W. K. 2005. Metabolism and detoxification of phytoalexins and analogs by phytopathogenic fungi. *Phytochemistry* 66(4), 391–411.

Pereira, D., McDonald, B. A. and Croll, D. 2020. The genetic architecture of emerging fungicide resistance in populations of a global wheat pathogen. *Genome Biology and Evolution* 12(12), 2231–44.

Radhakrishnan, G. V., Cook, N. M., Bueno-Sancho, V., Lewis, C. M., Persoons, A., Mitiku, A. D., Heaton, M., Davey, P. E., Abeyo, B., Alemayehu, Y., Badebo, A., Barnett, M., Bryant, R., Chatelain, J., Chen, X., Dong, S., Henriksson, T., Holdgate, S., Justesen, A. F., Kalous, J., Kang, Z., Laczny, S., Legoff, J. P., Lesch, D., Richards, T., Randhawa, H. S., Thach, T., Wang, M., Hovmøller, M. S., Hodson, D. P. and Saunders, D. G. O. 2019. Marple, a point-of-care, strain-level disease diagnostics and surveillance tool for complex fungal pathogens. *BMC Biology* 17(1), 65.

Samarakoon, U. C., Schnabel, G., Faust, J. E., Bennett, K., Jent, J., Hu, M. J., Basnagala, S. and Williamson, M. 2017. First report of resistance to multiple chemical classes of fungicides in Botrytis cinerea, the causal agent of gray mold from greenhouse-grown petunia in Florida. *Plant Disease* 101(6), 1052.

Sang, H., Popko, J. T., Jr. and Jung, G. 2019. Evaluation of a Sclerotinia homoeocarpa population with multiple fungicide resistance phenotypes under differing selection pressures. *Plant Disease* 103(4), 685–90.

Saunders, D. G. O. 2015. Hitchhiker's guide to multi-dimensional plant pathology. *New Phytologist* 205(3), 1028–33.

Tucker, M. A., Lopez-Ruiz, F., Cools, H. J., Mullins, J. G. L., Jayasena, K. and Oliver, R. P. 2020. Analysis of mutations in West Australian populations of Blumeria graminis f. sp. hordei CYP51 conferring resistance to DMI fungicides. *Pest Management Science* 76(4), 1265–72.

Van De Wouw, A. P., Elliott, V. L., Chang, S., López-Ruiz, F. J., Marcroft, S. J. and Idnurm, A. 2017. Identification of isolates of the plant pathogen Leptosphaeria maculans with resistance to the triazole fungicide fluquinconazole using a novel in planta assay. *PLoS ONE* 12(11), e0188106.

Van De Wouw, A. P., Howlett, B. J. and Idnurm, A. 2018. Changes in allele frequencies of avirulence genes in the blackleg fungus, Leptosphaeria maculans, over two decades in Australia. *Crop and Pasture Science* 69(1), 20–9.

Van De Wouw, A. P., Marcroft, S. J., Ware, A., Lindbeck, K., Khangura, R. and Howlett, B. J. 2014. Breakdown of resistance to the fungal disease, blackleg, is averted in commercial canola (Brassica napus) crops in Australia. *Field Crops Research* 166, 144–51.

van den Bosch, F., Oliver, R. P., Van Den Berg, F. and Paveley, N. 2014. Governing principles can guide fungicide-resistance management tactics. *Annual Review of Phytopathology* 52, 175–95.

Weber, R. W. S. and Entrop, A. P. 2016. Recovery of Botrytis strains with multiple fungicide resistance from raspberry nursery plants. *European Journal of Plant Pathology*, 1–4.

Yu, Q. and Powles, S. 2014. Metabolism-based herbicide resistance and cross-resistance in crop weeds: A threat to herbicide sustainability and global crop production. *Plant Physiology* 166(3), 1106–18.

Zhan, J. and McDonald, B. A. 2013. Experimental measures of pathogen competition and relative fitness. In: *Annual Review of Phytopathology* 51, 131–53.

Zulak, K. G., Cox, B. A., Tucker, M. A., Oliver, R. P. and Lopez-Ruiz, F. J. 2018. Improved detection and monitoring of fungicide resistance in Blumeria graminis f. sp. hordei with high-throughput genotype quantification by digital PCR. *Frontiers in Microbiology* 9, 706.

Chapter 2

Molecular evolution and mechanisms of fungicide resistance in plant pathogenic fungi

Laetitia Chartrain and James K. M. Brown, John Innes Centre, UK

1 Introduction

Chemicals have been used to reduce crop damage since Greek and Roman times. In the seventeenth century, lime and salt were used as seed treatment to prevent bunt on wheat. In the nineteenth century, sulphur and copper were used in foliar sprays and seed treatments to control a range of plant pathogens. In 1882, the famous Bordeaux mixture, a mixture of copper sulphate and hydrated lime, was developed to control grapevine mildew. Until the 1940s, chemical disease control continued to rely on inorganic chemical preparations which were prepared by the user and had to be applied in enormous amounts. After the war the agrochemical industry started to develop, and new chemicals such as dithiocarbamates and phthalimides appeared on the market. These compounds are still used as foliar sprays and seed treatments today. Fungicides used from the 1940s to the 1960s were multi-site inhibitors which affected

http://dx.doi.org/10.19103/AS.2022.0116.02

many fungal enzymes, had generally broad-spectrum effectiveness against many target species, were non-systemic because they did not move through the plant, and only had protectant activity, being ineffective against established infection. They had to be applied in large quantities and very frequently, often weekly, and they tended to be toxic not only to the crop or plant to which they were applied but also to the environment and the user.

In the 1960s and 1970s, the fungicide market expanded dramatically with the release of an entirely new type of fungicides. These were systemic fungicides which could be taken up by and travelled through plant tissues. These new formulations all had single-site modes of action which were highly specific to target pathogens but not host crop plants. They were more effective, less toxic to people and the natural environment and could be used as eradicants as well as protectants. This revolutionised crop protection and the main group of fungicides which are used today are all single-site fungicides. The history of fungicides has been reviewed in detail by Morton and Staub (2008) and Hollomon (2015).

The main disadvantage of single-site mode of action fungicides is that resistance in pathogen populations tends to evolve shortly after the introduction of a product. Resistance may be either qualitative or quantitative. Qualitative resistance is characterised by an abrupt loss of effectiveness of the fungicide, resulting in control failure. Isolates in the population tend to fall into two groups, resistant and sensitive. Quantitative resistance is characterised by a gradual decline in disease control with variable degrees of resistance in the pathogen population, resulting in a correspondingly gradual decline in fungicide performance.

This chapter reviews the groups of fungicides which have been most widely used in agriculture and horticulture over the last 50 years and summarises the mechanisms of resistance to each class of compounds. We address resistance to compounds active against fungi *sensu stricto* but not those used specifically to oomycete pathogens. Single-site fungicides target essential fungal cell processes. As fungal cells have many biochemical features and processes in common with animal cells the number of useful target proteins are limited. The main targets for fungicide development are the integrity of the fungal cell and fungal respiration. We focus on powdery mildew fungi because these pathogens are notorious for the speed with which they have evolved resistance to many groups of systemic fungicides. Why they are so prone to evolving resistance is not known but reasons may include the fact that they are haploid, so mutations to resistance are immediately exposed to natural selection by the fungicide, their short life cycle in their asexual phase, and the plasticity of their genomes (Spanu et al., 2010, Wicker et al., 2013, Bhosle et al., 2019, Bindschedler et al., 2016). Moreover, very unusually for a eukaryote, all reproductive cells of powdery mildew are exposed to the atmosphere and thus to mutagens including ultra-violet wavelengths of sunlight, chemical pollutants and cosmic rays (Arnold, 2018).

2 Methyl benzimidazole carbamate: fungicides which target the cytoskeleton

Three important classes of fungicides target structural features of the fungal cell. MBC fungicides target the cytoskeleton whereas the azole and amine groups affect damage the integrity of fungal membranes.

The cytoskeleton provides structure and shape to eukaryotic cells. An important part of the cytoskeleton consists of microtubules, which are polymers of tubulin proteins. There are two types of tubulin forming the microtubule, α-tubulin, encoded by the *Tub1* gene, and β-tubulin, encoded by *Tub2*. Methyl benzimidazole carbamate (MBC) or benzimidazole fungicides, which target β-tubulin, were introduced in the early 1970s and quickly became widely used in agriculture. They are effective against a broad range of ascomycete and basidiomycete pathogens of cereals, grapes, fruits and vegetables but are ineffective against oomycetes. They were the first single-site fungicides developed. Resistance to MBCs developed very soon after their introduction and has now been reported in over 100 species of plant pathogenic fungi (Hawkins and Fraaije, 2016, Vela-Corcía et al., 2018).

Resistance was first reported in cucurbit powdery mildew (Schroeder and Provvidenti, 1969) followed by *Botrytis* in grapevine (Ehrenhardt et al., 1973) then cereal powdery mildew (Vargas, 1972, 1973). As resistance started to emerge in the field, mutagenesis studies in model organisms such as brewer's yeast, *Neurospora crassa* and *Aspergillus nidulans* (teleomorph *Emericella nidulans*) were quickly set up to assess mechanisms and risks of resistance in advance of resistance becoming widespread in the field. These experiments led to the identification of the target of MBCs being β-tubulin and the identification of mutations responsible for resistance (Thomas et al., 1985, Fujimura et al., 1992, Fujimura et al., 1994, Orbach et al., 1986).

Several mutations in *Tub2* are associated with resistance, most of which are at amino acid 198 and 200. The most common substitutions are F200Y (phenylalanine mutated to tyrosine at residue 200) and E198A/K/G (glutamic acid mutated to alanine, lysine or glycine at residue 198); see Table 1 for amino acid codes (Hawkins and Fraaije, 2016, Vela-Corcía et al., 2018). In the cucurbit powdery mildew fungus *Podosphaera xanthii* (which may be synonymous with *P. fusca*; (Pérez-García et al., 2009)), resistance in the field was identified in the early 2000s (McGrath, 2001). No crystal structure for fungal β-tubulin is available yet but Vela-Corcía et al. (2018) were able to use the mammalian crystal structure from *Bos taurus* (cattle) and heterologous expression of β-tubulin in *E. coli* to explore the mechanism of resistance in *P. xanthii* to two

Table 1 Amino acid codes and properties

Amino acid	Code	Properties	Structure
Alanine	A	Hydrophobic (non-polar)	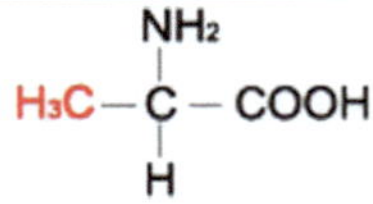
Arginine	R	Polar charged basic	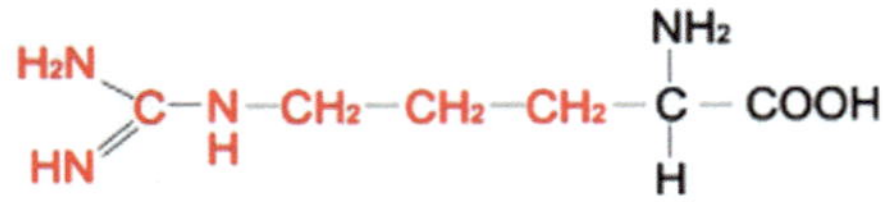
Asparagine	N	Polar uncharged	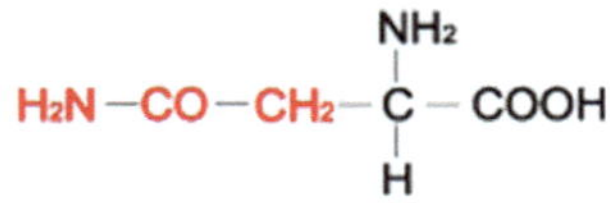
Aspartic acid	D	Polar charged acid	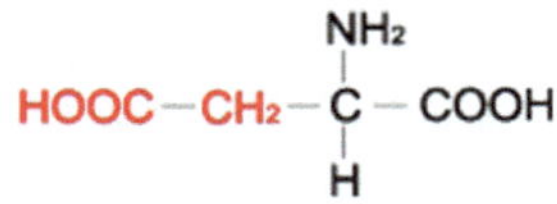
Cysteine	C	Polar uncharged	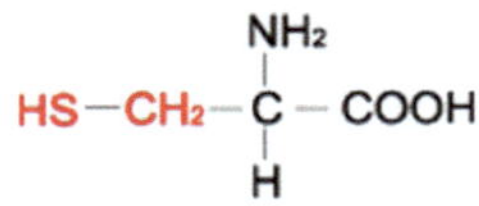
Glutamic acid	E	Polar charged acid	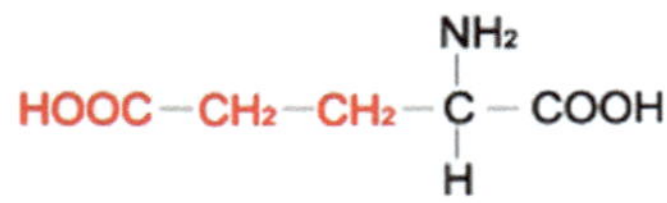
Glutamine	Q	Polar uncharged	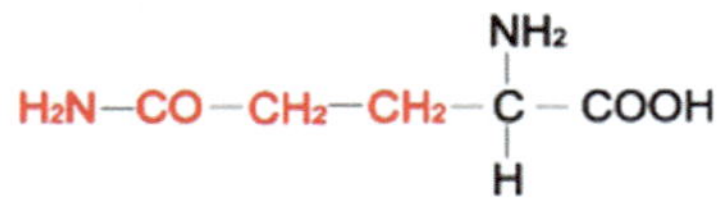
Glycine	G	Polar uncharged	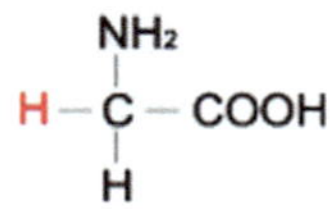
Histidine	H	Polar charged	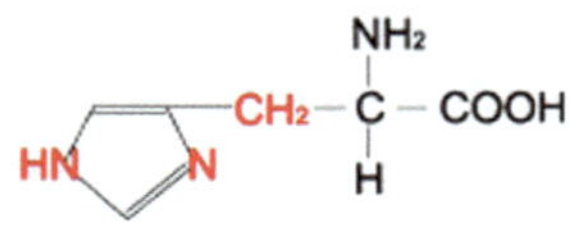
Isoleucine	I	Hydrophobic	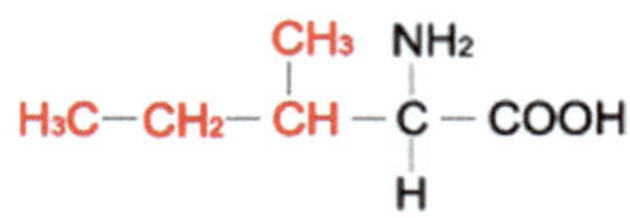
Leucine	L	Hydrophobic	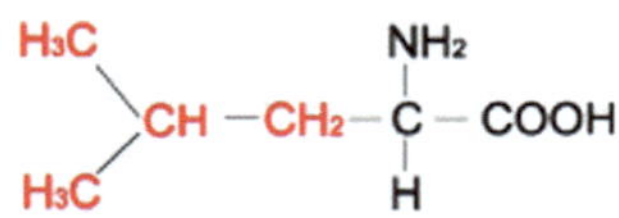

(*Continued*)

Table 1 (*Continued*)

Amino acid	Code	Properties	Structure
Lysine	K	Polar charged basic	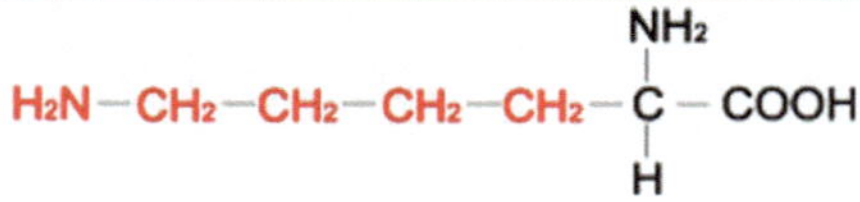
Methionine	M	Hydrophobic	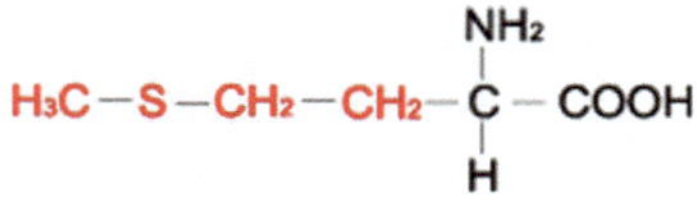
Phenylalanine	F	Hydrophobic	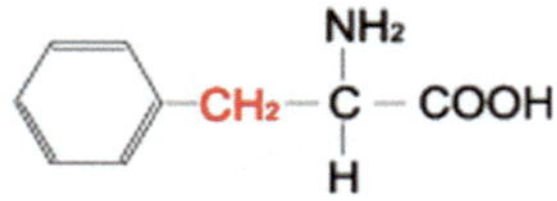
Proline	P	Hydrophobic	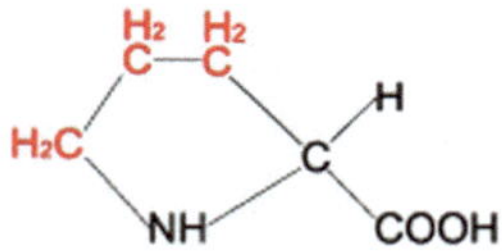
Serine	S	Polar uncharged	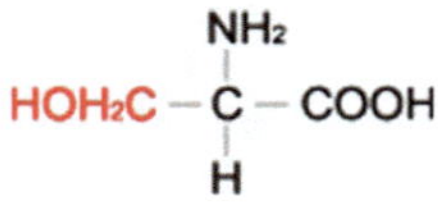
Threonine	T	Polar uncharged	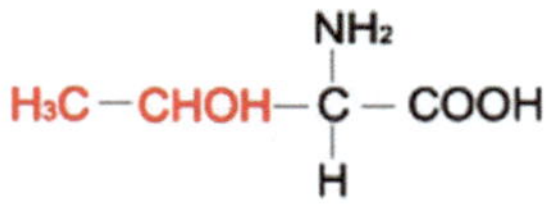
Tryptophan	W	Hydrophobic	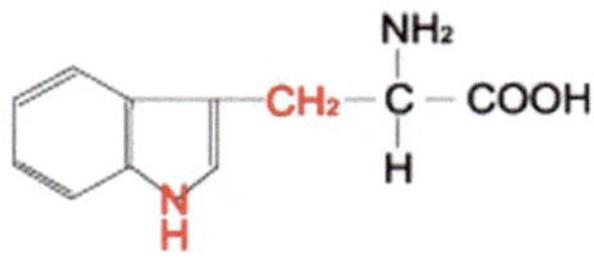
Tyrosine	Y	Hydrophobic	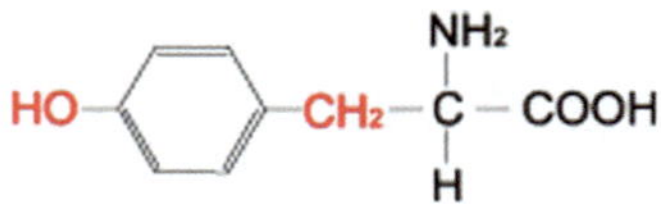
Valine	V	Hydrophobic	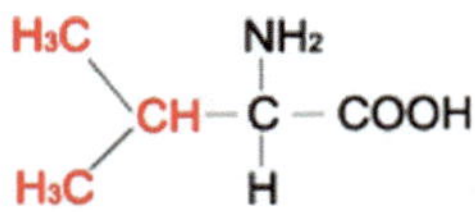

molecules, carbendazim, an MBC fungicide, and diethofencarb, an N-phenyl carbamate, which is a herbicide with additional activity against several fungi. In *P. xanthii*, resistance is conferred by the substitution of glutamic acid to alanine at position 198 (E198A). This study postulated that the E198A mutation induces subtle structural rearrangements of the entire protein, specifically that the resistant β-tubulin protein had four β-sheet strands which were not present in the sensitive protein. Furthermore, residues 198 and 200 did not appear to be part of the binding site.

Protein homology modelling in *P. xanthii* suggested that residue 198 is located deep inside the protein making in inaccessible to the fungicide; rather, the carbendazim binding site involves residues S138 and T178, while diethofencarb binds to S178 and Y222. It was predicted that carbendazim binds to β-tubulin via three hydrogen bonds, one at amino acid residue S138 and two at T178 and that the E198A mutation results in T178 being twisted out, precluding the establishment of two of the three hydrogen bonds between carbendazim and the β-tubulin target. A similar mechanism was predicted for diethofencarb, with the fungicide binding the protein via two hydrogen bonds at residues S138 and Y222. The modeling predicted that the E198A mutation results in the movement of Y222, preventing the formation of one of these hydrogen bonds. Site-directed mutagenesis of residues 138 and 178 was performed with both residues mutated to alanine. The resultant mutant proteins underwent dramatic misfolding in the absence of the fungicide and T178A was unable to bind to the fungicide (Fig. 1).

Resistance to MBC fungicides has been detected in several powdery mildew species. Indeed, resistance appeared in grapevine powdery mildew *Erysiphe necator* (anamorph *Uncinula necator*) in California vineyards in the late 1970s but the underlying mutations have not been investigated. This is probably due

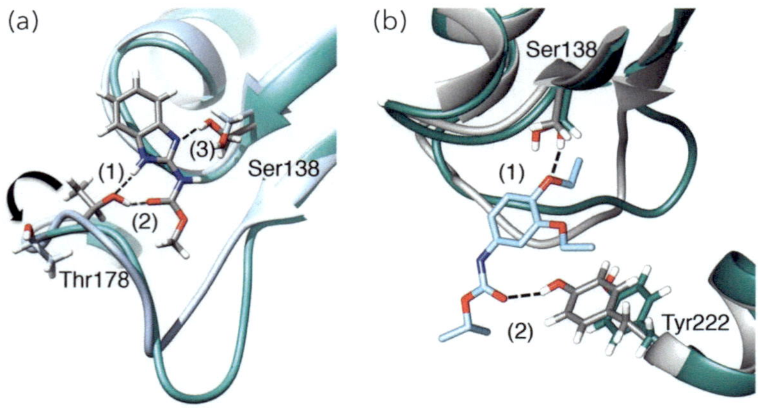

Figure 1 Molecular docking of MBC fungicides to superpositions of MBC-sensitive and resistant *Podosphaera xanthii* β-tubulin models. (a) Docking of carbendazim to MBC-sensitive *P. xanthii* β-tubulin (in cyan) via three hydrogen bonds (dashed lines) involving residues Threonine 178 (T178) and Serine 138 (S138). Thr178 is twisted out (arrow) in the MBC-resistant β-tubulin model (gray) preventing carbendazim binding to T178. (b) Docking of diethiofencarb to MBC-resistant β-tubulin model (cyan) via two hydrogen bonds that involve residues Ser138 and Tyr222 (Y222). Diethiofencarb cannot bind to MBC-sensitive β-tubulin (gray) due to movement of Tyr222. This figure is reproduced from Vela-Corcía et al. (2018) and is included in the article's Creative Commons CC-BY Licence.

Published by Burleigh Dodds Science Publishing Limited, 2023.

to the low importance of MBC use in grapevine mildew management (Pearson and Taschenberg, 1980, Kunova et al., 2021). In cucumber mildew *P. fusca* (syn. *P. xanthii*), resistance in the field was identified in the early 2000s (McGrath, 2001) and study of the β-tubulin gene sequence revealed that the E198A mutation conferred resistance to MBC fungicides in *P. fusca* (Vela-Corcía et al., 2014).

3 Azoles: inhibitors of sterol demethylation

The sterol synthesis pathway is the target of several classes of fungicide because ergosterol is specific to fungi whereas animals cells have cholesterol and fungi cannot survive without ergosterol. Like other sterols, ergosterol is an essential component of fungal cell membranes, determining the fluidity, permeability and activity of the membrane. The two main classes of agricultural fungicide which attack the sterol synthesis pathway are azoles and amines. In addition, terbinafine, used widely in medicine, inhibits the ERG1 squalene monooxygenase, an important step in the pathway to sterol, steroid and triterpenoid synthesis.

3.1 Activity of azoles

Since their introduction in the 1970s, azole antifungals have been the mainstay of antifungal treatments in human and veterinary medicine as well as arable agriculture and horticulture. They are the largest class of all fungicides. They have broad-spectrum activity and have remained largely effective against many medical and agricultural pathogens. Most agricultural azole fungicides fall into the triazole sub-class, in contrast to medical antifungals, most of which are imidazoles. A notable agricultural imidazole is the broad-spectrum fungicide prochloraz. Other, smaller groups of sterol C14 demethylation inhibitor fungicides have also been important, including the piperazine compound triforine, the pyrimidine fenarimol and the pyridine pyrifenox (Kuck et al., 1995), but they are not considered further here.

Although they are generally considered to be safe products, azole fungicides used in medicine and agriculture have been reported to have off-target effects on mammalian CYP enzymes other than CYP51, especially CYP19 and also CYP11 and CYP17 (Trösken et al., 2006); also see summaries in (Warrilow et al., 2013) and (Munkboel et al., 2019). Several of these CYP enzymes are involved in sex hormone synthesis; for example, CYP19A1, known as aromatase or oestrogen synthase, is involved in the conversion of androgens to oestrogens. At high doses, azole fungicides have generally small effects on mammalian reproduction (Zarn et al., 2003). As the European Union has banned agricultural (but not medical) fungicides with any endocrine disrupting effect at all, regardless of their value in food production, understanding the interaction of azoles with the active site of CYP51 in fungi is of critical importance to food security, particularly in Europe. Knowledge of the molecular mechanism and evolution of azole resistance is valuable in achieving this. An example of the fruits of applying molecular information to fungicide development is mefentrifluconazole (trade name Revysol), invented by BASF AG, which has high inhibition of fungal CYP51 and low inhibition of human CYP19 with no evidence of endocrine disruption (Tesh et al., 2019). This compound is particularly interesting because it has a flexible structure which allows it to bind to the binding pocket even when the shape of the pocket has been distorted by mutation, so the fungicide retains its efficacy (Strobel et al., 2020).

Azoles interfere with the fungal enzyme lanosterol 14α-demethylase, known as CYP51 or ERG11 (Van Den Bossche et al., 1984). Fungal CYP51 catalyzes a key step in the biosynthesis of ergosterol, a component of the fungal membrane that mediates membrane permeability and fluidity (Daum et al., 1998). CYP51 belongs to the cytochrome P450 monooxygenase (CYP) super-family. It is one of the most ancient families of proteins, exists in all biological kingdoms and has a narrow function: to remove the 14α methyl group from sterol precursors. CYP51 proteins have narrow specificity for their sterol substrates, so specificity of an azole fungicide for the enzyme in a fungus rather than its plant or animal host depends on interactions between side-chains of the fungicide and amino acids in the CYP51 protein (Parker et al., 2014).

3.2 Cyp51 sequence variation

Widespread use of azoles in both medicine and agriculture has resulted in reduced sensitivity in many fungi. The most common cause of reduced sensitivity is point mutations in the protein-coding sequence of the *Cyp51* gene. In pathogens such as *Candida albicans* and *Zymoseptoria tritici* (formerly *Mycosphaerella graminicola*), numerous mutations have been identified (Cools et al., 2011, Warrilow et al., 2019). Over 140 mutations have been reported in *C. albicans* (reviewed by Morio et al., (2010), most of which clustered in three hot-spot regions encoding amino acids 105 to 165, 266 to 287 and 405 to 488 (Marichal et al., 1999). At least 23 of these substitutions have been shown to be involved in reduced sensitivity to azoles (Morio et al., 2010). Similarly in *Z. tritici*, 22 substitutions associated with reduced sensitivity to azoles had been identified in field isolates by 2011 (Cools and Fraaije, 2008, Leroux et al., 2007, Stammler et al., 2008, Zhan et al., 2006, Jørgensen and Heick, 2021).

Many substitutions in *Z. tritici* are at amino acid residues homologous to those altered in *C. albicans*. In fact, the most common mutations are located at equivalent residues in many fungi. For example, Y137F in *Z. trictici* is homologous to Y132 in *C. albicans,* Y134 in *Puccinia triticina* (brown or leaf rust of wheat) and Y136 in *Blumeria graminis* (powdery mildew of cereals and grasses). To facilitate comparison between different fungi which are targets of the same fungicide, a system of unified nomenclature has been introduced (Mair et al., 2016). For each main class of fungicide, a well-studied archetype species was chosen, then orthologous amino acids in all species were assigned numerical labels based on the positions of the equivalent amino acids in the archetype protein. *Z. tritici* is the reference archetype for CYP51. Note that for β-tubulin, none of the fungal species discussed in Section 2 have insertions or deletions in amino acid sequences with respect to the archetype species, *Aspergillus nidulans*.

In contrast to *C. albicans*, *Z. tritici* and many other well-studied fungi, very few mutations have evolved in powdery mildews. The first mutation to be identified in powdery mildews was Y136F (Délye et al., 1998, Délye et al., 1997a, Délye et al., 1997b), corresponding to Y137F in the unified nomenclature (Mair et al., 2016), hereafter written [Y137F]. This mutation was first identified in grapevine mildew, *E. necator*, and was subsequently found in *B. graminis* (Wyand & Brown, 2005). This is the most common mutation for resistance to azoles and it is found in many fungi including the ascomycetes *C. albicans*, *Z. tritici* and powdery mildews and the basidiomycete rust fungi (Mair et al., 2016). [Y137] is conserved in all known wild-type CYP51 proteins in fungi and animals. It is located in the CR2 domain which is highly conserved and thought to be involved in substrate recognition (Fig. 2). This domain corresponds to Substrate Recognition Site 1 (SRS1) in human CYP51 (Podust et al., 2001, Leroux and Walker, 2011). In *E. necator*, Y136F is the only mutation found to be associated with azole resistance to date but in cereal powdery mildews other mutations have been discovered.

A larger study of barley powdery mildew expanded on the work of Délye et al. (1998), using more isolates with a wider range of resistance levels and identified a second mutation K147Q [K148] (Wyand and Brown, 2005). In plant pathogens, this mutation has only been detected in barley powdery mildew *Blumeria graminis* f.sp. *hordei* (*Bgh*) but has also been identified in *C. albicans* where it was associated with resistance to fluconazole in particular but had a limited effect on responses to itraconazole and voriconazole (Flowers et al., 2015).

Another four mutations were subsequently identified in Australia in *Bgh* isolates sampled between 2009 and 2013 (Tucker et al., 2020). These were K171E [K172], M301I [M304], R327G [R330] and S509T [S524T]. The latter three mutations are in regions conserved between *Bgh* and *Z. tritici*. The S509T mutation is of particular significance because it has spread very quickly in the Australian *Bgh* population and seems to confer the most significant increase in resistance to DMIs in Australia (Tucker et al., 2020), in particular to the triazole tebuconazole. This mutation is common in other fungi such as *Z. tritici* and is associated with decreased sensitivity to all commonly used azoles (Cools et al., 2011).

In *B. graminis*, as in *C. albicans and Z. tritici*, mutations in CYP51 have occurred sequentially, a process which was correlated with increasing resistance to DMI (Wyand and Brown, 2005, Tucker et al., 2020, Warrilow et al., 2019, Cools et al., 2011, Leroux et al., 2007). Resistance to DMI developed quickly but gradually in powdery mildews. Low levels of resistance first appeared in the late 1970s (Walmsley-Woodward et al., 1979). Subsequently, moderate resistance was recorded in 1982, followed by high resistance in 1984 and very high resistance in 1986 (Blatter et al., 1998, Brown et al., 1991a, Brown and Wolfe, 1991). In Australia in 2009, isolates showing moderate to high levels of resistance to tebuconazole were collected, all of which had the Y136F mutation. Higher resistance was associated with the combination of Y136F and a second mutation, S509T. The latter was first identified in *Bgh* in 2010 in Australia and by 2011 it had become universal there. Isolates carrying the doubly mutant genotype F136+T509 displayed increased resistance to all the formulations of DMIs tested including tebuconazole, suggesting that not only had resistance increased but the number of formulations affected had increased too (Tucker et al., 2015, Tucker et al., 2020). The same phenomenon has been observed in *Z. tritici* where isolates with the homologous genotype F137+T524 were identified in the early 2000 and quickly became universal. These strains showed reduced sensitivity to all azoles tested (Cools et al., 2011). The F137 genotype subsequently disappeared from the European population of *Z. tritici* (Stammler et al., 2008). The S524T mutation, on the other hand, has remained prominent in *Z. tritici*, in which this mutation appears to affect fungicide resistance most strongly when it is combined with another mutation such as L50S, D134G, V136A or Y461S (Cools and Fraaije, 2013). In a yeast heterologous expression system, the Y137F mutation alone (Y136F in *B. graminis*) caused a small decrease in DMI sensitivity but the presence of both Y137F and S524T (S509T in *B. graminis*) produced a much larger reduction in sensitivity to both tebuconazole and propoconazole (Tucker et al., 2020). Similarly, in *C. albicans* double substitutions have also been shown to greatly affect tolerance to tebuconazole compared to single mutations (Warrilow et al., 2019).

The CYP51 protein has been widely studied and crystal structures of the CYP51 in bacteria and fungi are now available (Podust et al., 2001, Hargrove et al., 2017, Hargrove et al., 2015, Keniya et al., 2018). All CYP51 proteins possess a typical P450 fold with a conserved core formed by α-helices and β-sheets. Helices E, I, J, K and L are located around the haem cavity (Figs 2 and 3). The active site which comprises the haem group is deeply embedded in the interior of the protein. In bacteria, Podust et al. (2001) have shown that azoles bind directly to the active site of the CYP51 with the azole ring of the fungicide perpendicular to the porphyrin group, such that the azole's nitrogen atom is coordinated to the haem iron. In bacteria, it is thought that the substrate

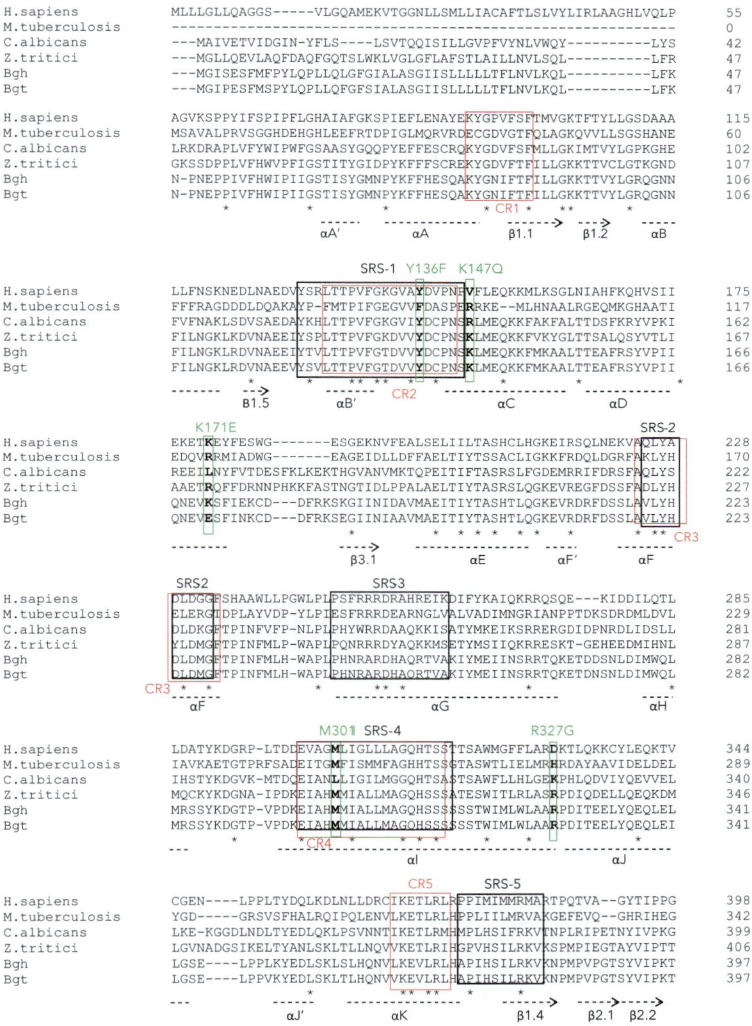

Figure 2 Amino acid sequence alignment of the *Cyp51* gene family. Sequences were downloaded as Fasta files from UniprotKB at https://www.uniprot.org. The alignment was carried out with ClustalOmega algorithm using the following Uniprot entries: *Homo sapiens* entry Q1658, *Mycobacterium tuberculosis* strain ATCC25618/H37Rv entry P9WPP9, *Candida albicans strain SC5314/ATCC MYA-2876 entry* P10613, *Zymoseptoria tritici* strain ST99CH_3D1 entry AOA1B1H4M, *Blumeria graminis* f.sp. *hordei (Bgh)* isolate DH14 entry N1J881 and *B. graminis* f.sp. *tritici* (*Bgt*) isolate JIW2 entry Q7Z7V5. Residues conserved across the *Cyp51* family are marked with an asterix *. Gotoh's predicted substrate recognition sites (SRSs) as are boxed in black (Gotoh, 1983). Conserved regions (CR) as identified by Délye et al. (1998) are boxed in red. Loci of *Bgh* mutations associated with resistance to azoles are boxed in green and the mutated amino acid highlighted in bold. α indicates α-helices and β indicates β-sheets as illustrated in Figure 2 of Leroux and Walker (2011).

```
H.sapiens          HQVCVSPTVNQRLKDSWVERLDFNPDRYLQDN-------------------------PAS   433
M.tuberculosis     DLVAASPAISNRIPEDFPDPHDFVPARYEQPRQEDL------------------------   378
C.albicans         HYVLVSPGYAHTSERYFDNPEDFDPTRWDTAAAKAN-----SVSFNSSDEVDYGFGKVSK   454
Z.tritici          HTLLAAPGTTSRMDEHFPDCLHWEPHRWDESPSEKYKHLSPTTALGSIAEEKEDYGLVSK   466
Bgh                HSLLAAPGWTSRDASYFPNPLKWDPHRWDTGSGGVIGTDME------DEKFDYGYGLIST   451
Bgt                HSLLAAPGWTSRDASYFPNPLKWDPHRWDTGSGGVIGTDME------DEKFDYGYGLIST   451
                         *                   * *
                   ---->  ---
                   β1.3   αK'

                                                                  SRS-6 S509T
H.sapiens          GEKFAYVPFGAGRHRCIGENFAYVQIKTIWSTMLRLYEFDLIDGYF--PTVNYTTMIHTP   491
M.tuberculosis     LNRWTWIPFGAGRHRCVGAAFAIMQIKAIFSVLLREYEFEMAQPP-ESYRNDHSKMVVQL   437
C.albicans         GVSSPYLPFGGGRHRCIGEQFAYVQLGTILTTFVYNLRWT-ID-GYKVPDPDYSSMVVLP   512
Z.tritici          GAASPYLPFGAGRHRCIGEQFAYVQLQTITATMVRDFKFYNVDGSDKVVGTDYSSLFSRP   526
Bgh                GAASPYLPFGAGRHRCIGEQFATVQLVTIMATMVRSFKFHNLDGRNSVAETDYSSMFSRP   511
Bgt                GAASPYLPFGAGRHRCIGEQFATVQLVTIMATMVRSFKFHNLDGRNSVAETDYSSMFSRP   511
                  CR6 *** ***** *  **  *   *
                                ---------------------   ---->      -->
                                         αL             β3.3       β4.1

H.sapiens          ENPV-IRYKRRSK------          503
M.tuberculosis     AQPACVRYRRRTGV-----          451
C.albicans         TEPAEIIWEKRETCMF---          528
Z.tritici          LSPAVVKWERREEKEEKN-          542
Bgh                MAPATIAWEKRDKKDKTEC          530
Bgt                MAPATIAWEKRDKKDKTEC          530
                     *        *
                   -->  ----->
                   β4.2 β3.2
```

Figure 2 (*Continued*)

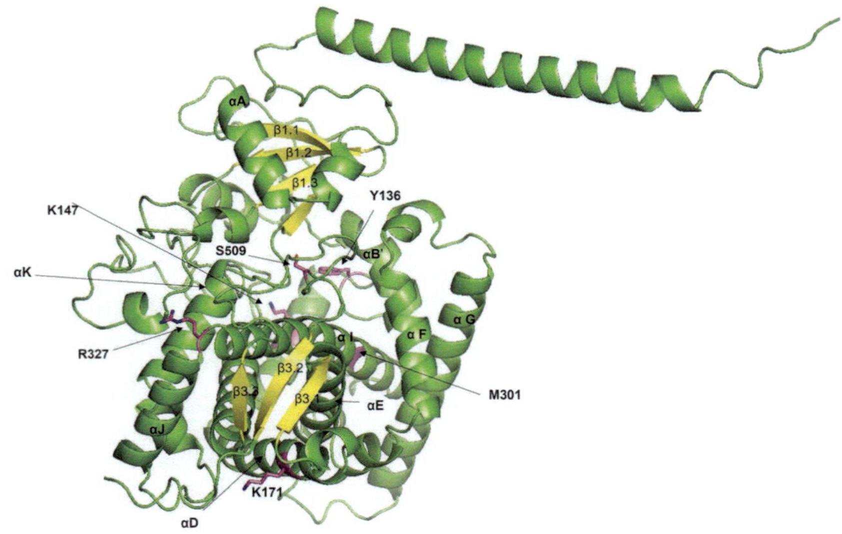

Figure 3 Alphafold prediction of the structure of the azole-sensitive form of the *Blumeria graminis* f.sp. *hordei* CYP51 protein. β-sheets are in yellow. α-helices are in green. Amino acid residues which are sites of known mutations associated with azole resistance (Section 3) are in purple. The Alphafold model was created by Oliver Powell.

enters the binding cavity through Entrance Channel 1 which runs parallel to the haem plane. Further characterisation of crystal structures in human have shown that eukaroytic and bacterial CYP51 proteins have different access channels (reviewed by Becher and Wirsel, 2012). Furthermore, some fungi such as C.

albicans have an additional external loop between residues 428 and 459. In *C. albicans*, substitutions associated with azole resistance have been mapped to the catalytic site (Y132F/H [Y137]), and the region below the haem (K143R [K148]) (Flowers et al., 2012).

The substitution of tyrosine by phenylalanine as Y136F [Y137F] in *B. graminis* results in a small but significant change. Tyrosine and phenylalanine have similar structures except that tyrosine has a hydroxyl group which is missing in phenylalanine (Table 1). The loss of the hydroxyl group is thought to increase the hydrophobicity at the border of the catalytic pocket without changing its configuration (Délye et al., 1997b). Tucker et al. (2020) showed that in *Bgh* the Y136F mutation resulted in an increased volume of the binding cavity and a reduction of the diameter of the access channel. It is thought that greater binding cavity volume may perturb the orientation of the fungicide molecule. The reduction of the diameter of the access channel is thought to affect the ability of larger fungicide molecules to access the haem cavity. In *Z. tritici*, the Y137F substitution was also shown to result in an overall increase in the size of the binding pocket. The F137 residue appeared to be in an obstructive position which may prevent the binding of triadimenol but not other azoles. This might explain why the frequency of F137 in *Z. tritici* has fallen as tebuconazole has largely been replaced by more modern triazoles for use on arable crops. F137 may also have an effect on Y459, causing it to move away from the position where it can bind to azole molecules. As the frequency of F137 in *Z. tritici* has declined in Europe, mutations which included a deletion around position Y459-Y461 have since become dominant in this fungus (Mullins et al., 2011).

K147Q [K148] in *Bgh* results in the substitution of glutamine for lysine. Both residues are similar (Table 1) but lysine is more polar. Similarly in *C. albicans*, the homologous mutation K143R results in the substitution of glutamine for arginine. Both arginine and lysine are thought to increase the net positive charge of the β loop where they are located. In *C. albicans*, K143R has been mapped to the other side of the haem group from the docking site of fluconazole (Flowers et al., 2015). It is thought a net positive charge may affect the orientation of the substrate in the binding cavity (Wyand and Brown, 2005).

Similarly, the change from serine to threonine at amino acid 509 [S524T] in *Bgh* is small because threonine only differs from serine by the addition of a methyl group (Table 1). S509T resulted in the greatest increase in cavity volume recorded by Tucker et al. (2020). More strikingly, the combination of F136 [F137] and T509 [T524] increased the volume of the cavity even further. In *Z. tritici*, the [S524T] mutation is thought to cause the loss of two ß-sheets in the vicinity of the binding cavity, leading to changes in the conformation of the cavity, making it more open and thereby reducing its affinity for the azole (Cools et al., 2011).

3.3 Expression of Cyp51

A second mechanism of resistance to azole involves increased expression of the *Cyp51* gene. In vitro, over-expression of *Cyp51* in *Saccharomyces cerevisiae* conferred resistance to fluconazole (Kontoyiannis et al., 1999). In vivo, increased expression of *Cyp51* has been shown to be linked to azole resistance in several fungi. In *Penicillum digitatum*, transcription was enhanced by the insertion of a 126-bp sequence tandemly repeated five times in the promoter of the *Cyp51* gene. Strains with the five repeats were resistant to a range of azoles whereas strains with only one copy of the enhancer sequence were sensitive. In the resistant strains, constitutive expression of *Cyp51* was about 100-fold higher than in azole-sensitive strains (Hamamoto et al., 2000). In *Venturia inaequalis*, a 553-bp insertion upstream of the *Cyp51A* gene was identified in myclobutanil-resistant isolates with high levels of *Cyp51* gene expression (Schnabel and Jones, 2001). Similarly, in the cherry leaf spot pathogen *Blumeriella jaappi*, the insertion of a retrotransposon sequence into the promoter was identified in azole-resistant isolates. This insertion was associated with a 5 to 12-fold increase in expression of the *Cyp51* gene (Ma et al., 2006). In *Z. tritici* a 120-bp insertion in the promoter of the *Cyp51* gene was also associated with over-expression of CYP51 and increased resistance to azoles (Cools et al., 2012).

A microarray experiment conducted by Dunkel et al. (2008) in *C. albicans* identified a transcription factor called *Upc2* as a regulator of *Cyp51*. This study showed that a gain of function mutation in *Upc2* resulted in over-expression of *Cyp51* and other genes in the ergosterol biosynthesis pathway and in higher resistance to azoles and other fungicides involved in the ergosterol pathway such as terbinafine.

In powdery mildews, enhanced expression of *Cyp51* has been associated with increased azole resistance. In *E. necator*, Frenkel et al. (2015) compared isolates from the US and Chile and found that those with the Y136F mutation had higher expression of *Cyp51*. A second mutation was identified in the *Cyp51* DNA sequence at base 1119 (*a1119c*). This synonymous mutation, which did not change the protein sequence, was associated with higher expression of the gene and increased resistance to azoles. This may have been the result of genetic hitch-hiking, with the (*a1119c)* mutation occurring by chance in a genotype carrying another mutation, as yet uncharacterised, which increased *Cyp51* expression, possibly in the promoter. Azole resistance in the *c1119* isolates was comparable to that of isolates with the F136 genotype.

3.4 Copy number variation

A second study by Jones et al. (2014) investigated isolates from seven vineyards in California and found that *Cyp51* copy number varied greatly. Isolates with high *Cyp51* copy number were significantly more likely to be present in azole-treated vineyards and all isolates with high copy number had the Y136F mutation suggesting that higher copy number only became advantageous once the resistance allele has been acquired. A further study by Rallos and Baudoin (2016) confirmed the existence of the silent mutation (*a1119c)* in American vineyards and its association with over-expression of *Cyp51* and azole resistance. Interestingly, this study found isolates with two alleles with either the Y or F amino-acids at residue 136, demonstrating gene duplication or increased copy number.

Duplication of the *Cyp51B* gene was also identified in a study of *B. graminis* f.sp. *tritici* (*Bgt*; powdery mildew of wheat) by Arnold (2018). Among US and UK isolates, those with both F and Y at residue 136 [137] were identified. In the UK population, these heteroallelic isolates were also heteroallelic at position 509 [524] with F136 always linked to T509 and Y136 to S509. In US isolates, the T509 mutation did not occur so heteroallelic isolates were either F136/S509 or Y136/S509. Heteroallelic isolates from the US had greater resistance to tebuconazole and prothioconazole on average than homoallelic isolates with Y136 only and similar resistance as homoallelic F136 isolates. In the UK population, heteroallelic Y136+S509 / F136+T509 isolates were significantly more resistant to both azoles than homoallelic F136 isolates, which in turn were more resistant than the Y136 isolates. *Cyp51* expression was similar in all heteroallelic isolates, whether from the US or the UK, to that of homoallelic F136 isolates. Copy number and gene expression was higher in UK isolates than in the US isolates, which was correlated with higher resistance to both triazoles. These findings indicate that in *Bgt*, increased resistance to azoles is associated with the presence of the Y136F mutation, increased *Cyp51* gene expression and higher copy number. Stronger resistance is associated with the combination of the Y136F and S509T mutations, especially in heteroallelic isolates with both Y136+S509 and F136+T509 copies of CYP51. Remarkably, no other coding sequence mutation was detected in these isolates. The small number of ORF mutations in *Cyp51* in *B. graminis* is in striking contrast to *Z. tritici* and *C. albicans*.

Copy number variation mediating resistance to azoles is not unique to powdery mildews. In *Candida glabrata*, increased expression of *Cyp51* was associated with increased resistance to azoles. In this instance, over-expression resulted from higher copy number in azole-resistant isolates in which the entire chromosome containing the *Cyp51* gene was duplicated (Marichal et al., 1997). Similarly, aneuploidy involving the formation of an isochromosome composed of the two left arms of chromosome 5, including the *Cyp51* locus, was associated with azole resistance in *C. albicans* (Selmecki et al., 2006).

3.5 Paralogue divergence

In Pezizomycotina, a subphylum of Ascomycota which comprises many plant and most human pathogens, Becher et al. (2011) found great variation not only in the copy number but also in the subtypes of *Cyp51* genes. Phylogenic analysis revealed three clades of CYP51 proteins, A, B and C encoded by paralogues genes *Cyp51A, B* and *C*. Clades A and B are the most common (Becher et al., 2011). Indeed, both CYP51A and CYP51B have been found in fungi such as *Aspergillus fumigatus* (Mellado et al., 2001), *A. nidulans* (Ferreira et al., 2005) and *Magnaporthe oryzae* (Yan et al., 2011). Other fungi such as *Aspergillus oryzae* (Ferreira et al., 2005) and *Fusarium* spp. have three paralogues of the CYP51 enzyme (Deng et al., 2007), CYP51A, B and C. There is striking variation within *Aspergillus*, in which some species such as *A. fumigatus* and *A. nidulans* have only one A- or B-type of CYP51 protein while other species have undergone recent duplication of one protein type. In *A. oryzae* and *A. terreus*, the *Cyp51A* gene has been duplicated and in *A. terreus* CYP51B has been duplicated. It is not known why some fungi have multiple paralogues and others do not, but knocking out one of the paralogues in *A. fumigatus* and *F. graminearum* resulted in increased sensitivity to azoles suggesting that additional paralogues may play a role in reducing sensitivity (Mellado et al., 2005, Fan et al., 2013). Furthermore, in *Rhyncosporium commune* (formerly *R. secalis*), *Cyp51A* had been lost from multiple lineages but re-emerged in the *R. commune* population when azoles where introduced in UK fields; this re-emergence was associated with increased resistance to azoles (Hawkins et al., 2014). In *A. flavus*, *A. oryzae* and *A. terreus*, phylogenetic studies have shown that the *Cyp51* gene (*Cyp51A* in the first two pathogens and *Cyp51B* in the latter) have recently been duplicated. Cereal powdery mildews have the *Cyp51B* paralogous gene (Mair et al., 2016).

3.6 Over-expression of efflux pumps

Finally, the third mechanism of resistance to azoles involves the over-expression of plasma membrane efflux pumps. Fungi can become resistant to multiple unrelated antifungal compounds with different modes of action by increasing the rate at which they transport them out of their cells. This phenomenon is called multi-drug resistance or MDR. It became an important clinical problem in the 1990s when cancer tumors suddenly became resistant to chemotherapeutic drugs (Gottesman and Pastan, 1993). There are two families of efflux pumps, the major facilitator super-family (MFS) and the ATB-binding cassette (ABC) pumps. MFS pumps have a limited range of substrates whereas ABC pumps have broader specificity. ABC transporters are found in all organisms. In prokaryotes they are involved in both the uptake and the efflux of compounds whereas in eukaryotes their function seems limited to efflux (Driessen et al., 2000). ABC transporters have been identified in many organisms and linked to resistance to drugs such as azoles. In clinical *Candida* isolates, over-expression of both ABC and MFS efflux pumps is the major mechanism for high-level resistance to azoles. Efflux pumps genes *Cdr1*, *Cdr2* (which encode ABC transporters) and *Md1* (encoding MFS transporters) are involved with azole resistance in *Candida* species (White, 1997). In vitro, some azole-resistant *C. albicans* strains expressed multiple pumps (Rogers and Barker, 2003). In *A. nidulans*, two genes, *atrA* and *atrB*, encoding ABC transporters were identified. Transcription of both was enhanced within a few minutes of treatment by several drugs including azole fungicides. Heterologous expression of *atrB* in an ABC-transporter-deficient yeast strain restored multi-drug resistance (Del Sorbo et al., 1997). A further study by deletion and over-expression mutants showed that *atrB* functions as a multi-drug transporter with affinity to substrates belonging to all major classes of fungicides including azoles (Andrade et al., (2000).

The role of ABC transporters in fungicide resistance has been documented in plant pathogens. In *Botrytis cinerea*, the *BcatrD* gene was cloned and its function characterised using gene-replacement and over-expression mutants. Replacement mutants where *BcatrD* had been replaced by the selectable marker hygromycin exhibited increased sensitivity to oxpoconazole and accumulated high amounts of the fungicide. Furthermore, mutants with the highest expression levels of *BcatrD* were the most resistant to oxpoconazole (Hayashi et al., 2002). Similarly in *Z. tritici*, ABC transporter encoding genes *MgAtr1* and *MgAtr2* have been cloned and characterised. Both were upregulated after treatment with cyproconazole (Zwiers and De Waard, 2000). Stergiopoulos et al. (2002) expanded on these experiments by cloning and characterising three more transporter genes, *Mgatr3*, *Mgatr4* and *Mgatr5* and showing that both *Mgatr4* and *Mgatr5* are upregulated by exposure to cyproconazole. Leroux and Walker (2011) tested the effect of known inhibitors of efflux pumps on resistance to azoles and other drugs, finding that changes in the sequence of *Cyp51* combined with over-expression of efflux pumps resulted in MDR in *Z. tritici*.

MFS pumps have not been as well-studied as ABC pumps but they have been shown to play a role in MDR in plant pathogens. In *Z. tritici* for example, *MgMfs1*, has been identified from an expressed sequenced tag (EST) screen. Heterologous over-expression of *Mgmfs1* in a hypersensitive strain of *S. cerevisiae* yeast resulted in increased resistance to several fungicidal compounds including azoles (Roohparvar et al., 2007).

In powdery mildews, the role of efflux transporters in mediating fungicide resistance has not been studied in depth. Frenkel et al. (2015) identified three ABC transporters and one MFS homologues in the genome of *E. necator* but found no evidence that they contributed to resistance to an azole fungicide, myclobutanil.

4 Amines: inhibitors of sterol reductase and sterol isomerase

Heterocyclic amine fungicides include morpholines, piperidines and spiroketalamines. They are especially effective against powdery mildews and rusts, and are used as both eradicants and protectants (Baloch and Mercer, 1987, Mercer, 1991). Amines are highly effective in inhibiting mycelial growth and often cause abnormal mycelium formation (Mercer, 1991). Like the azoles, morpholines target the ergosterol biosynthesis pathway but at two target enzymes: Δ^{14}-reductase (ERG24) (Lorenz and Parks, 1992) and Δ^{8} Δ^{7}-isomerase (ERG2) (Ashman et al., 1991) (Fig. 4). Inhibition of either Δ^{14}-reductase or Δ^{8} Δ^{7}-isomerase results in the depletion of ergosterol and the accumulation of toxic sterols such as ignosterol and fecosterol (Lorenz and Parks, 1991, Marcireau et al., 1990). Studies in *S. cerevisiae* and *Ustilago maydis* have shown that different morpholines act differently on each enzyme. For example, in yeast fenpropimorph mainly inhibits Δ^{14}-reductase whereas tridemorph mainly inhibits Δ^{8} Δ^{7}-isomerase (Baloch et al., 1984, Baloch and Mercer, 1987). In *C. albicans*, *Erg24* is involved in pathogenicity and resistance to fenpropimorph but not to tridemorph (Jia et al., 2002). Like *Cyp51,* some species have evolved to have two paralogues of *Erg24*. In Fusarium, *Egr24a* and *Erg24b* have been identified and disruption of the ERG24B protein results in hypersensitivity to fenpropidin and fenpropimorph (Liu et al., 2011).

In the UK, the main amine fungicides used have been the morpholines fenpropimorph, dodemorph and tridemorph and the piperidine fenpropidin. Dodemorph and tridemorph were introduced in the late 1960s for control of powdery mildew on diverse crops but the introduction of fenpropimorph and fenpropidin in the mid-1980s produced a step-change in the ability of farmers to control mildew even on the most susceptible varieties, such as Golden Promise spring barley. Within a few years, however, decreased sensitivity was identified in barley mildew in Scotland where these molecules had been used regularly. Indeed by 1988, Brown et al. (1991b) sampled isolates with reduced sensitivity to both fenpropimorph and fenpropidin in Scottish barley fields. Application of either of these fungicides selected isolates with reduced sensitivity to both of them, whereas application of tridemorph selected against resistance to fenpropidin (Brown and Evans, 1992). Resistance was incomplete and there was still moderately good field control of mildew by doses of fungicides lower than the recommended field application rate. To date this is still the case and fenpropimorph is still in use today to control cereal powdery mildew in the field.

In 2014, a population of *Bgt* with moderate resistance to fenpropimorph was identified in a glasshouse in the UK containing experimental populations of wheat (Arnold, 2018). Sequencing of the genes encoding the target enzymes of amine fungicides identified a mutation in the open reading frame of the *Erg24* gene in *Bgt*, V295L. A fenpropimorph-resistant *Bgh* isolate retained

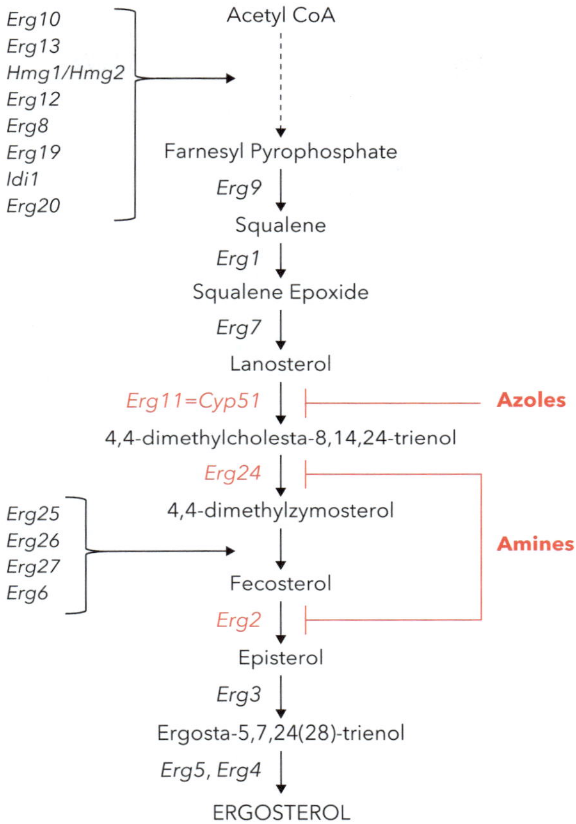

Figure 4 Summary of the ergosterol biosynthesis pathway in fungi. Fungicides which target the pathway are on the right-hand side and genes encoding key enzymes are on the left. Azoles and amines and genes for their respective target enzymes are highlighted in red.

from the study by Brown et al. (1992) had a D291N mutation in *Erg24*. No mutation was identified in *Erg2* in either *Bgh* or *Bgt*. To our knowledge, this is the only evidence of mutation in the ERG24 protein linked to fungal resistance. In *C. albicans*, over-expression of either *Erg24* or *Erg2* in the lab resulted in resistance to morpholines (Luna-Tapia et al., 2015) but this potential mechanism of resistance has not yet been studied in natural or clinical strains of the fungus.

5 Succinate dehydrogenase inhibitors: inhibitors of respiration at Complex II

Aerobic respiration happens all the time in animal, plants and fungi as a process by which organisms use oxygen to turn food into energy. This process involves the transfer of electrons through the mitochondrial membrane via a series of protein complexes. The respiration process starts with the oxidization of NADH into NAD by the first protein complex, which is at Complex I. This releases an electron which enters the respiration chain. Electrons then travel through protein Complexes II, III and IV until they reach the final acceptor, molecular oxygen (Fig. 5). There are three main classes of fungicides which target fungal respiration: the succinate dehydrogenase inhibitors (SDHIs) which target Complex II, and the Quinone-outside inhibitors (QoIs) and Quinone-inside inhibitors (QiI), which target different sites in Complex III.

SDHIs block the fungal respiration process by binding to the ubiquinone reduction site of Complex II of the respiratory chain also known as succinate dehydrogenase (SDH) or succinate:ubiquitine reductase (SQR). This group of fungicides have a carboxamide structure and are thus sometimes known as carboxamides. The first formulation, Carboxin, introduced in 1966 (Schmeling and Kulka, 1966) had a limited spectrum of action and it was mainly used to control smuts. Subsequently, new molecules were developed between 1971 and 1997 but the range remained limited until the release of boscalid in 2003 (Matheron and Porchas, 2004, Stammler et al., 2008, Stammler and Speakman, 2006). Boscalid was the first member of this class of fungicide to control a broad range of pathogens of fruit, vegetables and cereals, which was achieved by

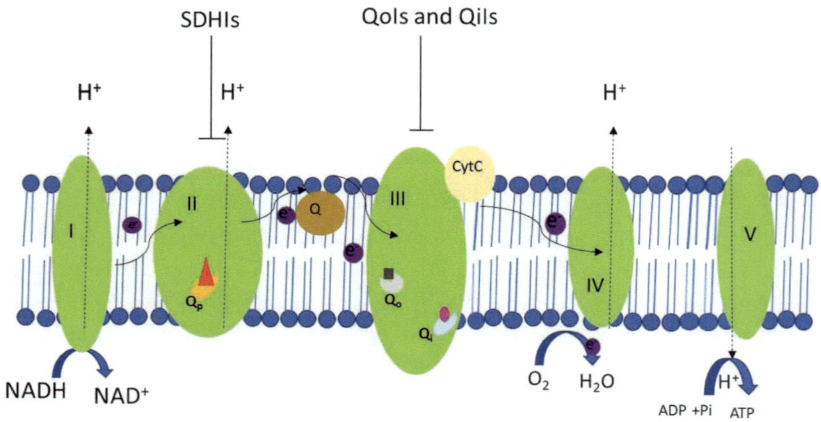

Figure 5 Schematic representation of the mitochondrial respiration chain. I, II, III and IV are the different protein complexes of the electron transfer chain. V is the ATP synthase complex. Q is the ubiquinone pool. CytC is the peripheral protein cytochrome c. Electron transfer route is shown with black arrows. The Qp binding site in Complex II is delineated by a red triangle representing Qp-inhibitor molecules (SDHIs). The Qo and Qi binding sites in Complex III are delineated by a dark gray square and a pink circle respectively representing Qo and Qi-inhibitor molecules respectively (QoI and QiIs).

significant modifications of the carboxamide molecule (Glättli et al., 2010). The discovery of boscalid triggered a renaissance of SDHIs and a new generation of SDHIs with broader specificity were released from 2010 onwards. SDHIs have been on the market for over 40 years and over the last 20 years, have become the class of fungicide with the most compounds released onto the market for crop disease control.

SDHIs are a structurally diverse class of fungicide with a broad spectrum of action. They all have in common an amide bond, which is strictly conserved among SDHIs and defines the orientation of the molecule. It is essential for the hydrogen-bond interactions in the ubiquinone-binding site of SDH. The SDH enzyme is a mitochondrial heterodimer formed by four subunits: SDHA (flavoprotein), SDHB (iron sulfur subunit), SDHC (cytochrome b560 subunit) and SDHD (cytochrome b small subunit). SDHC and Dare membrane proteins which anchor the SDHAB dimer to the membrane of the mitochondria. Amino acid sequences of the *SdhA* and *SdhB* subunits are highly conserved among fungal species in contrast to the *SdhC* and *SdhD* units which are diverse. The SDH enzyme transfers succinate-derived electrons to the ubiquinone pool of the respiratory chain and as such SDH is an essential component of the respiratory chain (Cecchini, 2003, Huang and Millar, 2013, Hägerhäll, 1997). All crop protection SDHIs target the hydrophobic ubiquinone-binding pocket (Qp) formed by residues of the subunits B, C and D. The fungicide binds to the Qp site, thus blocking access for the substrate and preventing further cycling of succinate oxidation. Crystal structures of SDH have been determined (Yankovskaya et al., 2003, Sun et al., 2005, Huang et al., 2006) and several homology models and docking predictions have been drawn for common plant pathogens such as *Z. tritici* and *B. cinerea* (Fraaije et al., 2012, Scalliet et al., 2012, Ruprecht et al., 2009, Amiri et al., 2020). Key amino acids which form the Qp pocket have been identified in *SdhB*, *SdhC* and *SdhD* (Horsefield et al., 2006, Hägerhäll, 1997, Sierotzki and Scalliet, 2013). The reference species for numbering of *Sdh* gene sequences in relation to fungicide resistance mutations is *Pyrenophora teres* (Mair et al., 2016). In *Z. tritici* for example, W224 [W234] in *SdhB*, Y130 [Y146] in *SdhD* and S83[S72] in *SdhC* all establish hydrogen bonds with the central amid bond and are directly involved with ubiquinone binding and catalysis while A87 [A87] in *SdhC* and H267 [H277] in *SdhB* are both buried deep in the binding pocket and are thought to interact with the core section of SDHI fungicides (Sierotzki and Scalliet, 2013).

More than 40 amino acid mutations encoded by *SdhB*, *SdhC* and *SdhD* have been linked to decreased sensitivity to SDHIs in a wide range of pathogenic fungi, notably *B. cinerea*, *Z. tritici*, *Sclerotinia sclerotiorum (S. sclerotiorum)* and *Alternaria alternata*. The most frequent mutations have been reported in *SdhB* at 277 [H277R/Y/N/L] (Sang and Lee, 2020), where H227R and H227Y are the most common changes. Less frequent are mutations at codon 230 [P230F/L/T] and 235 [N235 I/T] in *SdhB*. *SdhC* is the subunit where the most mutations have been identified across fungi as a whole with 14 mutations, while 6 mutations have been identified in *SdhD* (reviewed by Sang and Lee (2020) and Vielba-Fernández (2021)).

In powdery mildews, decreased sensitivity to boscalid has been observed sporadically in grapevine mildew *E. necator* since 2014. Cherrad et al. (2018) studied isolates from French vineyards resistant to boscalid in 2015 and identified two mutations in *SdhB* associated with resistance homologous to those reported in other fungi, H242R/Y [H277R/Y]. Resistance was specific to boscalid and did not affect the other SDHIs tested, fluxapyroxad and fluopyram. No other mutations were identified in *SdhC* although, in Germany (Graf, 2017) identified two mutations in *SdhC,* both at residue 169 in *E. necator* [G169 D/S]. In cucurbit mildew *P. xanthii*, decreased sensitivity to a range of SDHIs including boscalid has been reported in the US, Japan and Spain (Vielba-Fernández et al., 2021, Miyamoto et al., 2020, McGrath, 2008, Keinath et al., 2018). In Japan, boscalid was introduced to control pathogens such as gray mold *(B. cinerea*) and sclerotinia rot (*S. sclerotiorum*) on cucumber and was not specifically targeted at the control of powdery mildews. Nonetheless, resistance to boscalid occurred in *P. xanthii* in 2008 and 2009, probably because of the selection pressure from fungicide applications to control other diseases. Four new generation SDHIs have been registered specifically to control cucurbit powdery mildew in Japan from 2010 onwards and various levels of resistance have developed to all of them, from moderate to very high (Miyamoto et al., 2020). Gene sequencing showed that mutations in *SdhD* S121P [S118], H137R [H134R] and *SdhC* A84V [S73 P/V], G172D [G159] and G151R [G138] were associated with decreased sensitivity to SDHIs in Japan and Spain (Miyamoto et al., 2020, Vielba-Fernández et al., 2021). It is worth noting that, in these studies, less than 50% of the isolates tested showed decreased sensitivity to any of the SDHIs used in the experiments. Even so, polymorphism in the level of sensitivity would still be detrimental to crop management if the more resistant genotypes were selected by continued use of the fungicide.

To our knowledge, there has been no report of resistance to SDHIs in cereal powdery mildews to date. Cereal powdery mildews are not the primary target for this class of fungicide (AHDB, 2022). Nevertheless, given that SDHIs are widely used on cereals against *Z. tritici* of wheat, *Rhynchosporium commune* of barley and other diseases, *B. graminis* is being exposed to them. Future studies of responses of cereal powdery mildew to SDHIs will need to compare current isolates with those in historic collections sampled before the introduction of boscalid.

6 Quinone-outside inhibitors: inhibitors of respiration at Complex III

A second group of fungicides that target fungal respiration is QoIs, also known as the strobilurin group. QoIs were introduced to the market in 1996 and quickly became one of the most important classes of agricultural fungicide, accounting for 20% of the global fungicide market within the first 10 years of introduction (Bartlett et al., 2002). QoIs are active against all major classes of fungal pathogen although control levels vary. They have been of particular value against powdery and downy mildews and brown rusts (Balba, 2007). They revolutionised disease control of grapevine mildew because, for the first time, both powdery and downy mildews could be controlled using one fungicide even though the causal organisms are in different kingdoms of eukaryote. In cereals, QoIs are associated with increased yield and grain quality. This is called the 'greening effect', which seems to be caused by delayed senescence (Bartlett et al., 2002). It has also been shown that azoxystrobin (one of the first QoIs, released in 1996) inhibits the production of ascospores and oospores of grapevine powdery and downy mildews, potentially reducing the evolutionary potential of these pathogens in vineyards (Godwin and Cortesi, 1999, Vercesi et al., 2001).

The QoI group binds to the quinol oxidation (Qo) site of the cytochrome *bc1* protein complex, also known as Complex III of the respiration chain. Complex III is formed by 10 to 11 polypeptides in eukaryotes with the catalytic site of the protein being formed by 3 subunits: cytochrome *b*, cytochrome *c1* and the Rieske iron sulfur protein (ISP) (Becker et al., 1981). The catalytic mechanism is called the Q-cycle and it requires two distinct binding sites, *Qo*, the ubiquinol oxidation site, and Qi, the ubiquinone reduction site. Both sites are located within the cytochrome b subunit but on opposite sides of the mitochondrial membrane, *Qo* on the outside of the mitochondrion and *Qi* on the inside (Fig. 5). The QoI fungicides target the Qo site whereas some natural products such as Antimycin A and a new fungicide, fenpicoxamid, target the Qi site. All the Complex III subunits are encoded by the nucleus except cytochrome b, which is encoded by the mitochondrial genome. QoIs bind tightly to the Qo pocket preventing binding of ubiquinol. This inhibition blocks the transfer of electron between cytochrome *b* and cytochrome *c1*, leading to energy deficiency in the fungal cell (Fisher et al., 2020, Vielba-Fernández et al., 2021).

Resistance to QoIs emerged very quickly in many pathogens including powdery mildews. In Europe, QoIs were introduced in 1996 to control cereal powdery mildews and by 1998 resistant wheat isolates of *Bgt* were identified in Germany with high levels of resistance (resistance factor >50) (Chin et al., 2001). In Japan, a rapid decline of QoI efficacy against cucumber powdery mildew was reported within two years of QoI introduction to the market (Grasso et al., 2006). In the US, QoIs were introduced in 1998 to control grapevine downy and powdery mildews (*Plasmopora viticola* and *E. necator* respectively). In 1999, Wong and Wilcox (2002) found that two per cent of the *E. necator* population was resistant to low doses of azoxystrobin. Within a few years, a much higher proportion of the pathogen population was resistant (90% in Virginia) to much higher doses of azoxystrobin (resistance factor >100) resulting in control failure (Baudoin et al., 2008). Resistance persisted for at least 4 years after QoIs were no longer applied to control *E. necator* (Rallos et al., 2014). Evidence for similar strong selection for resistance has been established in other species of powdery mildews such as sugar beet and apple powdery mildews (reviewed by Vielba-Fernández et al. (2020).

In *E. necator*, the resistance described earlier was caused by the substitution of glycine by alanine at position 143 in the Cytochrome b protein (G143A) which corresponds to G143A in the reference archetype species for *Z. tritici*. Mutations in the *Cytb* gene have been found in two regions corresponding to amino acids 120-155 and 255-280 of the Qo site (Degli Esposti et al., 1993, Sierotzki et al., 2000, Fisher and Meunier, 2001). In the folded cytochrome b protein, these domains are close to each other and important for substrate binding (Brasseur et al., 1996, Geier et al., 1994). The G143A substitution is by far the most common to QoI-resistance, having been found in over 25 species of phytopathogenic ascomycetes and oomycetes including *Mycosphaerella* (including *Zymoseptoria*), *Alternaria, Blumeria* and other fungal species (reviewed by Fernández-Ortuño et al. (2010). Interestingly, A143 is the wild-type residue in basidiomycetes that naturally produce QoIs, such as *Mycena galopoda* (Kraiczy et al., 1996). G143A is associated with complete or qualitative resistance. It is thought to result in steric hindrance between the fungicide and the cytochrome b but does not hinder interaction of the protein with the ubiquinol substrate (Esser et al., 2004). (Steric hindrance at a given atom in a molecule is the congestion caused by the physical presence of surrounding ligands which slow down or prevent a reaction at the atom.) It is thought that G143A has spread very rapidly among pathogen populations because it confers a high level of resistance without a negative effect on enzyme activity (Rallos et al., 2014).

Not all pathogen species have evolved the G143A mutation, however. In all rust species, resistance has never developed. Grasso et al. (Grasso et al., 2006) identified an intron located immediately after the G143 codon which was predicted to alter splicing and thus to prevent this mutation from occurring. In some powdery mildew species such as *P. xanthii*, both the G143 and A143 alleles of *Cytb* co-exist in the same cell, a phenomenon called heteroplasmy (Vielba-Fernández et al., 2018). In sexual crosses, *B. graminis*, which is a hermaphrodite, inherits the maternal genotype at *Cytb* and in a cross between a QoI-sensitive isolate with G143 and a resistant one with A143, all progeny of any one chasmothecium, the sexual structure of powdery mildews, had either G143 or A143 (Robinson et al., 2002).

Other common substitutions in pathogenic fungi and oomycetes including downy mildews and rusts are F129L (phenylalanine to leucine) and G137R/S (glycine to arginine or serine) (Sierotzki et al., 2007, Gisi et al., 2002), which both confer less resistance than G143A. The side chain of F129 is oriented toward the hydrophobic cavity that facilitates substrate access to Qo. It is thought that F129 is involved with van der Walls and aromatic-aromatic interactions between the QoI fungicide and the bc1 complex proteins. The F129L substitution is thought to destabilise both types of interactions between the bound fungicide and the bc1 proteins (Esser et al., 2004). Like G143A, it is thought that F129L does not affect vitality or fitness of isolates carrying the mutation (Sierotzki et al., 2007, Esser et al., 2004). This is in contrast with mutations at G137, located at the N-terminal region of helix cd1. It is thought that the G137R/S mutation results in the distortion of the loop connecting helixes C and cd1 by arginine or serine, which are bulkier than glycine. In yeast, the G137R mutation caused a significant reduction in respiratory growth (Fisher and Meunier, 2001). In phytopathogenic fungi, mutations at G137 have been identified in the lab but have remained rare in the field (Sierotzki et al., 2007, Standish et al., 2016, Standish et al., 2018).

7 Conclusion

Despite improvements in agronomy and breeding (Brown 2021), fungicides continue to play an integral role in sustaining food production. The challenges of climate change and an ever-growing human population means that fungicides will still be needed for the foreseeable future. Yet resistance to modern single-target fungicides is now common, particularly in agriculture where many important pathogens have developed reduced sensitivity to many of the major classes of systemic fungicides. These compounds now need to be used carefully to maintain efficacy. The most common mechanism for resistance to all the major classes of fungicide used in agriculture today is mutation in the target gene. A single amino acid substitution can alter the conformation of the binding site and thus reduce affinity for the fungicide. These mutations are often conserved between pathogens. Some fungi have evolved more complex mechanisms of resistance to the most commonly used fungicides such as the azoles, where pathogens have evolved resistance through a combination of mechanisms such as mutation of the target gene, copy number variation and over-expression of the target gene, as well as changes in transport processes. There is a constant arms race between pathogen evolution and fungicide discovery, and new fungicides are needed urgently, both for agriculture and medicine (Fisher et al., 2018). The advent of new technologies, such as protein modelling, prediction of substrate and inhibitor docking, and real-time pathogen detection may help the chemical industry develop compounds with safer chemistries and new, more durable modes of action.

8 Acknowledgements

The writing of this chapter was supported by BBSRC through the Plant Health Institute Strategic Programme. We thank Oliver Powell (University of East Anglia and John Innes Centre) for generating the Alphafold model in Figure 3.

9 References

AHDB. 2022. Fungicide management in cereals. Available at: https://ahdb.org.uk/knowledge-library/fungicide-resistance-management-in-cereals (Accessed 29 March 2022).

Amiri, A., Zuniga, A. I. and Peres, N. A. 2020. Mutations in the membrane-anchored SDHC subunit affect fitness and sensitivity to succinate dehydrogenase inhibitors in *Botrytis cinerea* populations from multiple hosts. *Phytopathology* 110(2): 327–335.

Andrade, A. C., Del Sorbo, G., Van Nistelrooy, J. G. M. and Waard, M. A. 2000. The ABC transporter AtrB from *Aspergillus nidulans* mediates resistance to all major classes of fungicides and some natural toxic compounds. *Microbiology* 146(8): 1987–1997.

Arnold, C. 2018. Molecular evolution of fungicide resistance in *Blumeria graminis*. PhD Doctoral, University of East Anglia.

Ashman, W. H., Barbuch, R. J., Ulbright, C. E., Jarrett, H. W. and Bard, M. 1991. Cloning and disruption of the yeast C-8 sterol isomerase gene. *Lipids* 26(8): 628–632.

Balba, H. 2007. Review of strobilurin fungicide chemicals. *Journal of Environmental Science and Health: Part B, Pesticides, Food Contaminants, and Agricultural Wastes* 42(4): 441–451.

Baloch, R. I. and Mercer, E. I. 1987. Inhibition of sterol delta-8-delta-7-isomerase and delta-14-reductase by fenpropimorph, tridemorph and fenpropidin in cell-free enzyme-systems from *Saccharomyces cerevisiae*. *Phytochemistry* 26(3): 663–668.

Baloch, R. I., Mercer, E. I., Wiggins, T. E. and Baldwin, B. C. 1984. Inhibition of ergosterol biosynthesis in *Saccharomyces cerevisiae* and *Ustilago maydis* by tridemorph, fenpropimorph and fenpropidin. *Phytochemistry* 23(10): 2219–2226.

Bartlett, D. W., Clough, J. M., Godwin, J. R., Hall, A. A., Hamer, M. and Parr-Dobrzanski, B. 2002. The strobilurin fungicides. *Pest Management Science* 58(7): 649–662.

Baudoin, A., Olaya, G., Delmotte, F., Colcol, J. F. and Sierotzki, H. 2008. QoI resistance of *Plasmopara viticola* and *Erysiphe necator* in the mid-Atlantic united states. *Plant Health Progress* 9(1).

Becher, R., Weihmann, F., Deising, H. B. and Wirsel, S. G. R. 2011. Development of a novel multiplex DNA microarray for *Fusarium graminearum* and analysis of azole fungicide responses. *BMC Genomics* 12: 52.

Becher, R. and Wirsel, S. G. 2012. Fungal cytochrome P450 sterol 14α-demethylase (CYP51) and azole resistance in plant and human pathogens. *Applied Microbiology and Biotechnology* 95(4): 825–840.

Becker, W. F., Von Jagow, G., Anke, T. and Steglich, W. 1981. Oudemansin, strobilurin A, strobilurin B and myxothiazol: New inhibitors of the BC1 segment of the respiratory chain with an E-beta-methoxyacrylate system as common structural element. *FEBS Letters* 132(2): 329–333.

Bhosle, S. M., Marathe, N., Bheri, M. and Makandar, R. 2019. Detection of putative pathogenicity and virulence genes of *Erysiphe pisi* using genome-wide in-silico search and their suppression by er2 mediated resistance in garden pea. *Microbial Pathogenesis* 136: 103680.

Bindschedler, L. V., Panstruga, R. and Spanu, P. D. 2016. Mildew-omics: How global analyses aid the understanding of life and evolution of powdery mildews. *Frontiers in Plant Science* 7: 123.

Blatter, R. H. E., Brown, J. K. M. and Wolfe, M. S. 1998. Genetic control of the resistance of *Erysiphe graminis f. sp. hordei* to five triazole fungicides. *Plant Pathology* 47(5): 570–579.

Brasseur, G., Saribaş, A. S. and Daldal, F. 1996. A compilation of mutations located in the cytochrome b subunit of the bacterial and mitochondrial BC1 complex. *Biochimica et Biophysica Acta* 1275(1–2): 61–69.

Brown, J. K. M. 2021. Achievements in breeding cereals with durable disease resistance in Northwest Europe. In: *Achieving Durable Disease Resistance in Cereals* edited by R.P. Oliver, published by BDS Publishing, Cambridge, UK. 825–871. http://dx.doi.org/10.19103/AS.2021.0092.39.

Brown, J. K. M. and Evans, N. N. 1992. Selection on responses of barley powdery mildew to morpholine and piperidine fungicides. *Crop Protection* 11(5): 449–457.

Brown, J. K. M., Jessop, A. C. and Rezanoor, H. N. 1991a. Genetic uniformity in barley and its powdery mildew pathogen. *Proceedings of the Royal Society of London. Series B* 246(1315): 83–90.

Brown, J. K. M., Slater, S. E. and See, K. A. 1991b. Sensitivity of *Erysiphe graminis f sp hordei* to morpholine and piperidine fungicides. *Crop Protection* 10(6): 445–454.

Brown, J. K. M., Jessop, A. C., Thomas, S. and Rezanoor, H. N. 1992. Genetic-control of the response of *Erysiphe graminis f sp hordei* to ethirimol and triadimenol. *Plant Pathology* 41(2): 126–135.

Brown, J. K. M. and Wolfe, M. S. 1991. Levels of resistance of *Erysiphe graminis f sp hordei* to the systemic fungicide triadimenol. *Netherlands Journal of Plant Pathology* 97(4): 251–263.

Cecchini, G. 2003. Function and structure of complex II of the respiratory chain: Annual review. *Annual Review of Biochemistry* 72: 77–109.

Cherrad, S., Charnay, A., Hernandez, C., Steva, H., Belbahri, L. and Vacher, S. 2018. Emergence of Boscalid-resistant strains of *Erysiphe necator* in French vineyards. *Microbiological Research* 216: 79–84.

Chin, K. M., Chavaillaz, D., Kaesbohrer, M., Staub, T. and Felsenstein, F. G. 2001. Characterizing resistance risk of *Erysiphe graminis f.sp tritici* to strobilurins. *Crop Protection* 20(2): 87–96.

Cools, H. J., Bayon, C., Atkins, S., Lucas, J. A. and Fraaije, B. A. 2012. Overexpression of the sterol 14 alpha-demethylase gene (MgCYP51) in *Mycosphaerella graminicola* isolates confers a novel azole fungicide sensitivity phenotype. *Pest Management Science* 68(7): 1034–1040.

Cools, H. J. and Fraaije, B. A. 2008. Are azole fungicides losing ground against Septoria wheat disease? Resistance mechanisms in *Mycosphaerella graminicola*. *Pest Management Science* 64(7): 681–684.

Cools, H. J. and Fraaije, B. A. 2013. Update on mechanisms of azole resistance in *Mycosphaerella graminicola* and implications for future control. *Pest Management Science* 69(2): 150–155.

Cools, H. J., Mullins, J. G. L., Fraaije, B. A., Parker, J. E., Kelly, D. E., Lucas, J. A. and Kelly, S. L. 2011. Impact of recently emerged sterol 14 alpha-demethylase (CYP51) variants of *Mycosphaerella graminicola* on azole fungicide sensitivity. *Applied and Environmental Microbiology* 77(11): 3830–3837.

Daum, G., Lees, N. D., Bard, M. and Dickson, R. 1998. Biochemistry, cell biology and molecular biology of lipids of *Saccharomyces cerevisiae*. *Yeast* 14(16): 1471–1510.

Degli Esposti, M. D., Devries, S., Crimi, M., Ghelli, A., Patarnello, T. and Meyer, A. 1993. Mitochondrial cytochrome-b evolution and structure of the protein. *Biochimica et Biophysica Acta* 1143(3): 243–271.

Del Sorbo, G., Andrade, A. C., Vannistelrooy, J. G. M., Vankan, J. A. L., Balzi, E. and Dewaard, M. A. 1997. Multidrug resistance in *Aspergillus nidulans* involves novel ATP-binding cassette transporters. *Molecular and General Genetics* 254(4): 417–426.

Délye, C., Bousset, L. and Corio-Costet, M. F. 1998. PCR cloning and detection of point mutations in the eburicol 14 alpha-demethylase (CYP51) gene from *Erysiphe graminis f. sp. hordei*, a "recalcitrant" fungus. *Current Genetics* 34(5): 399–403.

Délye, C., Laigret, F. and Coriocostet, M. F. 1997a. Cloning and sequence analysis of the eburicol 14 alpha-demethylase gene of the obligate biotrophic grape powdery mildew fungus. *Gene* 195(1): 29–33.

Délye, C., Laigret, F. and Coriocostet, M. F. 1997b. A mutation in the 14 alpha-demethylase gene of *Uncinula necator* that correlates with resistance to a sterol biosynthesis inhibitor. *Applied and Environmental Microbiology* 63(8): 2966–2970.

Deng, J. X., Carbone, I. and Dean, R. A. 2007. The evolutionary history of cytochrome P450 genes in four filamentous Ascomycetes. *BMC Evolutionary Biology* 7: 30.

Driessen, A. J. M., Rosen, B. P. and Konings, W. N. 2000. Diversity of transport mechanisms: Common structural principles. *Trends in Biochemical Sciences* 25(8): 397–401.

Dunkel, N., Liu, T. T., Barker, K. S., Homayouni, R., Morschhäuser, J. and Rogers, P. D. 2008. A gain-of-function mutation in the transcription factor *Upc2*p causes upregulation of ergosterol biosynthesis genes and increased fluconazole resistance in a clinical *Candida albicans* isolate. *Eukaryotic Cell* 7(7): 1180–1190.

Ehrenhardt, H., Eichhorn, K. and Thate, R. 1973. *Zur Frage der Resistenzbildung von Botrytis cinerea Gegenuber Systemischen Fungiziden. Nachrichtenbl Deut Pflanzenschutzdienstes*.

Esser, L., Quinn, B., Li, Y. F., Zhang, M. Q., Elberry, M., Yu, L., Yu, C. A. and Xia, D. 2004. Crystallographic studies of quinol oxidation site inhibitors: A modified classification of inhibitors for the cytochrome bc(1) complex. *Journal of Molecular Biology* 341(1): 281–302.

Fan, J., Urban, M., Parker, J. E., Brewer, H. C., Kelly, S. L., Hammond-Kosack, K. E., Fraaije, B. A., Liu, X. and Cools, H. J. 2013. Characterization of the sterol 14α-demethylases of *Fusarium graminearum* identifies a novel genus-specific CYP51 function. *New Phytologist* 198(3): 821–835.

Fernández-Ortuño, D., Torés, J. A., Vicente, A. D. and Pérez-García, A. 2010. The QoI fungicides, the rise and fall of a successful class of agricultural fungicides. In: Carisse, O. (Ed.), *Fungicides*. London: IntechOpen.

Ferreira, M. E., Colombo, A. L., Paulsen, I., Ren, Q., Wortman, J., Huang, J., Goldman, M. H. and Goldman, G. H. 2005. The ergosterol biosynthesis pathway, transporter genes, and azole resistance in *Aspergillus fumigatus*. *Medical Mycology* 43(Suppl. 1): S313–S319.

Fisher, M. C., Hawkins, N. J., Sanglard, D. and Gurr, S. J. 2018. Worldwide emergence of resistance to antifungal drugs challenges human health and food security. *Science* 360(6390): 739–742. doi: 10.1126/science.aap7999. PMID: 29773744 (Accessed 18 May 2018).

Fisher, N. and Meunier, B. 2001. Effects of mutations in mitochondrial cytochrome b in yeast and man - Deficiency, compensation and disease. *European Journal of Biochemistry* 268(5): 1155–1162.

Fisher, N., Meunier, B. and Biagini, G. A. 2020. The cytochrome bc(1) complex as an antipathogenic target. *FEBS Letters* 594(18): 2935–2952.

Flowers, S. A., Barker, K. S., Berkow, E. L., Toner, G., Chadwick, S. G., Gygax, S. E., Morschhauser, J. and Rogers, P. D. 2012. Gain-of-function mutations in *Upc2* are a frequent cause of *Erg11* upregulation in azole-resistant clinical isolates of *Candida albicans*. *Eukaryotic Cell* 11(10): 1289–1299.

Flowers, S. A., Colon, B., Whaley, S. G., Schuler, M. A. and Rogers, P. D. 2015. Contribution of clinically derived mutations in *Erg11* to azole resistance in *Candida albicans*. *Antimicrobial Agents and Chemotherapy* 59(1): 450–460.

Fraaije, B. A., Bayon, C., Atkins, S., Cools, H. J., Lucas, J. A. and Fraaije, M. W. 2012. Risk assessment studies on succinate dehydrogenase inhibitors, the new weapons in the battle to control Septoria leaf blotch in wheat. *Molecular Plant Pathology* 13(3): 263–275.

Frenkel, O., Cadle-Davidson, L., Wilcox, W. F. and Milgroom, M. G. 2015. Mechanisms of resistance to an azole fungicide in the grapevine powdery mildew fungus, *Erysiphe necator*. *Phytopathology* 105(3): 370–377.

Fujimura, M., Kamakura, T., Inoue, H., Inoue, S. and Yamaguchi, I. 1992. Sensitivity of *Neurospora crassa* to benzimidazoles and N-phenylcarbamates: Effect of amino acid substitutions at position 198 in β-tubulin. *Pesticide Biochemistry and Physiology* 44(3): 165–173.

Fujimura, M., Kamakura, T., Inoue, H. and Yamaguchi, I. 1994. Amino-acid alterations in the beta-tubulin gene of *Neurospora crassa* that confer resistance to carbendazim and diethofencarb. *Current Genetics* 25(5): 418–422.

Geier, B. M., Haase, U. and Von Jagow, G. 1994. Inhibitor binding to the Qp-site of BC1 complex: Comparative studies of yeast mutants and natural inhibitor resistant fungi. *Biochemical Society Transactions* 22(1): 203–209.

Gisi, U., Sierotzki, H., Cook, A. and McCaffery, A. 2002. Mechanisms influencing the evolution of resistance to Qo inhibitor fungicides. *Pest Management Science* 58(9): 859–867.

Glättli, A., Grote, T. and Stammler, G. 2010. SDH-inhibitors: History, biological performance and molecular mode of action. In: Dehne, H. W., Deising, H. B., Gisi, U., Kuck, K. H., Russell, P. E. and Lyr, H. (Eds.), *Modern Fungicides and Antifungal Compounds VI*.

16th International Reinhardsbrunn Symposium. Friedrichroda, Germany: Deutsche Phytomedizinische Gesellschaft e.V. Selbstverlag.

Godwin, J. R. and Cortesi, P. Effects of azoxystrobinon sexual stages of *Uncinula necator*. In: *First International Powdery Mildew Conference*, Avignon, France, Vol. 43, 1999.

Gotoh, O. 1983. Prediction of melting profiles and local helix stability for sequenced DNA. *Advances in Biophysics* 16: 1–52.

Gottesman, M. M. and Pastan, I. 1993. Biochemistry of multidrug resistance mediated by the multidrug transporter. *Annual Review of Biochemistry* 62: 385–427.

Graf, S. B. 2017. Characterisation of metrafenone and succinate dehydrogenase inhibitor resistant isolates of the grapevine powdery mildew *Erysiphe necator*. PhD Technische Universitat Kaiserslautern.

Grasso, V., Palermo, S., Sierotzki, H., Garibaldi, A. and Gisi, U. 2006. Cytochrome b gene structure and consequences for resistance to Qo inhibitor fungicides in plant pathogens. *Pest Management Science* 62(6): 465–472.

Hägerhäll, C. 1997. Succinate: Quinone oxidoreductases: Variations on a conserved theme. *Biochimica et Biophysica Acta* 1320(2): 107–141.

Hamamoto, H., Hasegawa, K., Nakaune, R., Lee, Y. J., Makizumi, Y., Akutsu, K. and Hibi, T. 2000. Tandem repeat of a transcriptional enhancer upstream of the sterol 14 alpha-demethylase gene (CYP51) in *Penicillium digitatum*. *Applied and Environmental Microbiology* 66(8): 3421–3426.

Hargrove, T. Y., Garvey, E. P., Hoekstra, W. J., Yates, C. M., Wawrzak, Z., Rachakonda, G., Villalta, F. and Lepesheva, G. I. 2017. Crystal structure of the new investigational drug candidate vt-1598 in complex with *Aspergillus fumigatus* sterol 14 alpha-demethylase provides insights into its broad-spectrum antifungal activity. *Antimicrobial Agents and Chemotherapy* 61(7): e00570-17.

Hargrove, T. Y., Wawrzak, Z., Lamb, D. C., Guengerich, F. P. and Lepesheva, G. I. 2015. Structure-functional characterization of cytochrome P450 sterol 14 alpha-demethylase (CYP51B) from *Aspergillus fumigatus* and molecular basis for the development of antifungal drugs. *Journal of Biological Chemistry* 290(39): 23916–23934.

Hawkins, N. J., Cools, H. J., Sierotzki, H., Shaw, M. W., Knogge, W., Kelly, S. L., Kelly, D. E. and Fraaije, B. A. 2014. Paralog re-emergence: A novel, historically contingent mechanism in the evolution of antimicrobial resistance. *Molecular Biology and Evolution* 31(7): 1793–1802.

Hawkins, N. J. and Fraaije, B. A. 2016. Predicting resistance by mutagenesis: Lessons from 45 years of MBC resistance. *Frontiers in Microbiology* 7: 1814.

Hayashi, K., Schoonbeek, H.-J. and De Waard, M. A. 2002. Expression of the ABC transporter BcatrD from *Botrytis cinerea* reduces sensitivity to sterol demethylation inhibitor fungicides. *Pesticide Biochemistry and Physiology* 73(2): 110–121.

Hollomon, D. W. 2015. Fungicide resistance: Facing the challenge. *Plant Protection Science* 51: 170–176.

Horsefield, R., Yankovskaya, V., Sexton, G., Whittingham, W., Shiomi, K., Ōmura, S., Byrne, B., Cecchini, G. and Iwata, S. 2006. Structural and computational analysis of the quinone-binding site of complex II (succinate-ubiquinone oxidoreductase): A mechanism of electron transfer and proton conduction during ubiquinone reduction. *Journal of Biological Chemistry* 281(11): 7309–7316.

Huang, L. S., Sun, G., Cobessi, D., Wang, A. C., Shen, J. T., Tung, E. Y., Anderson, V. E. and Berry, E. A. 2006. 3-Nitropropionic acid is a suicide inhibitor of mitochondrial respiration that, upon oxidation by complex II, forms a covalent adduct with a catalytic base arginine in the active site of the enzyme. *Journal of Biological Chemistry* 281(9): 5965–5972.

Huang, S. and Millar, A. H. 2013. Succinate dehydrogenase: The complex roles of a simple enzyme. *Current Opinion in Plant Biology* 16(3): 344–349.

Jia, N., Arthington-Skaggs, B., Lee, W., Pierson, C. A., Lees, N. D., Eckstein, J., Barbuch, R. and Bard, M. 2002. *Candida albicans* sterol C-14 reductase, encoded by the *Erg24* gene, as a potential antifungal target site. *Antimicrobial Agents and Chemotherapy* 46(4): 947–957.

Jones, L., Riaz, S., Morales-Cruz, A., Amrine, K. C. H., Mcguire, B., Gubler, W. D., Walker, M. A. and Cantu, D. 2014. Adaptive genomic structural variation in the grape powdery mildew pathogen, *Erysiphe necator*. *BMC Genomics* 15(1): 1081.

Jørgensen, L. N. and Heick, T. M. 2021. Azole use in agriculture, horticulture, and wood preservation – Is it indispensable? *Frontiers in Cellular and Infection Microbiology* 11: 730297.

Keinath, A. P., Rennberger, G. and Kousik, C. S. 2018. First report of resistance to Boscalid in *Podosphaera xanthii*, cucurbit powdery mildew, in South Carolina. *Plant Health Progress* 19(3): 220–221.

Keniya, M. V., Sabherwal, M., Wilson, R. K., Woods, M. A., Sagatova, A. A., Tyndall, J. D. A. and Monk, B. C. 2018. Crystal structures of full-length lanosterol 14 alpha-demethylases of prominent fungal pathogens *Candida albicans* and *Candida glabrata* provide tools for antifungal discovery. *Antimicrobial Agents and Chemotherapy* 62(11): e01134-18.

Kontoyiannis, D. P., Sagar, N. and Hirschi, K. D. 1999. Overexpression of *Erg11*p by the regulatable GAL1 promoter confers fluconazole resistance in *Saccharomyces cerevisiae*. *Antimicrobial Agents and Chemotherapy* 43(11): 2798–2800.

Kraiczy, P., Haase, U., Gencic, S., Flindt, S., Anke, T., Brandt, U. and Vonjagow, G. 1996. The molecular basis for the natural resistance of the cytochrome bc(1) complex from strobilurin-producing basidiomycetes to center Q(P) inhibitors. *European Journal of Biochemistry* 235(1–2): 54–63.

Kuck, K. H., Scheinpflug, H. and Pontzen, R. 1995. DMI fungicides. In: *Modern Selective Fungicides*: *Properties, Applications, Mechanisms of Action*. Verlag, NY: Gustav Fisher.

Kunova, A., Pizzatti, C., Saracchi, M., Pasquali, M. and Cortesi, P. 2021. Grapevine powdery mildew: Fungicides for its management and advances in molecular detection of markers associated with resistance. *Microorganisms* 9(7): 1541.

Leroux, P., Albertini, C., Gautier, A., Gredt, M. and Walker, A. S. 2007. Mutations in the *Cyp51* gene correlated with changes in sensitivity to sterol 14 alpha-demethylation inhibitors in field isolates of *Mycosphaerella graminicola*. *Pest Management Science* 63(7): 688–698.

Leroux, P. and Walker, A. S. 2011. Multiple mechanisms account for resistance to sterol 14 alpha-demethylation inhibitors in field isolates of *Mycosphaerella graminicola*. *Pest Management Science* 67(1): 44–59.

Liu, X., Fu, J., Yun, Y., Yin, Y. and Ma, Z. 2011. A sterol C-14 reductase encoded by *FgErg24b* is responsible for the intrinsic resistance of *Fusarium graminearum* to amine fungicides. *Microbiology (Reading)* 157(6): 1665–1675.

Lorenz, R. T. and Parks, L. W. 1991. Physiological effects of fenpropimorph on wild-type *Saccharomyces cerevisiae* and fenpropimorph-resistant mutants. *Antimicrobial Agents and Chemotherapy* 35(8): 1532–1537.

Lorenz, R. T. and Parks, L. W. 1992. Cloning, sequencing, and disruption of the gene encoding sterol C-14 reductase in *Saccharomyces cerevisiae*. *DNA and Cell Biology* 11(9): 685–692.

Luna-Tapia, A., Peters, B. M., Eberle, K. E., Kerns, M. E., Foster, T. P., Marrero, L., Noverr, M. C., Fidel, P. L. and Palmer, G. E. 2015. ERG2 and ERG24 are required for normal vacuolar physiology as well as *Candida albicans* pathogenicity in a murine model of disseminated but not vaginal candidiasis. *Eukaryotic Cell* 14(10): 1006–1016.

Ma, Z. H., Proffer, T. J., Jacobs, J. L. and Sundin, G. W. 2006. Overexpression of the 14 alpha-demethylase target gene (CYP51) mediates fungicide resistance in *Blumeriella jaapii*. *Applied and Environmental Microbiology* 72(4): 2581–2585.

Mair, W., Lopez-Ruiz, F., Stammler, G., Clark, W., Burnett, F., Hollomon, D., Ishii, H., Thind, T. S., Brown, J. K. M., Fraaije, B., Cools, H., Shaw, M., Fillinger, S., Walker, A. S., Mellado, E., Schnabel, G., Mehl, A. and Oliver, R. P. 2016. Proposal for a unified nomenclature for target-site mutations associated with resistance to fungicides. *Pest Management Science* 72(8): 1449–1459.

Marcireau, C., Guilloton, M. and Karst, F. 1990. Invivo effects of fenpropimorph on the yeast *Saccharomyces cerevisiae* and determination of the molecular-basis of the antifungal property. *Antimicrobial Agents and Chemotherapy* 34(6): 989–993.

Marichal, P., Koymans, L., Willemsens, S., Bellens, D., Verhasselt, P., Luyten, W., Borgers, M., Ramaekers, F. C. S., Odds, F. C. and Van Den Bossche, H. 1999. Contribution of mutations in the cytochrome P450 14 alpha-demethylase (*Erg11p*, *Cyp51p*) to azole resistance in Candida albicans. *Microbiology* 145(10): 2701–2713.

Marichal, P., Van Den Bossche, H., Odds, F. C., Nobels, G., Warnock, D. W., Timmerman, V., Van Broeckhoven, C., Fay, S. and Moselarsen, P. 1997. Molecular-biological characterization of an azole-resistant *Candida glabrata* isolate. *Antimicrobial Agents and Chemotherapy* 41(10): 2229–2237.

Matheron, M. E. and Porchas, M. 2004. Activity of boscalid, fenhexamid, fluazinam, fludioxonil, and vinclozolin on growth of *Sclerotinia minor* and *S. sclerotiorum* and development of lettuce drop. *Plant Disease* 88(6): 665–668.

McGrath, M. 2008. Fungicide sensitivity in *Podosphaera xanthii* and efficacy for cucurbit powdery mildew in NY, USA, in 2003–2006. *Journal of Plant Pathology* 90: 90.

McGrath, M. T. 2001. Fungicide resistance in cucurbit powdery mildew: Experiences and challenges. *Plant Disease* 85(3): 236–245.

Mellado, E., Diaz-Guerra, T. M., Cuenca-Estrella, M. and Rodriguez-Tudela, J. L. 2001. Identification of two different 14-alpha sterol demethylase-related genes (*Cyp51A* and *Cyp51B*) in *Aspergillus fumigatus* and other aspergillus species. *Journal of Clinical Microbiology* 39(7): 2431–2438.

Mellado, E., Garcia-Effron, G., Buitrago, M. J., Alcazar-Fuoli, L., Cuenca-Estrella, M. and Rodriguez-Tudela, J. L. 2005. Targeted gene disruption of the 14-alpha sterol demethylase (*Cyp51A*) in *Aspergillus fumigatus* and its role in azole drug susceptibility. *Antimicrobial Agents and Chemotherapy* 49(6): 2536–2538.

Mercer, E. I. 1991. Morpholine antifungals and their mode of action. *Biochemical Society Transactions* 19(3): 788–793.

Miyamoto, T., Hayashi, K., Okada, R., Wari, D. and Ogawara, T. 2020. Resistance to succinate dehydrogenase inhibitors in field isolates of *Podosphaera xanthii* on cucumber: Monitoring, cross-resistance patterns and molecular characterization. *Pesticide Biochemistry and Physiology* 169: 104646.

Morio, F., Loge, C., Besse, B., Hennequin, C. and Le Pape, P. 2010. Screening for amino acid substitutions in the *Candida albicans* ERG11 protein of azole-susceptible and azole-resistant clinical isolates: New substitutions and a review of the literature. *Diagnostic Microbiology and Infectious Disease* 66(4): 373–384.

Morton, V. and Staub, T. 2008. A short history of fungicides. *APSnet Feature Articles*.

Mullins, J. G. L., Parker, J. E., Cools, H. J., Togawa, R. C., Lucas, J. A., Fraaije, B. A., Kelly, D. E. and Kelly, S. L. 2011. Molecular modelling of the emergence of azole resistance in *Mycosphaerella graminicola*. *PLoS ONE* 6(6): e20973.

Munkboel, C. H., Rasmussen, T. B., Elgaard, C., Olesen, M.-L. K., Kretschmann, A. C. and Styrishave, B. 2019. The classic azole antifungal drugs are highly potent endocrine disruptors in vitro inhibiting steroidogenic CYP enzymes at concentrations lower than therapeutic C_{max}. *Toxicology* 425: 152247.

Orbach, M. J., Porro, E. B. and Yanofsky, C. 1986. Cloning and characterization of the gene for beta-tubulin from a benomyl-resistant mutant of *Neurospora crassa* and its use as a dominant selectable marker. *Molecular and Cellular Biology* 6(7): 2452–2461.

Parker, J. E., Warrilow, A. G., Price, C. L., Mullins, J. G., Kelly, D. E. and Kelly, S. L. 2014. Resistance to antifungals that target CYP51. *Journal of Chemical Biology* 7(4): 143–161.

Pearson, R. and Taschenberg, E. 1980. Benomyl-resistant strains of *Uncinula necator* on grapes. *Plant Diseases (USA)*.

Pérez-García, A., Romero, D., Fernández-Ortuño, D., López-Ruiz, F., De Vicente, A. and Torés, J. A. 2009. The powdery mildew fungus *Podosphaera fusca* (synonym Podosphaera xanthii), a constant threat to cucurbits. *Molecular Plant Pathology* 10(2): 153–160.

Podust, L. M., Poulos, T. L. and Waterman, M. R. 2001. Crystal structure of cytochrome P450 14 alpha-sterol demethylase (CYP51) from *Mycobacterium tuberculosis* in complex with azole inhibitors. *Proceedings of the National Academy of Sciences of the United States of America* 98(6): 3068–3073.

Rallos, L. E. and Baudoin, A. B. 2016. Co-occurrence of two allelic variants of *Cyp51* in *Erysiphe necator* and their correlation with over-expression for dmi resistance. *PLoS ONE* 11(2): e0148025.

Rallos, L. E. E., Johnson, N. G., Schmale, D. G., 3rd, Prussin, A. J., 2nd and Baudoin, A. B. 2014. Fitness of *Erysiphe necator* with G143A-based resistance to quinone outside inhibitors. *Plant Disease* 98(11): 1494–1502.

Robinson, H. L., Ridout, C. J., Sierotzki, H., Gisi, U. and Brown, J. K. M. 2002. Isogamous, hermaphroditic inheritance of mitochondrion-encoded resistance to Qo inhibitor fungicides in *Blumeria graminis f. sp tritici*. *Fungal Genetics and Biology: FG and B* 36(2): 98–106.

Rogers, P. D. and Barker, K. S. 2003. Genome-wide expression profile analysis reveals coordinately regulated genes associated with stepwise acquisition of azole resistance in *Candida albicans* clinical isolates. *Antimicrobial Agents and Chemotherapy* 47(4): 1220–1227.

Roohparvar, R., De Waard, M. A., Kema, G. H. J. and Zwiers, L. H. 2007. MgMFS1, a major facilitator superfamily transporter from the fungal wheat pathogen *Mycosphaerella graminicola*, is a strong protectant against natural toxic compounds and fungicides. *Fungal Genetics and Biology: FG and B* 44(5): 378–388.

Ruprecht, J., Yankovskaya, V., Maklashina, E., Iwata, S. and Cecchini, G. 2009. Structure of *Escherichia coli* succinate:quinone oxidoreductase with an occupied and empty quinone-binding site. *Journal of Biological Chemistry* 284(43): 29836–29846.

Sang, H. and Lee, H. B. 2020. Molecular mechanisms of succinate dehydrogenase inhibitor resistance in phytopathogenic fungi. *Radiation Protection Dosimetry* 26: 1–7.

Scalliet, G., Bowler, J., Luksch, T., Kirchhofer-Allan, L., Steinhauer, D., Ward, K., Niklaus, M., Verras, A., Csukai, M., Daina, A. and Fonné-Pfister, R. 2012. Mutagenesis and functional studies with succinate dehydrogenase inhibitors in the wheat pathogen *Mycosphaerella graminicola*. *PLoS ONE* 7(4): e35429.

Schmeling, B. V. and Kulka, M. 1966. Systemic fungicidal activity of 1,4-oxathiin derivatives. *Science* 152(3722): 659–660.

Schnabel, G. and Jones, A. L. 2001. The 14alpha-demethylase (*Cyp51a1*) gene is overexpressed in *Venturia inaequalis* strains resistant to myclobutanil. *Phytopathology* 91(1): 102–110.

Schroeder, W. and Provvidenti, R. 1969. Resistance to benomyl in powdery mildew of cucurbits. *Plant Disease Reporter* 53: 271–275.

Selmecki, A., Forche, A. and Berman, J. 2006. Aneuploidy and isochromosome formation in drug-resistant *Candida albicans*. *Science* 313(5785): 367–370.

Sierotzki, H., Frey, R., Wullschleger, J., Palermo, S., Karlin, S., Godwin, J. and Gisi, U. 2007. Cytochrome b gene sequence and structure of *Pyrenophora teres* and *P. tritici-repentis* and implications for QoI resistance. *Pest Management Science* 63(3): 225–233.

Sierotzki, H., Parisi, S., Steinfeld, U., Tenzer, I., Poirey, S. and Gisi, U. 2000. Mode of resistance to respiration inhibitors at the cytochrome bc1 enzyme complex of *Mycosphaerella fijiensis* field isolates. *Pest Management Science* 56(10): 833–841.

Sierotzki, H. and Scalliet, G. 2013. A review of current knowledge of resistance aspects for the next-generation succinate dehydrogenase inhibitor fungicides. *Phytopathology* 103(9): 880–887.

Spanu, P. D., Abbott, J. C., Amselem, J., Burgis, T. A., Soanes, D. M., Stüber, K., Ver Loren Van Themaat, E., Brown, J. K., Butcher, S. A., Gurr, S. J., Lebrun, M. H., Ridout, C. J., Schulze-Lefert, P., Talbot, N. J., Ahmadinejad, N., Ametz, C., Barton, G. R., Benjdia, M., Bidzinski, P., Bindschedler, L. V., Both, M., Brewer, M. T., Cadle-Davidson, L., Cadle-Davidson, M. M., Collemare, J., Cramer, R., Frenkel, O., Godfrey, D., Harriman, J., Hoede, C., King, B. C., Klages, S., Kleemann, J., Knoll, D., Koti, P. S., Kreplak, J., López-Ruiz, F. J., Lu, X., Maekawa, T., Mahanil, S., Micali, C., Milgroom, M. G., Montana, G., Noir, S., O'Connell, R. J., Oberhaensli, S., Parlange, F., Pedersen, C., Quesneville, H., Reinhardt, R., Rott, M., Sacristán, S., Schmidt, S. M., Schön, M., Skamnioti, P., Sommer, H., Stephens, A., Takahara, H., Thordal-Christensen, H., Vigouroux, M., Wessling, R., Wicker, T. and Panstruga, R. 2010. Genome expansion and gene loss in powdery mildew fungi reveal tradeoffs in extreme parasitism. *Science* 330(6010): 1543–1546.

Stammler, G., Carstensen, M., Koch, A., Semar, M., Strobel, D. and Schlehuber, S. 2008. Frequency of different CYP51-haplotypes of *Mycosphaerella graminicola* and their impact on epoxiconazole-sensitivity and -field efficacy. *Crop Protection* 27(11): 1448–1456.

Stammler, G. and Speakman, J. 2006. Microtiter method to test the sensitivity of *Botrytis cinerea* to Boscalid. *Journal of Phytopathology* 154(7–8): 508–510.

Standish, J. R., Avenot, H. F., Brenneman, T. B. and Stevenson, K. L. 2016. Location of an intron in the cytochrome b gene indicates reduced risk of QoI fungicide resistance in *Fusicladium effusum*. *Plant Disease* 100(11): 2294–2298.

Standish, J. R., Brenneman, T. B. and Stevenson, K. L. 2018. Dynamics of fungicide sensitivity in *Venturia effusa* and fungicide efficacy under field conditions. *Plant Disease* 102(8): 1606–1611.

Stergiopoulos, I., Gielkens, M. M. C., Goodall, S. D., Venema, K. and De Waard, M. A. 2002. Molecular cloning and characterisation of three new ATP-binding cassette transporter genes from the wheat pathogen *Mycosphaerella graminicola*. *Gene* 289(1–2): 141–149.

Strobel, D., Bryson, R., Semar, M., Stammler, G., Kienle, M. and Smith, J. 2020. Mefentrifluconazole (Revysol®). The first Isopropanol-azole. In: Deising, H. B., Fraaije, B., Mehl, A., Oerke, E. C., Sierotzki, H. and Stammler, G. (Eds.), *Modern Fungicides and Antifungals Compounds IX, April 7–11 2019 Friedrichroda, Germany. DPG Spectrum Phytomedizin*, pp. 259–264.

Sun, F., Huo, X., Zhai, Y., Wang, A., Xu, J., Su, D., Bartlam, M. and Rao, Z. 2005. Crystal structure of mitochondrial respiratory membrane protein complex II. *Cell* 121(7): 1043–1057.

Tesh, S. A., Tesh, J. M., Fegert, I., Buesen, R., Schneider, S., Mentzel, T., Van Ravenzwaay, B. and Stinchcombe, S. 2019. Innovative selection approach for a new antifungal agent mefentrifluconazole (Revysol®) and the impact upon its toxicity profile. *Regulatory Toxicology and Pharmacology* 106: 152–168.

Thomas, J. H., Neff, N. F. and Botstein, D. 1985. Isolation and characterization of mutations in the beta-tubulin gene of *Saccharomyces cerevisiae*. *Genetics* 111(4): 715–734.

Trösken, E. R., Fischer, K., Völkel, W. and Lutz, W. K. 2006. Inhibition of human CYP19 by azoles used as antifungal agents and aromatase inhibitors, using a new LC-MS/MS method for the analysis of estradiol product formation. *Toxicology* 219(1–3): 33–40.

Tucker, M. A., Lopez-Ruiz, F., Cools, H. J., Mullins, J. G. L., Jayasena, K. and Oliver, R. P. 2020. Analysis of mutations in West Australian populations of *Blumeria graminis f. sp. hordei* CYP51 conferring resistance to DMI fungicides. *Pest Management Science* 76(4): 1265–1272.

Tucker, M. A., Lopez-Ruiz, F., Jayasena, K. and Oliver, R. P. 2015. Origin of fungicide-resistant barley powdery mildew in Western Australia: Lessons to be learned. In: Ishii, H. and Hollomon, D. W. (Eds.), *Fungicide Resistance in Plant Pathogens: Principles and a Guide to Practical Management*. Tokyo: Springer Japan.

Van Den Bossche, H. V., Lauwers, W., Willemsens, G., Marichal, P., Cornelissen, F. and Cools, W. 1984. Molecular-basis for the antimycotic and antibacterial activity of N-substituted imidazoles and triazoles - the inhibition of isoprenoid biosynthesis. *Pesticide Science* 15(2): 188–198.

Vargas, J. M. 1972. Benzimidazole-resistant strain of *Erysiphe graminis*. *Phytopathology* 62: 795–795.

Vargas, J. M. 1973. A benzimidazole resistant strain of *Erysiphe graminis*. *Phytopathology* 63(11): 1366–1368.

Vela-Corcía, D., Bellón-Gómez, D., López-Ruiz, F., Torés, J. A. and Pérez-García, A. 2014. The *Podosphaera fusca Tub2* gene, a molecular "Swiss Army knife" with multiple applications in powdery mildew research. *Fungal Biology* 118(2): 228–241.

Vela-Corcía, D., Romero, D., De Vicente, A. and Pérez-García, A. 2018. Analysis of β-tubulin-carbendazim interaction reveals that binding site for MBC fungicides does not include residues involved in fungicide resistance. *Scientific Reports* 8(1): 7161.

Vercesi, A., Vavassori, A., Faoro, F. and Bisiach, M. 2001. Effect of azoxystrobin on the oospores of *Plasmopara viticola*. In: Spencer-Phillips, P., Gisi, U. and Lebeda, A. (Eds.), *Advances in Downy Mildew Research*. Dordrecht: Kluwer Academic Publishers.

Vielba-Fernández, A., Bellón-Gómez, D., Torés, J. A., De Vicente, A., Pérez-García, A. and Fernández-Ortuño, D. 2018. Heteroplasmy for the cytochrome b gene in *Podosphaera xanthii* and its role in resistance to Qoi fungicides in Spain. *Plant Disease* 102(8): 1599–1605.

Vielba-Fernández, A., Polonio, Á., Ruiz-Jiménez, L., De Vicente, A., Pérez-García, A. and Fernández-Ortuño, D. 2020. Fungicide resistance in powdery mildew fungi. *Microorganisms* 8(9): 1431.

Vielba-Fernández, A., Polonio, Á., Ruiz-Jiménez, L., De Vicente, A., Pérez-García, A. and Fernández-Ortuño, D. 2021. Resistance to the SDHI fungicides Boscalid and fluopyram in *Podosphaera xanthii* populations from commercial cucurbit fields in Spain. *Journal of Fungi* 7(9): 733.

Walmsley-Woodward, D. J., Laws, F. A. and Whittington, W. J. 1979. Studies on the tolerance of *Erysiphe graminis f. sp hordei* to systemic fungicides. *Annals of Applied Biology* 92(2): 199–209.

Warrilow, A. G., Nishimoto, A. T., Parker, J. E., Price, C. L., Flowers, S. A., Kelly, D. E., Rogers, P. D. and Kelly, S. L. 2019. The evolution of azole resistance in *Candida albicans* sterol 14 alpha-demethylase (CYP51) through incremental amino acid substitutions. *Antimicrobial Agents and Chemotherapy* 63(5): e02586-18.

Warrilow, A. G., Parker, J. E., Kelly, D. E. and Kelly, S. L. 2013. Azole affinity of sterol 14α-demethylase (CYP51) enzymes from *Candida albicans* and *Homo sapiens*. *Antimicrobial Agents and Chemotherapy* 57(3): 1352–1360.

White, T. C. 1997. Increased mRNA levels of ERG16, CDR, and MDR1 correlate with increases in azole resistance in *Candida albicans* isolates from a patient infected with human immunodeficiency virus. *Antimicrobial Agents and Chemotherapy* 41(7): 1482–1487.

Wicker, T., Oberhaensli, S., Parlange, F., Buchmann, J. P., Shatalina, M., Roffler, S., Ben-David, R., Doležel, J., Šimková, H., Schulze-Lefert, P., Spanu, P. D., Bruggmann, R., Amselem, J., Quesneville, H., Ver Loren Van Themaat, E., Paape, T., Shimizu, K. K. and Keller, B. 2013. The wheat powdery mildew genome shows the unique evolution of an obligate biotroph. *Nature Genetics* 45(9): 1092–1096.

Wong, F. P. and Wilcox, W. F. 2002. Sensitivity to azoxystrobin among isolates of *Uncinula necator*: Baseline distribution and relationship to myclobutanil sensitivity. *Plant Disease* 86(4): 394–404.

Wyand, R. A. and Brown, J. K. M. 2005. Sequence variation in the *Cyp51* gene of *Blumeria graminis* associated with resistance to sterol demethylase inhibiting fungicides. *Fungal Genetics and Biology: FG and B* 42(8): 726–735.

Yan, X., Ma, W. B., Li, Y., Wang, H., Que, Y. W., Ma, Z. H., Talbot, N. J. and Wang, Z. Y. 2011. A sterol 14α-demethylase is required for conidiation, virulence and for mediating sensitivity to sterol demethylation inhibitors by the rice blast fungus *Magnaporthe oryzae*. *Fungal Genetics and Biology: FG and B* 48(2): 144–153.

Yankovskaya, V., Horsefield, R., Törnroth, S., Luna-Chavez, C., Miyoshi, H., Léger, C., Byrne, B., Cecchini, G. and Iwata, S. 2003. Architecture of succinate dehydrogenase and reactive oxygen species generation. *Science* 299(5607): 700–704.

Zarn, J. A., Brüschweiler, B. J. and Schlatter, J. R. 2003. Azole fungicides affect mammalian steroidogenesis by inhibiting sterol 14 alpha-demethylase and aromatase. *Environmental Health Perspectives* 111(3): 255–261.

Zhan, J., Stefanato, F. L. and McDonald, B. A. 2006. Selection for increased cyproconazole tolerance in *Mycosphaerella graminicola* through local adaptation and in response to host resistance. *Molecular Plant Pathology* 7(4): 259–268.

Zwiers, L. H. and De Waard, M. A. 2000. Characterization of the ABC transporter genes *MgAtr1* and *MgAtr2* from the wheat pathogen *Mycosphaerella graminicola*. *Fungal Genetics and Biology: FG and B* 30(2): 115–125.

Chapter 3

Tracking the development of fungicide resistance

Francisco J. Lopez-Ruiz, Curtin University, Australia

1 Introduction

Since the introduction of the dithiocarbamate fungicide class in the 1940s, fungicides have gained increasing importance in the control of crop diseases and remain a basic component of existing integrated disease management (IDM) strategies in developed countries for most crops. In recent years, the increased affordability of some key actives, together with better training and support in disease management, has also contributed to the growing adoption of fungicides in many low- and middle-income countries. With a protection benefit comparable to that of genetic resistance, fungicides have contributed to the steady yield increases seen over the last few decades. They, therefore, remain an integral part of medium-term strategies to ensure worldwide food security. In addition to their role in directly controlling disease, fungicides also support the effectivity of other IDM tools. It has been suggested, for example, that fungicides play an important role in protecting genetic resistance in crops from rapid evolution and spread of virulent pathotypes by reducing pathogen growth rates (Carolan et al., 2017).

Unfortunately, the effectiveness of fungicides in controlling diseases has led to their widespread application sometimes without following the best practice in avoiding problems such as the development of resistance. As a

http://dx.doi.org/10.19103/AS.2022.0116.03

result, resistance to fungicides has emerged rapidly in some cases, challenging IDM programmes and increasing pressure on other disease management tools. To date, a total of 259 field resistance cases to 23 fungicide modes of action (MoA) have been reported for a large number of crop diseases (Fungicide Resistance Action Committee (FRAC), 2020).

When fungicide resistance develops, management strategies need to be rapidly adjusted to avoid fungicide control failure. One of the key elements required in any sound IDM strategy is the ability to monitor pathogen populations so that disease control methods can be tailored to take into account and manage patterns of resistance (e.g. by combining fungicides with different MoA). Significant efforts have therefore been devoted to the monitoring of fungicide resistance populations in many different crops and countries (R4P network, 2021). This chapter reviews relevant strategies aimed at tracking the emergence and evolution of fungicide resistance.

2 Key objectives and requirements in tracking fungicide resistance

Traditionally, the vast majority of practical fungicide resistance detections have derived from field observations. Generally, growers and agronomists are the first ones to observe abnormal fungicide performance in the field, for example, by spotting the development of disease symptoms in treated crops. Unfortunately, this typically occurs when the disease is already well established, is difficult to control and will almost certainly result in yield or quality losses. In addition, such observations cannot identify the nature/cause of any emerging pattern of resistance.

The inability to conduct adequate in-season management of fungicide resistance continues to be an important challenge for agriculture. Where growers and agronomists have access to adequate extension support in dealing with fungicide resistance, the lack of near real-time resistance data is, arguably, the main reason why ineffective chemistries are still applied later in the season in those situations where fungicide failure has already been experienced. Typically, by the time that the results from the analysis of disease samples collected after a failed spray become available, the window for spraying an alternative effective fungicide has closed.

Fungicide resistance monitoring plays a central role in the stewardship of existing and new fungicides. This activity, which is conducted by both research organisations and chemical manufacturers, often involves the annual analysis of large numbers of samples collected from fields, trials and the environment (e.g. airborne spores) across multiple growing regions and countries. The aim of these analyses is to establish if sensitivity shifts to particular fungicides have occurred. This information is then compared to crop practices, including fungicide application regimes, to determine potential associations between resistance patterns and disease management. As an example, in 2021 an excess of 1500 samples were analysed in Australia for the presence of demethylase inhibitor (DMI) resistance in the barley pathogens *Pyrenophora teres* f. *teres* and *Pyrenophora teres* f. *maculata* after a resistance outbreak had been reported.

As noted, a fundamental requirement for any analytical technique is the ability to rapidly monitor pathogen populations so that disease control methods can be tailored in sufficient time to take into account and manage patterns of resistance. As the previous discussion shows, key requirements for any analytical technique in meeting these objectives include the following:

- Specificity (ability to correctly identify specific targets);
- Accuracy and reliability/reproducibility (e.g. minimal false positives or negatives);
- Ability to process large volumes of samples;
- Speed of analysis;
- Cost;
- Footprint, that is, the type/amount of equipment/resources and level of expertise required for efficient operation;
- Ideally, the ability to operate in the field (to allow rapid results which can then be used to tailor an IDM programme in sufficient time for early (and thus more effective) disease control; and
- Ability to detect new, previously unknown resistant phenotypes.

The development of genotypic methods for fungicide resistance detection allows analysis of high volumes of infected plant material or fungal isolates, given the high-throughput power that molecular biology offers. Figure 1 identifies current technologies against two of the criteria listed earlier: speed and cost.

As Fig. 1 suggests, analysis cost is an important factor that limits the practical application and future industry adoption of any given molecular detection methodology. This is often due to the relatively large number of samples that need to be tested to generate robust results at the field level. *In vitro* analysis remains the most expensive approach to the detection and/or quantification of fungicide resistance once labour costs have been factored in. The continuous reduction in sequencing costs over the last 15 or so years has made possible the implementation of different next-generation sequencing (NGS) approaches. However, despite the advent of the NGS era, the use of labour-intensive *in vitro* and *in vivo* phenotypic fungal characterisation methods continues to be an essential tool for monitoring and management of fungicide resistance. This means that, despite the relatively low cost of *in vitro* analyses and the low-to-medium costs of molecular approaches, labour remains the

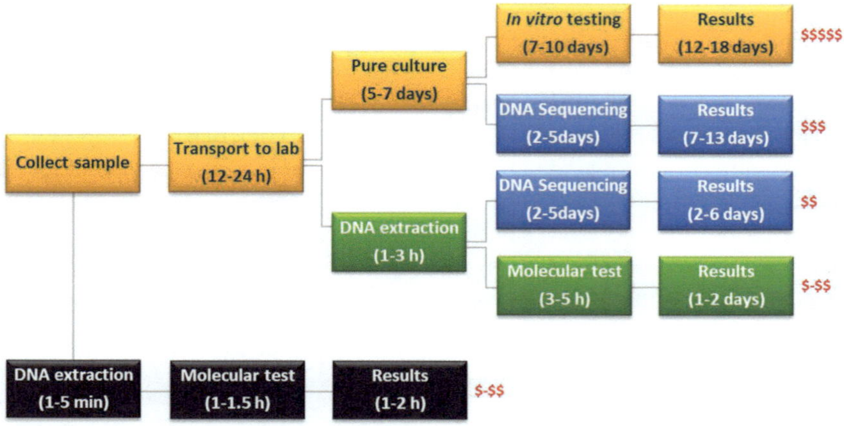

Figure 1 Diagram showing the different fungicide resistance analysis streams that a disease field sample can follow. Orange: *in vitro*; Blue: DNA sequencing from pure cultures; Green: molecular test (i.e. qPCR) conducted on field samples directly; Black: in-field detection. A time range for each of the steps is given in brackets. A cost estimate is provided at the end of each stream ($). The time required for the interpretation of results has not been factored in. Source: Modified from: Dodhia et al. (2019). qPCR, quantitative polymerase chain reaction.

biggest budget component in any resistance test. Hence, the reduction in labour associated with rapid molecular assays able to be conducted in the field should contribute to improving the adoptability of resistance analytical techniques by the agricultural industry.

3 Phenotypic tracking of fungicide resistance

Baseline sensitivity analyses are an essential component in any sound stewardship management package aimed at maximising the lifespan of new fungicide molecules (Russell, 2022). Once a new fungicide has been released, the availability of data on unexposed populations allows for tracking sensitivity shifts within fungal populations that are being exposed to the new fungicide over a number of seasons. This information is then used by product developers as part of the stewardship process and by agronomists for the adjustment of disease management practices, specifically those directly related to the performance of existing fungicide molecules and chemical control programmes.

When resistance to a fungicide molecule within a particular MoA group develops, other molecules within the same group are also often affected (Brent and Hollomon, 2007). This phenomenon, known as cross-resistance, can impact a number of fungicides within an MoA, such as the majority of target site mutations affecting the DMIs, or the entire fungicide class as occurs with mutation G143A in the case of quinone outside inhibitors (Ishii et al., 2021; FRAC, 2022). The analysis of fungicide panels, both *in vivo* and *in vitro*, makes it possible to determine cross-sensitivity patterns when fungicide resistance first emerges. As with baselines, the information generated is essential for the development of robust fungicide stewardship strategies and effective field resistance management practices where only the molecules that retain activity against the fungicide-resistant mutants are incorporated within chemical control programmes.

As well as identifying resistance patterns, the use of analytical techniques such as bioassays also provides researchers with the added capacity to investigate other aspects in understanding of fungicide resistance and its management. These aspects are as follows:

1 Establishment of baselines for new fungicides and fungal naïve populations;
2 Understanding of sensitivity levels of specific pathogen genotypes to panels of different fungicides;
3 Discovery of resistant pathogen phenotypes that could potentially be associated with novel genotypes; and
4 Quantification of sensitivity shifts associated with particular genotypes when molecular markers are not available.

Traditional DNA-based methods such as polymerase chain reaction (PCR), quantitative PCR (qPCR), digital PCR (dPCR), loop-mediated isothermal amplification (LAMP) and cleaved amplified polymorphic sequences have the limitation of requiring pre-existing knowledge on the genetic modifications associated with any given fungicide resistant phenotype in order to be able to detect it. Resistance monitoring based on these detection methods can then become problematic with the emergence of new resistant phenotypes associated with novel target site mutations or other genetic modifications for which molecular detection tests have not yet been developed. In these situations, researchers run the risk of underestimating the actual frequency of resistance or not being able to detect it at all, which can lead to inappropriate management.

The implementation of phenotypic testing approaches, like effective concentration 50 (EC_{50}) analysis, or the use of fungicide discriminatory concentrations, in combination with high-throughput molecular detection methods, can contribute to developing more effective fungicide resistance monitoring strategies. The use of these two approaches can ensure that the emergence of new resistant genotypes is detected by phenotypic methods ahead of any potential fungicide resistance outbreaks resulting from the increase in the frequency of these genotypes due to the detection limitations of molecular methods.

Molecular markers might not be available for the analysis of fungicide resistance for a number of reasons, that is, lack of a known resistance mechanism or resistance emerging in a new disease for which genetic resources such as genome information have not been developed yet. In these situations, resistance level quantification and frequency monitoring depend on the availability of suitable phenotypic methods. An example of this can be the use of fungicide discriminatory doses for the detection and frequency quantification of resistant pathogens (Russell, 2002; Weber and Hahn, 2011; Lehner et al., 2015). This approach, which is often used in parallel with molecular detection, makes it possible not only to categorise isolates based on their resistance level but also to filter them so that only those exhibiting a particular level of resistance are investigated further (i.e. sequencing of the target site or transporter expression analysis).

Table 1 Different methodologies used for the calculation of EC_{50} data in studies of *Botrytis cinerea* isolated from wine grape, table grape and non-grape hosts

Fungicide	EC_{50} phenotyping methods[a]	Reference
Azoxystrobin	AMG	Yin et al. (2015); Banno et al. (2009)
	AGT	Leroux et al. (2010); Weber and Hahn (2011)
	MIC	Harper et al. (2022)
Boscalid	MIC	Campia et al. (2017); Stammler and Speakman (2006); Harper et al. (2022); Toffolatti et al. (2020)
	MIC, AMG	Alberoni et al. (2011)
Fenhexamid	AMG	Albertini and Leroux (2004); Latorre and Torres (2012); Beresford et al. (2017)
	AMG, AGT, AGE	Leroux et al. (1999)
	AMG, AGE	Fillinger et al. (2008)
	MIC	Harper et al. (2022); Toffolatti et al. (2020)
Fludioxonil	AGT	Weber and Hahn (2011)
	AGE	Fernández-Ortuño et al. (2015)
	AMG	Latorre and Torres (2012)
	MIC	Harper et al. 2022; Toffolatti et al. (2020)
Iprodione	AMG, AGT, AGE	Leroux et al. (1999)
	MIC	Harper et al. (2022)
Pyrimethanil or cyprodinil	AMG	Avenot et al. (2018); Gullino et al. (2000); Sergeeva et al. (2002); Latorre and Torres (2012); Beresford et al. (2017)
	MIC	Campia et al. (2017); Harper et al. (2022); Toffolatti et al. (2020)

[a]MIC, microtitre; AMG, agar-mycelial growth; AGT, agar-germ tube elongation; AGE, agar-germination; EC_{50}, effective concentration 50. *Source*: Modified from Harper et al. (2022).

It is important to note that results obtained using this kind of approach can vary due to factors like ambient temperature fluctuations, the ability of the operator or the growth rate of specific isolates (Russell, 2002). The use of different phenotypic assays for the sensitivity analysis of a given fungus to a particular fungicide can contribute to generating significant data variability among laboratories (Table 1). In the absence of genotypic evidence, it is reasonable to assume that differences in phenotypic data can be attributed to the variability generated by the available analysis techniques. Is for this reason that genetic differences, such as alternative target site mutations, need to be easily identified to avoid data bias. To deal with issues of data variability, efforts have been made by organisations like the Fungicide Resistance Action Committee (FRAC; https://www.frac.info/) to standardise fungicide resistance testing methodologies, enabling researchers to produce readily comparable data (see FRAC-approved bioassay methods at https://www.frac.info/knowledge-database/monitoring-methods).

4 Developments in genotyping techniques

One of the main goals of most studies aimed at characterising fungicide resistance is the development of molecular markers for the detection of mutations associated with specific resistant phenotypes (Banno et al., 2009; Chen et al., 2019; Liu et al., 2019). These methods, which in many cases can be scaled up to increase the results throughput, are faster, have a relatively low footprint in the laboratory, are generally cheaper than traditional phenotyping approaches due to lower labour requirements and provide results that are less prone to interpretation bias since phenotypes can be ascertained based on previously established correlations between genetic modifications and specific sensitivity levels. The use of molecular methods for the detection of fungicide resistance has been reviewed in a number of studies (Ma and Michailides, 2005; Capote et al., 2012; Sierotzki et al., 2019).

PCR-based assays are by far the most widely used genotyping target amplification-based methods in fungicide resistance research. While PCR-based methodologies are limited by the available knowledge on the existing genetic modifications associated with fungicide resistance, the possibility of utilising different types of polymerases, probes and dyes (i.e. qPCR) and different platforms, together with the ability to modify assay conditions to suit multiple different targets and mutations, make PCR-based methods very adaptable tools for the monitoring of resistance. This versatility, combined with the relatively small investment required, reduced footprint and high-throughput potential, has allowed PCR-based assays to remain an essential tool for the detection and quantification of fungicide resistance in many crops (Hollomon, 2015).

Some of these target amplification-based methods such as qPCR, dPCR or LAMP can be 'multiplexed' to simultaneously detect several different genotypes and, due to their high specificity, can be used for the detection of specific genotypes directly from infected tissue and from samples containing mixed genotypes and pathogen populations (Fraaije et al., 2002; Duan et al., 2018b; Hellin et al., 2020).

Although molecular biology cannot be used to establish fungicide sensitivity baselines, it can provide an indirect idea as to whether a pathogen population is shifting in sensitivity to a fungicide without the need of having to conduct phenotypic assays. Monitoring frequency variation of alleles associated with specific sensitivity levels can reveal if a pathogen population is decreasing their overall sensitivity to a given fungicide due to, for example, repeated use, or increasing it as a result of the introduction of more dynamic crop and fungicide rotations, among other crop protection strategies. This can be particularly relevant in those situations where fungicides are known to underperform when resistant alleles in a population reach a certain frequency threshold (McCartney et al., 2007; Walker et al., 2013; Blake et al., 2018; Heick et al., 2020).

Molecular biology can also be used to detect the emergence of alleles associated with fungicide resistance in new regions or to quantify the frequency of resistant genotypes so that chemical control programmes can be adapted accordingly allowing growers to avoid the use of ineffective chemicals, which in turn will benefit the environment and will increase the quality of the produce.

Resistance is generally selected from existing pathogen genetic variation, which can already be known to the researcher, or from *de novo* mutations (Hawkins et al., 2019). It is important to make a distinction between molecular methods that require pre-existing knowledge on fungicide resistance-associated genetic changes that are likely to be found in the pathogen population and those methods able to provide information about potential new mutations before these have been characterised by researchers.

In recent years, there have been several reports on the detection, and potential for quantification, of plant pathogens through new sensing technologies (Dyussembayev et al., 2021). Nanomaterial-, DNA-, antibody- and enzymatic-based biosensors, among others, have been explored for the rapid and accurate detection of several viruses and some bacterial pathogens (see Fang and Ramasamy (2015) for a comprehensive review of these methods). However, the adoption of these technologies for the detection and quantification of pathogens and fungicide resistance seems to be still far from being able to initiate a paradigm change in field management of fungicide resistance. It remains to be seen whether molecular methodologies can be simplified enough so that industry adoption becomes a reality and fungicide resistance management strategies can be implemented during the growing season.

5 Developments in in-field detection and quantification techniques

Although PCR-based target quantification technology has made considerable advances in such areas as reducing instrument footprint and increasing target sensitivity and specificity by improving probe design flexibility, quantification of mutations associated with fungicide resistance remains mainly laboratory based (Dodhia et al., 2021). *In-situ* detection and quantification of resistant alleles offer the potential to inform the choice of optimum fungicides for crop disease management in a timely manner.

Until the adoption of pyrosequencing in the last decade, quantitative allele-specific PCR (ASqPCR) has traditionally been the main molecular tool used for the frequency analysis of specific genotypes associated with fungicide resistance (Fig. 2). This can probably be attributed to the generally higher sensitivity of ASqPCR at mutation frequencies below 10% compared to pyrosequencing (Sierotzki et al., 2020). Different probe and intercalating dye-based assays have been used for the quantification of gene alterations leading to fungicide resistance such as the die SYBR Green for the quantification of cytochrome b (Cytb) mutation G143A in the wheat pathogen *Blumeria graminis* f.sp. *tritici* and the probe-based detection of Cyp51 mutation S524T in the wheat pathogen *Zymoseptoria tritici* (Fraaije et al., 2002; Hellin et al., 2020). Up until recently, the use of ASqPCR had been limited to laboratory settings probably

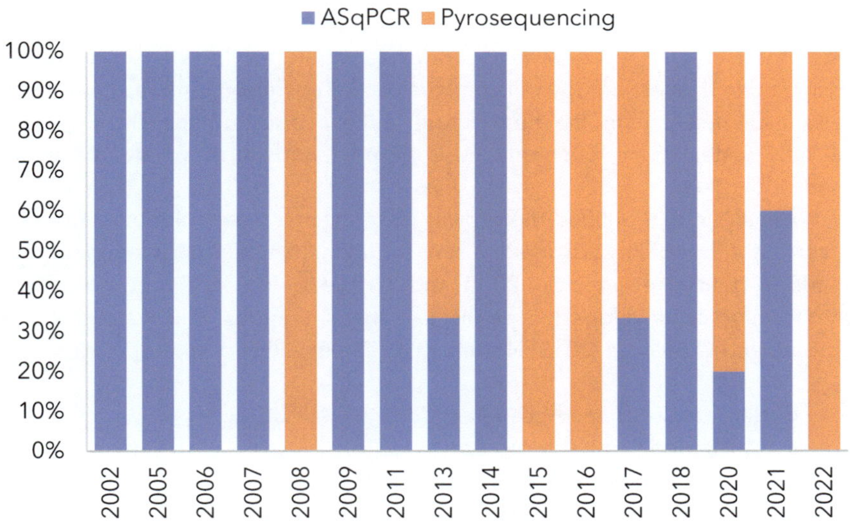

Figure 2 Percentage of 50 random publications listed in Web of Science between 2002 and 2022 describing the use of quantitative allele-specific polymerase chain reaction (ASqPCR) (blue) or pyrosequencing (orange) for quantification of genetic changes associated with fungicide resistance.

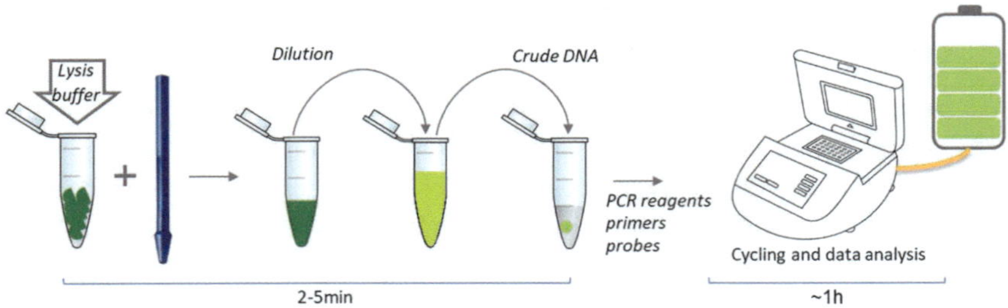

Figure 3 Schematic representation of the in-field pipeline for the quantification of cytochrome b mutation G143A in *Blumeria graminis* f.sp. *tritici*. Source: Modified from Dodhia et al. (2021). PCR tubes and thermocycler images: https://www.labicons.net. PCR, polymerase chain reaction.

due to the lack of light and field-deployable thermocyclers and robust assay pipelines that could cope with conditions in the field.

A study by Dodhia et al. (2021) describes a fungicide resistance survey based on the development of an in-field ASqPCR pipeline for the detection of the Cytb mutation G143A in *B. graminis* f.sp. *tritici*. This study combined a simple and effective quick DNA extraction method with the use of a robust Taq polymerase and a fast magnetic induction cycler connected to a battery (Fig. 3). The results showed no difference in the frequency of G143A between the runs conducted in the field and those performed in the laboratory on a highly sensitive digital PCR platform and using pure genomic DNA.

Despite these improvements, in-field ASqPCR still requires expensive equipment and trained experts to produce accurate results. It is reasonable to assume that this technology will continue to play an important role in fungicide resistance research in the near future while supporting the development of new faster, more economic, user-friendly and sensitive quantification methodologies.

Vrana (1996) identified the growing interest in isothermal methodologies given a number of key advantages over traditional amplification technologies. LAMP was first reported in 2000 and is one of the different isothermal methodologies developed over the last three decades which also include (Lobato and O'Sullivan, 2018; Notomi et al., 2000; Gill and Amree, 2020):

- nucleic acid sequence-based amplification (NASBA);
- self-sustained sequence replication (3SR);
- recombinase polymerase amplification (RPA);
- multiple displacement amplification (MDA);
- rolling circle amplification (RCA);
- Helicase-dependent amplification (HDA); and
- strand displacement amplification (SDA).

LAMP and other isothermal techniques have particular potential as in-field diagnostic tools given their simplified approach to target amplification (a simple heater block replaces a thermocycler), generally reduced time to result and lower cost compared to traditional amplification approaches. The advantages and disadvantages of these techniques have been reviewed in a number of studies (Craw and Balachandran, 2012; Yan et al., 2014; Deng and Gao, 2015).

LAMP is an affordable, highly-specific and sensitive method with the capacity to be deployed in low-resource settings. However, advantages of LAMP such as versatility, reduced per-sample cost and analysis speed must be offset by the number of false positives it gives, lack of quantification capacity and a fairly complex primer design and optimisation process (Suleman et al., 2016; Hardinge and Murray, 2019). Nevertheless, LAMP is, to our knowledge, the only isothermal detection tool that is widely used in the field. The use of LAMP has been facilitated by the array of detection protocols developed to date and the existence of fairly inexpensive colorimetric tests and battery-powered portable devices that allow the *in-situ* analysis of samples (Fukuta et al., 2006; Hsu et al., 2012; Kogovšek et al., 2015; Harrison et al., 2017).

Since its inception, LAMP has been modified to increase both sensitivity and specificity, playing a central role in fungicide resistance research and monitoring (Liu et al., 2017). Allele-specific LAMP combines different evolutions of the LAMP technique such as restriction fragment length polymorphism LAMP, in which the digestion of LAMP products leads to the identification of mutations of interest. Another example is peptide nucleic acid-locked nucleic acid (PNA-LNA)-mediated LAMP, which uses a PNA probe to avoid amplification of the wildtype target (Itonaga et al., 2016; Shao et al., 2011). In addition to the detection of different plant pathogens, the capacity of LAMP to genotype samples has enabled successful detection of pesticide-resistant alleles in several plant pathogens (Table 2) and weeds (Pan et al., 2015), in both laboratory and field settings. LAMP protocols have been developed for the detection of single nucleotide polymorphisms (SNPs) associated with fungicide resistance, such as the benzimidazole resistance mutation F200Y in *Fusarium asiaticum* and the succinate dehydrogenase inhibitor resistance mutation H272R in *Botrytis cinerea* (Duan et al., 2016; Fan et al., 2018), and larger genetic changes like promoter insertions in the target gene *Cyp51* of DMI-resistant *Z. tritici* isolates (King et al., 2017).

LAMP detection of mutations associated with fungicide resistance in the laboratory and in the field represents a considerable step forward for the monitoring of particular fungicide-resistant genotypes. This approach also allows, to some degree, for a more proactive management of the resistance problem as it can provide critical information on the pathogen resistance status if, for example, specific mutations associated with high levels of resistance to a given fungicide active are found in a field. However, the lack of a simple

Table 2 LAMP protocols developed for the detection of alleles associated with fungicide resistance in plant pathogens

Target[a]	Pathogen	Reference[b]
β-tubulin F167Y	*Fusarium graminearum*	Duan et al. (2014)
β-tubulin E198A	*Sclerotinia sclerotiorum*	Duan et al. (2015)
β-tubulin E198A	*Botrytis cinerea*	Liu et al. (2019)
β-tubulin E198A	*Podosphaera xanthii*	Vielba-Fernandez et al. (2019)
β-tubulin E198A	*Colletotrichum gloeosporioides*	Yuan et al. (2022)
β-tubulin F200Y	*Fusarium asiaticum*	Duan et al. (2016)
β-tubulin F200Y	*Botrytis cinerea*	Duan et al. (2018b)
β-tubulin E198K, F200Y	*Fusarium fujikuroi*	Li et al. (2022)
β-tubulin F167Y, E198Q, F200Y	*Fusarium graminearum* spp.	Komura et al. (2018)
β-tubulin E198A, E198K, E198V	*Botrytis cinerea*	Duan et al. (2018a)
Cyp51 120 bp promoter insert	*Zymoseptoria tritici*	King et al. (2017)
Cyp51 65 bp promoter insert	*Monilinia fructicola*	Chen et al. (2019)
Cytb G143A	*Botrytis cinerea*	Hu et al. (2017)
Cytb G143A	*Colletotrichum gloeosporioides*	Wu et al. (2019)
Cytb G143A	*Cercospora beticola*	Shrestha et al. (2020)
Cytb G143A	*Corynespora cassiicola*	Li et al. (2021)
Cytb G143A	*Botryosphaeria dothidea*	Wang et al. (2022)
Cytb G143A	*Glomerella cingulata*	Ren et al. (2022)
Cytb F129L	*Pyricularia oryzae*	Li et al. (2022)
MFS1 519 bp promoter insert	*Zymoseptoria tritici*	King et al. (2017)
SdhB H272R	*Botrytis cinerea*	Fan et al. (2018)
SdhC N75S	*Corynespora cassiicola*	Zhu et al. (2020)

[a]Cytb, cytochrome b; *Cyp51*, sterol 14α-demethylase; *MFS1*, major facilitator superfamily transporter 1; *SdhB*, succinate dehydrogenase subunit B.
[b]Only references published in peer-reviewed journals and specifying the molecular targets were included.

target quantification strategy limits the management impact of this technique at the field level. Although turbidimetry- and fluorescence-based quantitative LAMP methods were developed shortly after LAMP became available (Zhang et al., 2014), to our knowledge, no studies involving the use of these LAMP alternatives for the quantification of fungicide resistance mutations have been reported to date.

6 Developments in sequencing techniques

It is generally accepted that the adoption of amplification-based molecular detection technologies has contributed significantly to improving the understanding and management of fungicide resistance (Henson and French, 1993; Schena et al., 2004; Sierotzki and Gisi, 2002). An example is the provision of more precise resistance distribution and frequency maps. The speed, affordability, relative simplicity and high-throughput capacity of some molecular methods compared to existing phenotyping alternatives are key advantages. However, in order to ensure effective surveillance of resistance, molecular methods need to be flexible enough to detect new, previously unknown resistant genotypes.

Since the release of the first NGS instrument in 2005, this technology has proven revolutionary in many areas (Escobar-Zepeda et al., 2015). NGS has not only allowed researchers to sequence higher volumes more rapidly and at comparatively lower cost but also led to the development of new approaches where resistance questions can now be answered from a genome-wide scale perspective. The evolution of sequencing technology has led to the recent application of two noteworthy approaches to the study of fungicide resistance:

- Genome-wide association studies (GWASs); and
- Nanopore sequencing.

These are discussed in Section 6.1 and Section 6.2.

6.1 Developments in genome-wide association studies

The polygenic nature of some fungicide resistance phenotypes is a common phenomenon that increases the complexity of their molecular characterisation and monitoring. As a consequence, the contribution to the phenotype of additional genes is often established once the traditional target sites have been investigated. For example, the role of the major facilitator superfamily (MFS) drug efflux pump MgMFS1 in highly DMI-resistant *Z. tritici* strains was determined in 2015, several years after these strains had already been reported as carrying multiple different mutations in the Cyp51 target (Leroux and Walker, 2011; Omrane et al., 2015). This, however, is a direct consequence of the relatively limited information available at the genome level for the majority of pathogens in relation to fungicide resistance.

GWAS was born in the early 2000s as a response to the need for establishing links between specific phenotypic traits and potential modifications in genes at the full genome scale through the sequencing of large numbers of individuals (Klein et al., 2005). GWAS is perhaps to sequencing what sequencing is to amplification technologies, in that it accelerates the discovery of new loci and strongly supports their link to particular phenotypes. About 10 years after the publication of the first GWAS, this technique was applied to the identification of mutations in novel genes associated with fungicide resistance, also known as quantitative trait nucleotides or QTNs. In 2016, Talas et al. reported the use of GWAS for the identification of QTNs in 51 genes associated with sensitivity to the DMI fungicide propiconazole in *Fusarium graminearum*. Their analysis revealed genes known to be involved in fungicide sensitivity response such as DNA mismatch repair protein, ATP binding cassette (ABC) and MFS transporters. They also found other novel genes not previously known to play a role in the sensitivity to fungicides. Among the 51 genes found, QTNs in 3 of them explained 51% of the observed genetic variance (17% each) associated with propiconazole sensitivity in this pathogen. Two of these genes, the transcription factor TFIID and a hypothetical protein with potential metal ion binding, ATP binding or RNA splicing functions, had never been associated with changes in fungicide sensitivity before in pathogenic fungi (Talas et al., 2016).

GWAS have highlighted the role played by other novel genes in fungicide sensitivity response in different pathogens (Table 3). These associations could also be established with changes in intergenic regions, and with synonymous mutations in the fungicide target gene *Cyp51*. Spanner et al. (2021) found a strong association between Cyp51 silent mutation E170 and resistance to the DMI tetraconazole in the sugar beet pathogen *Cercospora beticola*. Significantly higher levels of resistance were observed when mutation E170 appeared in combination with other Cyp51 nonsynonymous mutations. The same correlation was found when the authors sequenced the *Cyp51* gene in a group of isolates collected several years after the *C. beticola* population was used in their GWAS analysis. This example clearly illustrates how this genome association approach can contribute not only to the discovery of new mutations that later can be used for developing improved monitoring strategies but also to the characterisation of the complex genetics behind particular fungicide sensitivity responses.

The existence of QTNs associated with decreased fungicide sensitivity in intergenic regions has also been observed by various authors (Mohd-Assaad

Table 3 Examples of novel genes associated with sensitivity to fungicides from the demethylase inhibitor group and identified by genome-wide association in different plant pathogens

Pathogen	Fungicide	Novel genes	Reference
Rhynchosporium commune	Cyproconazole	Calcium-transporter YVC1 Transcription activator (TA) Saccharopine dehydrogenase (SDH)	Mohd-Assaad et al. (2016)
Fusarium graminearum	Propiconazole	Transcription factor TFIID Acyl-CoA thioesterase Hypothetical protein FGSG_0693	Talas et al. (2016)
Parastagonospora nodorum	Propiconazole	MFS transporter SNOG_14185	Pereira et al. (2020)
Cercospora beticola	Tetraconazole	Tripeptidyl-peptidase Sed2-like protein	Spanner et al. (2021)

et al., 2016; Spanner et al., 2021; Talas et al., 2016). To our knowledge, the specific role played by any of these QTNs has not yet been characterised, but it has been suggested that these changes could be associated with regulatory regions (either directly or indirectly through linkage disequilibrium) or with poorly annotated genes (Talas et al., 2016). The fact that only a small fraction of mutations are likely to play a regulatory role (Rockman, 2012; Stenson et al., 2017) supports the current approach followed by authors to target fungicide resistance QTNs in genes.

The use of GWAS for the analysis of large fungal populations has indirectly contributed to improve our understanding of the relationship between the evolution of fungicide resistance mutations and the potential impact on the fitness of isolates harbouring these mutations. Some studies could not find any relevant trade-offs associated with isolates resistant to different DMI fungicides and the analysed traits. This was the case for the wheat pathogen *Parastagonospora nodorum* and propiconazole (Pereira et al., 2020) and for *C. beticola* and tetraconazole (Spanner et al., 2021). However, the analyses conducted by Mohd-Assaad et al. (2016) in the barley pathogen *Rhynchosporium commune* and the fungicide cyproconazole showed that novel loci associated with decreased fungicide sensitivity were also involved in reduced growth rates. These apparently contradictory results are to be expected considering the large number of novel loci found in these studies and their potential contribution to regulatory processes that probably transcend their role in fungicide sensitivity variation in resistant isolates. Moreover, the fitness penalty associated with the resistance mechanism could also explain the observed growth reduction.

6.2 Developments in nanopore and other long-read sequencing approaches

The long-sought capacity to deploy fast, accurate and affordable molecular approaches that can inform disease management practices in near real-time is currently a reality for a number of mutations associated with fungicide resistance. Several of these mutations can be easily detected and quantified from complex field samples using a range of amplification-based and sequencing methods (Fraaije et al., 2002; Dodhia et al., 2021). These and similar tools are likely to be adopted more broadly by the industry in the near future. However, one could argue that this process is still in its infancy for a number of reasons. We have discussed earlier in this chapter the limitations of these methodologies and how sequencing approaches are necessary for the detection of recently emerged mutations for which we have no prior knowledge. Moreover, the fact that mutation combinations in some fungicide target genes can lead to a variety of fungicide sensitivity responses implies that

Table 4 Examples of *Cyp51* haplotypes associated with varying levels of fungicide resistance in different plant pathogens that cannot be resolved by short-read or Sanger sequencing in complex field samples

Pathogen	Gene	Haplotype	Reference
Zymoseptoria tritici	*Cyp51*	*L50S, I381V, Y461H*	Huf et al. (2018)
		L50S, D134G, V136A, I381V, Y461H	
		L50S, S188N, I381V, ΔY459/G460, N513K	
Pyrenophora teres f. *maculata*	*Cyp51*	-46 PtTi-5 promoter insertion, F489L	Mair et al. (2020)
		-90 PtTi-2 promoter insertion, F489L	
Blumeria graminis f.sp. *hordei*	*Cyp51*	Y137F, M304I, R330G, S524T	Tucker et al. (2020)

haplotypes associated with these changes need to be adequately resolved if the expected outcome is to produce meaningful management advice from complex field samples.

Short-read and Sanger sequencing approaches have played a central role in this regard by providing the capacity to identify combinations of existing mutations and also new ones. The problem arises when changes in fungicide sensitivity levels are associated with specific haplotypes which carry multiple mutations across longer gene regions (Table 4). In these cases, short-read and Sanger sequencing approaches lack the capacity to profile such haplotypes unless the analyses are conducted on pure isolates, which dramatically delays turnaround time for the majority of agriculturally relevant pathogens and introduces a potential bias towards haplotypes harboured by easier-to-culture isolates.

Third-generation sequencing (TGS) came into light when Pacific Biosciences (PacBio) commercialised the single-molecule real-time (SMRT) DNA sequencing system in 2011 (McCarthy, 2010; Weirather et al., 2017). The key difference between second-generation sequencing (also known as NGS) and TGS is the capacity of the latter to generate very long reads (1–100 kb), which overcomes the challenges associated with the sequencing of highly complex or repetitive genomes.

In 2021, Samilis et al. reported the first attempt to profile fungal isolates carrying target site mutations associated with fungicide resistance using long-read sequencing. The authors of this work developed a PacBio pipeline (SMRT technology and circular consensus sequencing (Eid et al., 2009)) that involved the multiplex amplification and sequencing of the *Z. tritici* fungicide target genes *Cyp51*, *Cytb* and the *Sdh* subunits *B*, *C* and *D*, in several groups of isolates for which target site sequences and fungicide sensitivity levels had been previously characterised. In addition to the long-read capacity, the assay throughput was considerable as it could sequence the targets for up to 96 isolates, including nine additional housekeeping genes, which enhanced the ability to genotype *Z. tritici* populations. The analysis produced reads for the majority of isolates/targets and these made it possible to successfully profile isolates for all fungicide resistance mutations when compared to Sanger sequencing.

Although PacBio shows overall high data quality and scalability, another TGS approach that has drawn a reasonable amount of attention in recent years has been nanopore sequencing. This long-read sequencing technology has the capacity to provide higher data yields and the added advantage of being conducted in-house due to the low setup costs compared to existing PacBio platforms. The release in 2014 of the first MinION device marks a key milestone in the development of a TGS technology that not only has been used for the improvement of complex genomes (such as those high in repetitive regions) or the assembly of new ones but also for the sequencing of epigenomes, transcriptomes and epitranscriptomes, among other applications (Deamer et al., 2016; Wang et al., 2021).

However, a key drawback with the use of nanopore sequencing technology is its relatively low base-calling accuracy compared to other short-read technologies such as Illumina (Delahaye and Nicolas, 2021; Noakes et al., 2019). Unfortunately, this is an important impediment that may deter the adoption of this technology for profiling of fungicide-resistant pathogen populations and the use of this information to develop disease management approaches tailored to the control of resistant populations.

In recent years, the improvements in the accuracy of nanopore sequencing through chemistry and flow cell evolution, the use of specifically developed molecular methods to reduce error rate and the implementation of new base-calling and error correction algorithms (Boza et al., 2017; David et al., 2017; Karst et al., 2021; Sereika et al., 2022) have opened the door to the application of this technology to the detection of mutations as part of point of care rapid clinical diagnoses. The potential for this technology to be incorporated into other diagnostic areas is considerable with rapid development likely in the next few years.

The use of nanopore for the detection of fungicide resistance mutations is perhaps one of the most exciting applications of sequencing technology in plant pathology in recent years. This technology represents a considerable step forward for the monitoring of known mutations and for the discovery of novel ones. Recently, some studies have developed specific nanopore sequencing pipelines for the analysis of the genetic composition of pathogen populations showing different levels of resistance to DMI fungicides. This is particularly relevant given the high genetic complexity underpinning resistance in some key pathogens.

In a *Z. tritici* study, Gutierrez Vazquez et al. (2022) successfully used nanopore sequencing to investigate the genotype composition of pathogen populations. They identified different *Cyp51* genotypes associated with fungicide resistance in *Z. tritici* mock populations and field samples. Genotypes were successfully resolved at frequencies above 5%, which is a significant step forward in the development of better approaches for the surveillance of fungicide resistance. However, this study also highlights how the relatively high sequencing error rate associated with this technology can be a limitation for profiling relevant genotypes present at low frequencies during the emergence phase of fungicide resistance (van den Bosch et al., 2011).

Following a similar approach, Zulak et al. (2023) reported a substantial improvement in the error rate when mock communities including different DMI-resistant haplotypes of the barley pathogen *P. teres* f. *maculata*, *P. teres* f. *teres* and a hybrid of these two were profiled. In this study, an alternative error-proofing strategy based on the use of short sequences that uniquely tag amplicons called unique molecular identifiers (Karst et al., 2021) allowed for reads to be grouped based on their original templates and consensus sequences generated on a per molecule basis. This approach managed to produce high-density clusters identical to the reference sequences allowing for a very accurate profiling of a diverse *P. teres* spp. haplotype range. Sequencing errors were only limited to the lower density clusters which had some SNPs and indels mainly in homopolymer regions.

Recently, a number of nanopore SNP and insertion target profiling approaches were presented at the 20th International Reinhardsbrunn Symposium which showcases recent research in fungicide resistance. The fully *in-situ* pipeline reported by Kleemann et al. (2023) for the profiling of *Cyp51*, *Cytb* and the *Sdh* subunits *B*, *C* and *D*, in *Z. tritici* mixed field samples, allowed to produce real-time data within 2–3 h and with an error rate of about 6%. Although the error rate shown by this pipeline is considerably higher than that reported by Zulak et al. (2023), this is probably compensated by the *in-situ* capacity and the speed at which results can be generated across a range of major fungicide resistance targets.

While nanopore technology has the potential to improve the management of fungicide resistance by providing access to faster and accurate monitoring, it still cannot resolve samples containing mixed multi-resistant isolates in which specific haplotypes carry multiple mutations across more than one gene. For this reason, it is highly unlikely that fungicide resistance monitoring will transition solely to the use of genotypic analysis approaches in the short term, except for those mutations or haplotypes that can be sequenced or profiled using a simple combination of target amplification-based tests.

7 Conclusion and future trends

Over the last few decades, considerable efforts have been made to develop phenotypic and genotypic methods able to characterise fungicide sensitivity shifts as well as identify and quantify the presence of mutations or haplotypes in both fungal isolates and complex populations. Despite these efforts, researchers continue to investigate technologies that are faster, simpler and sufficiently affordable to allow their adoption and deployment to have a practical impact on the management of fungicide resistance.

Typically, detecting resistant phenotypes when frequencies are low requires extensive surveys in growing regions where conducive conditions for the selection of resistance are known to occur. Often, when resistance has been confirmed, growers have already started to experience its adverse effects. This approach cannot deliver in-season intelligence fast enough for effective adjustment of existing disease management programmes.

The current development of more sensitive detection technologies occurs essentially in the molecular space. Many of these techniques have the limitation of requiring pre-existing knowledge of the genetic modifications associated with any fungicide-resistant phenotype in order to be able to detect it. It is for this reason that emphasis is being placed on approaches able to cope with an increasingly complex resistance mutation landscape.

The development of long- and short-read sequencing methods and target amplification-based assays for the genotyping of fungicide resistance will continue to capitalise on the vast phenotypic and genotypic knowledge generated to date and produce practical approaches for the fast and accurate diagnosis of major fungicide resistance mutations in near real-time conditions.

The capacity to generate more data from the screening of existing resources, such as high-throughput sequencing of the aerobiome collected by networks of spore traps, will allow for the development of more accurate models that will inform better resistance management tactics. Similarly, the use of in-field simple tests for the detection of specific resistant genotypes similar to those that have been widely distributed around the world for the detection of COVID-19 can aid in the targeted adjustment of chemical disease control programmes and improve outbreak surveillance.

Resistance will continue to evolve in a process fuelled by the successful role played by fungicides in the management of crop diseases. On the other hand, research will continue to develop fungicide resistance phenotypic and genotypic detection approaches in an attempt to flatten this curve so that the industry has the time to adapt to the ever-changing chemical management landscape.

8 Acknowledgements

I would like to thank Katherine Zulak and Kejal Dodhia for their constructive suggestions that greatly improved this chapter. I have been fortunate enough to enjoy many discussions with very inquisitive minds such as Richard Oliver, James Brown, Chris Ridout, Frank van den Bosch, Andreas Mehl, Gerd Stammler, Omer Frenkel and others. The ideas emerging from these discussions continue to have a profound impact on the direction of my research.

9 References

Alberoni, G., Ciriani, A., Banorri, M., Collina, M. and Brunelli, A. (2011). Sensitivity to Boscalid of *Stemphylium vesicarium* and *Botrytis cinerea* in Italy. In: *Modern Fungicides and Antifungal Compounds*. 16th International Reinhardsbrunn Symposium, Friedrichroda, Germany (vol. VI), pp. 273–277.

Albertini, C. and Leroux, P. (2004). A *Botrytis cinerea* putative 3-keto reductase gene (ERG27) that is homologous to the mammalian 17β-hydroxysteroid dehydrogenase type 7 gene (17β-HSD7). *European Journal of Plant Pathology* 110(7), 723–733. https://doi.org/10.1023/B:EJPP.0000041567.94140.05.

Avenot, H. F., Quattrini, J., Puckett, R. and Michailides, T. J. (2018). Different levels of resistance to cyprodinil and iprodione and lack of fludioxonil resistance in *Botrytis cinerea* isolates collected from pistachio, grape, and pomegranate fields in California. *Crop Protection* 112, 274–281. https://doi.org/10.1016/j.cropro.2018.06.005.

Banno, S., Yamashita, K., Fukumori, F., Okada, K., Uekusa, H., Takagaki, M., Kimura, M. and Fujimura, M. (2009). Characterization of QoI resistance in *Botrytis cinerea* and identification of two types of mitochondrial *cytochrome b* gene. *Plant Pathology* 58(1), 120–129. https://doi.org/10.1111/j.1365-3059.2008.01909.x.

Beresford, R. M., Wright, P. J., Middleditch, C. L., Vergara, M. J., Hasna, L., Wood, P. N. and Agnew, R. H. (2017). Sensitivity of *Botrytis cinerea* to fungicides used in New Zealand wine grape spray programmes. *New Zealand Plant Protection* 70, 285–294. https://doi.org/10.30843/nzpp.2017.70.62.

Blake, J. J., Gosling, P., Fraaije, B. A., Burnett, F. J., Knight, S. M., Kildea, S. and Paveley, N. D. (2018). Changes in field dose-response curves for demethylation inhibitor (DMI) and quinone outside inhibitor (QoI) fungicides against *Zymoseptoria tritici*, related to laboratory sensitivity phenotyping and genotyping assays. *Pest Management Science* 74(2), 302–313. https://doi.org/10.1002/ps.4725.

Boza, V., Brejova, B. and Vinar, T. (2017). DeepNano: Deep recurrent neural networks for base calling in MinION nanopore reads. *PLoS ONE* 12(6), e0178751. https://doi.org/10.1371/journal.pone.0178751.

Brent, K. J. and Hollomon, D. W. (2007). *Fungicide Resistance in Crop Pathogens: How Can It Be Managed (FRAC Monograph No. 1)*. Fungicide Resistance Action Committee, Brussels, Belgium.

Campia, P., Venturini, G., Moreno-Sanz, P., Casati, P. and Toffolatti, S. L. (2017). Genetic structure and fungicide sensitivity of *Botrytis cinerea* populations isolated from grapevine in northern Italy. *Plant Pathology* 66(6), 890–899. https://doi.org/10.1111/ppa.12643.

Capote, N., Pastrana, A. M., Aguado, A. and Sánchez-Torres, P. (2012). Molecular tools for detection of plant pathogenic fungi and fungicide resistance. In: Cumagun, C. J. R. (Ed.). *Plant Pathology*. IntechOpen.

Carolan, K., Helps, J., van den Berg, F., Bain, R., Paveley, N. and van den Bosch, F. (2017). Extending the durability of cultivar resistance by limiting epidemic growth rates. *Proceedings. Biological Sciences* 284(1863), 20170828. http://doi.org/10.1098/rspb.2017.0828.

Chen, S., Schnabel, G., Yuan, H. and Luo, C. (2019). LAMP detection of the genetic element 'Mona' associated with DMI resistance in *Monilinia fructicola*. *Pest Management Science* 75(3), 779–786. https://doi.org/10.1002/ps.5178.

Craw, P. and Balachandran, W. (2012). Isothermal nucleic acid amplification technologies for point-of-care diagnostics: A critical review. *Lab on a Chip* 12(14), 2469–2486. https://doi.org/10.1039/C2LC40100B.

David, M., Dursi, L. J., Yao, D., Boutros, P. C. and Simpson, J. T. (2017). Nanocall: An open source basecaller for Oxford nanopore sequencing data. *Bioinformatics* 33(1), 49–55. https://doi.org/10.1093/bioinformatics/btw569.

Deamer, D., Akeson, M. and Branton, D. (2016). Three decades of nanopore sequencing. *Nature Biotechnoly* 6, 518–524. https://doi.org/10.1038/nbt.3423.

Delahaye, C. and Nicolas, J. (2021). Sequencing DNA with nanopores: Troubles and biases. *PLoS ONE* 16(10), e0257521. https://doi.org/10.1371/journal.pone.0257521.

Deng, H. and Gao, Z. (2015). Bioanalytical applications of isothermal nucleic acid amplification techniques. *Analytica Chimica Acta* 853, 30–45. https://doi.org/10.1016/j.aca.2014.09.037.

Dodhia, K. N., Oliver, R. P. and Lopez-Ruiz, F. J. (2019). Faster than a football match: Towards in situ disease and fungicide resistance quantification [Oral presentation]. 20th International Reinhardsbrunn Symposium Modern Fungicides and Antifungal Compounds, Friedrichroda, Germany.

Dodhia, K. N., Cox, B. A., Oliver, R. P. and Lopez-Ruiz, F. J. (2021). Rapid in situ quantification of the strobilurin resistance mutation G143A in the wheat pathogen *Blumeria graminis* f. sp. *tritici*. *Scientific Reports* 11(1), 4526. https://doi.org/10.1038/s41598-021-83981-9.

Duan, Y., Zhang, X., Ge, C., Wang, Y., Cao, J., Jia, X., Wang, J. and Zhou, M. (2014). Development and application of loop-mediated isothermal amplification for detection of the F167Y mutation of carbendazim-resistant isolates in *Fusarium graminearum*. *Scientific Reports* 4, 7094. https://doi.org/10.1038/srep07094.

Duan, Y. B., Yang, Y., Wang, J. X., Liu, C. C., He, L. L. and Zhou, M. G. (2015). Development and application of loop-mediated isothermal amplification for detecting the highly benzimidazole-resistant isolates in *Sclerotinia sclerotiorum*. *Scientific Reports* 5, 17278. https://doi.org/10.1038/srep17278.

Duan, Y., Yang, Y., Li, T., Zhao, D., Cao, J., Shi, Y., Wang, J. and Zhou, M. (2016). Development of a rapid and high-throughput molecular method for detecting the F200Y mutant genotype in benzimidazole-resistant isolates of *Fusarium asiaticum*. *Pest Management Science* 72(11), 2128–2135. https://doi.org/10.1002/ps.4243.

Duan, Y. B., Yang, Y., Wang, J. X., Chen, C. J., Steinberg, G., Fraaije, B. A. and Zhou, M. G. (2018a). Simultaneous detection of multiple benzimidazole-resistant β-tubulin variants of *Botrytis cinerea* using loop-mediated isothermal amplification. *Plant Disease* 102(10), 2016–2024. https://doi.org/10.1094/PDIS-03-18-0542-RE.

Duan, Y. B., Yang, Y., Li, M. X., Li, T., Fraaije, B. A. and Zhou, M. G. (2018b). Development and application of a simple, rapid and sensitive method for detecting moderately carbendazim-resistant isolates in *Botrytis cinerea*. *Annals of Applied Biology* 172(3), 355–365. https://doi.org/10.1111/aab.12426.

Dyussembayev, K., Sambasivam, P., Bar, I., Brownlie, J. C., Shiddiky, M. J. A. and Ford, R. (2021). Biosensor Technologies for Early Detection and Quantification of Plant Pathogens. *Frontiers in Chemistry* 9, 636245. https://doi.org/10.3389/fchem.2021.636245.

Eid, J., Fehr, A., Gray, J., Luong, K., Lyle, J., Otto, G., Peluso, P., Rank, D., Baybayan, P., Bettman, B., Bibillo, A., Bjornson, K., Chaudhuri, B., Christians, F., Cicero, R., Clark, S., Dalal, R., Dewinter, A., Dixon, J., Foquet, M., Gaertner, A., Hardenbol, P., Heiner, C., Hester, K., Holden, D., Kearns, G., Kong, X., Kuse, R., Lacroix, Y., Lin, S., Lundquist, P., Ma, C., Marks, P., Maxham, M., Murphy, D., Park, I., Pham, T., Phillips, M., Roy, J., Sebra, R., Shen, G., Sorenson, J., Tomaney, A., Travers, K., Trulson, M., Vieceli, J., Wegener, J., Wu, D., Yang, A., Zaccarin, D., Zhao, P., Zhong, F., Korlach, J. and Turner, S. (2009). Real-time DNA sequencing from single polymerase molecules. *Science* 323(5910), 133–138. https://doi.org/10.1126/science.1162986.

Escobar-Zepeda, A., Vera-Ponce de Leon, A. and Sanchez-Flores, A. (2015). The road to metagenomics: From microbiology to dna sequencing technologies and bioinformatics. *Frontiers in Genetics* 6, 348. https://doi.org/10.3389/fgene.2015.00348.

Fan, F., Yin, W. X., Li, G. Q., Lin, Y. and Luo, C. X. (2018). Development of a LAMP method for detecting SDHI fungicide resistance in *Botrytis cinerea*. *Plant Disease* 102(8), 1612–1618.

Fang, Y. and Ramasamy, R. P. (2015). Current and prospective methods for plant disease detection. *Biosensors* 5(3), 537–561. https://doi.org/10.3390/bios5030537.

Fernández-Ortuño, D., Grabke, A., Li, X. and Schnabel, G. (2015). Independent emergence of resistance to seven chemical classes of fungicides in *Botrytis cinerea*. *Phytopathology* 105(4), 424–432. https://doi.org/10.1094/PHYTO-06-14-0161-R.

Fillinger, S., Leroux, P., Auclair, C., Barreau, C., Al Hajj, C. and Debieu, D. (2008). Genetic analysis of fenhexamid-resistant field isolates of the phytopathogenic fungus *Botrytis cinerea*. *Antimicrobial Agents and Chemotherapy* 52(11), 3933–3940. https://doi.org/10.1128/AAC.00615-08.

Fraaije, B. A., Butters, J. A., Coelho, J. M., Jones, D. R. and Hollomon, D. W. (2002). Following the dynamics of strobilurin resistance in *Blumeria graminis* f. sp. *tritici* using quantitative allele-specific real-time PCR measurements with the fluorescent dye SYBR Green I. *Plant Pathology* 51, 45–54. https://doi.org/10.1046/j.0032-0862.2001.00650.x.

Fukuta, S., Mizukami, Y., Ishida, A. and Kanbe, M. (2006). Development of loop-mediated isothermal amplification (LAMP)-based SNP markers for shelf-life in melon (*Cucumis melo* L.). *Journal of Applied Genetics* 47(4), 303–308. https://doi.org/10.1007/BF03194639.

Fungicide Resistance Action Committee (2020). List of first confirmed cases of plant pathogenic organisms resistant to disease control agents. Available at: https://www.frac.info/docs/default-source/publications/list-of-resistant-plant-pathogens/list-of-first-confirmed-cases-of-plant-pathogenic-organisms-resistant-to-disease-control-agents_05_2020.pdf?sfvrsn=7073499a_2.

Fungicide Resistance Action Committee (2022). FRAC recommendations for QoI fungicides. Available at: https://www.frac.info/frac-teams/working-groups/qol-fungicides/recommendations-for-qoi.

Gill, P. and Amree, A. H. (2020). AS-LAMP: A new and alternative method for genotyping. *Avicenna Journal of Medical Biotechnology* 12(1), 2–8.

Gullino, M. L., Bertetti, D., Monchiero, M. and Garibaldi, A. (2000). Sensitivity to anilinopyrimidines and phenylpyrroles in *Botrytis cinerea* in north-Italian vineyards. *Phytopathologia Mediterranea* 39, 433–446. Available at: https://www.jstor.org/stable/26456575.

Gutierrez Vazquez, Y., Adams, I. P., McGreig, S., Walshaw, J., van den Berg, F., Sanderson, R., Pufal, H., Conyers, C., Langton, D., Broadhead, R., Harrison, C. and Boonham, N. (2022). Profiling azole resistant haplotypes within Zymoseptoria tritici populations using nanopore sequencing. *Frontiers in Agronomy* 4, 943440. https://doi.org/10.3389/fagro.2022.943440.

Hardinge, P. and Murray, J. A. H. (2019). Reduced false positives and improved reporting of loop-mediated isothermal amplification using quenched fluorescent primers. *Scientific Reports* 9(1), 7400. https://doi.org/10.1038/s41598-019-43817-z.

Harper, L. A., Paton, S., Hall, B., McKay, S., Oliver, R. P. and Lopez-Ruiz, F. J. (2022). Fungicide resistance characterized across seven modes of action in *Botrytis cinerea* isolated from Australian vineyards. *Pest Management Science* 78(4), 1326–1340. https://doi.org/10.1002/ps.6749.

Harrison, C., Tomlinson, J., Ostoja-Starzewska, S. and Boonham, N. (2017). Evaluation and validation of a loop-mediated isothermal amplification test kit for detection of *Hymenoscyphus fraxineus*. *European Journal of Plant Pathology* 149(2), 253–259. https://doi.org/10.1007/s10658-017-1179-8.

Hawkins, N. J., Bass, C., Dixon, A. and Neve, P. (2019). The evolutionary origins of pesticide resistance. *Biological Reviews of the Cambridge Philosophical Society* 94(1), 135-155 https://doi.org/10.1111/brv.12440.

Heick, T. M., Matzen, N. and Jørgensen, L. N. (2020). Reduced field efficacy and sensitivity of demethylation inhibitors in the Danish and Swedish *Zymoseptoria tritici* populations. *European Journal of Plant Pathology* 157(3), 625–636. https://doi.org/10.1007/s10658-020-02029-2.

Hellin, P., Duvivier, M., Clinckemaillie, A., Bataille, C., Legrève, A., Heick, T. M., Jørgensen, L. N., Andersson, B., Samils, B., Rodemann, B., Berg, G. and Kildea, S. (2020). Multiplex qPCR assay for simultaneous quantification of CYP51-S524T and SdhC-H152R substitutions in European populations of *Zymoseptoria tritici*. *Plant Pathology* 69(9), 1666–1677. https://doi.org/10.1111/ppa.13252.

Henson, J. M. and French, R. C. (1993). The polymerase chain reaction and plant disease diagnosis. *Annual Review of Phytopathology* 31, 81–109. https://doi.org/10.1146/annurev.py.31.090193.000501.

Hu, X. R., Dai, D. J., Wang, H. D. and Zhang, C. Q. (2017). Rapid on-site evaluation of the development of resistance to quinone outside inhibitors in *Botrytis cinerea*. *Scientific Reports* 7(1), 13861. https://doi.org/10.1038/s41598-017-13317-z.

Hollomon, D. W. (2015). Fungicide resistance: 40 years on and still a major problem. In: Ishii, H. and Hollomon, D. W. (Eds). *Fungicide Resistance in Plant Pathogens: Principles and a Guide to Practical Management*. Springer, Tokyo, pp. 3–11.

Hsu, T. H., Gwo, J. C. and Lin, K. H. (2012). Rapid sex identification of papaya (*Carica papaya*) using multiplex loop-mediated isothermal amplification (mLAMP). *Planta* 236(4), 1239–1246. https://doi.org/10.1007/s00425-012-1681-3.

Huf, A., Rehfus, A., Lorenz, K. H., Bryson, R., Voegele, R. T. and Stammler, G. (2018). Proposal for a new nomenclature for CYP51 haplotypes in *Zymoseptoria tritici* and analysis of their distribution in Europe. *Plant Pathology* 67(8), 1706–1712. https://doi.org/10.1111/ppa.12891.

Ishii, H., Bryson, P. K., Kayamori, M., Miyamoto, T., Yamaoka, Y. and Schnabel, G. (2021). Cross-resistance to the new fungicide mefentrifluconazole in DMI-resistant fungal pathogens. *Pesticide Biochemistry and Physiology* 171, 104737. https://doi.org/10.1016/j.pestbp.2020.104737.

Itonaga, M., Matsuzaki, I., Warigaya, K., Tamura, T., Shimizu, Y., Fujimoto, M., Kojima, F., Ichinose, M. and Murata, S. (2016). Novel methodology for rapid detection of KRAS mutation using PNA-LNA mediated loop-mediated isothermal amplification. *PLoS ONE* 11(3), e0151654. https://doi.org/10.1371/journal.pone.0151654.

Karst, S. M., Ziels, R. M., Kirkegaard, R. H., Sørensen, E. A., McDonald, D., Zhu, Q., Knight, R. and Albertsen, M. (2021). High-accuracy long-read amplicon sequences using unique molecular identifiers with nanopore or PacBio sequencing. *Nature Methods* 18(2), 165–169. https://doi.org/10.1038/s41592-020-01041-y.

King, K., Kirikyali, N., West, J. and Fraaije, B. (2017). Rapid LAMP Assays to Detect MgCYP51 and/or MgMFS1 overexpressing Strains of *Zymoseptoria tritici* in Leaf Samples. In: Deising, H. B., Fraaije, B., Mehl, A., Oerke, E. C., Sierotzki, H. and Stammler, G. (Eds). *Modern Fungicides and Antifungal Compounds* (vol. VIII), Deutsche Phytomedizinische Gesellschaft e.V pp. 67–72.

Klein, R. J., Zeiss, C., Chew, E. Y., Tsai, J. Y., Sackler, R. S., Haynes, C., Henning, A. K., SanGiovanni, J. P., Mane, S. M., Mayne, S. T., Bracken, M. B., Ferris, F. L., Ott, J., Barnstable, C. and Hoh, J. (2005). Complement factor H polymorphism in age-related macular degeneration. *Science* 308(5720), 385–389. https://doi.org/10.1126/science.1109557.

Kleemann, J., Ordonez-Roman, N., Mehl, A., Buer, B. and Al-Dilaimi, A. (2023). Portable Genotyping Analysis Platform (PGAP) – A novel approach to monitor fungicide resistance mutations from anywhere using Oxford nanopore Sequencing. In: 20th International Reinhardsbrunn Symposium Modern Fungicides and Antifungal Compounds, Friedrichroda, Germany.

Kogovšek, P., Hodgetts, J., Hall, J., Prezelj, N., Nikolić, P., Mehle, N., Lenarčič, R., Rotter, A., Dickinson, M., Boonham, N., Dermastia, M. and Ravnikar, M. (2015). LAMP assay and rapid sample preparation method for on-site detection of *Flavescence dorée* Phytoplasma in grapevine. *Plant Pathology* 64(2), 286–296. http://doi.org/10.1111/ppa.12266.

Komura, R., Kawakami, T., Nakajima, K., Suzuki, H. and Nakashima, C. (2018). Simultaneous detection of benzimidazole-resistant strains of Fusarium head blight using the loop-mediated isothermal amplification-fluorescent loop primer method. *Journal of General Plant Pathology* 84, 247–253. https://doi.org/10.1007/s10327-018-0788-1.

Latorre, B. A. and Torres, R. (2012). Prevalence of isolates of *Botrytis cinerea* resistant to multiple fungicides in Chilean vineyards. *Crop Protection* 40, 49-52. https://doi.org/10.1016/j.cropro.2012.03.022.

Lehner, M. S., Paula Júnior, T. J., Silva, R. A., Vieira, R. F., Carneiro, J. E. S., Schnabel, G. and Mizubuti, E. S. G. (2015). Fungicide sensitivity of *Sclerotinia sclerotiorum*: A thorough assessment using discriminatory dose, EC50, high-resolution melting analysis, and description of new point mutation associated with thiophanate-methyl resistance. *Plant Disease* 99(11), 1537–1543. https://doi.org/10.1094/PDIS-11-14-1231-RE.

Leroux, P., Chapeland, F., Desbrosses, D. and Gredt, M. (1999). Patterns of cross-resistance to fungicides in *Botryotinia fuckeliana* (*Botrytis cinerea*) isolates from French vineyards. *Crop Protection* 18(10), 687–697. https://doi.org/10.1016/S0261-2194(99)00074-5.

Leroux, P., Gredt, M., Leroch, M. and Walker, A. S. (2010). Exploring mechanisms of resistance to respiratory inhibitors in field strains of *Botrytis cinerea*, the causal agent of gray mold. *Applied and Environmental Microbiology* 76(19), 6615–6630. https://doi.org/10.1128/AEM.00931-10.

Leroux, P. and Walker, A. S. (2011). Multiple mechanisms account for resistance to sterol 14α-demethylation inhibitors in field isolates of *Mycosphaerella graminicola*. *Pest Management Science* 67(1), 44–59. https://doi.org/10.1002/ps.2028.

Li, X., Li, C., Li, G., Zhu, J., Liu, F., Jiang, L., Mu, W. and Liu, X. (2021). Detection of a point mutation (G143A) in cyt b of *Corynespora cassiicola* that confers pyraclostrobin resistance. *Horticulturae* 7, 155. https://doi.org/10.3390/horticulturae7060155.

Li, C., Wang, K., Zhang, H., Yang, D., Deng, Y., Wang, Y. and Qi, Z. (2022). Development of a LAMP method for detecting F129L mutant in azoxystrobin-resistant *Pyricularia oryzae*. *Fungal Biology* 126(1), 47–53. https://doi.org/10.1016/j.funbio.2021.10.002.

Li, F. J., Komura, R., Nakashima, C., Shimizu, M., Kageyama, K. and Suga, H. (2022). Molecular diagnosis of thiophanate-methyl-resistant strains of *Fusarium fujikuroi* in Japan. *Plant Disease* 106(2), 634–640. https://doi.org/10.1094/PDIS-07-21-1501-RE.

Liu, W., Huang, S., Liu, N., Dong, D., Yang, Z., Tang, Y., Ma, W., He, X., Ao, D., Xu, Y., Zou, D. and Huang, L. (2017). Establishment of an accurate and fast detection method using molecular beacons in loop-mediated isothermal amplification assay. *Scientific Reports* 7, 40125. https://doi.org/10.1038/srep40125.

Liu, Y. H., Yuan, S. K., Hu, X. R. and Zhang, C. Q. (2019). Shift of sensitivity in *Botrytis cinerea* to benzimidazole fungicides in strawberry greenhouse ascribing to the rising-lowering of E198A subpopulation and its visual, on-site monitoring by Loop-mediated Isothermal amplification. *Scientific Reports* 9(1), 11644. https://doi.org/10.1038/s41598-019-48264-4.

Lobato, I. M. and O'Sullivan, C. K. (2018). Recombinase polymerase amplification: Basics, applications and recent advances. *Trends in Analytical Chemistry* 98, 19–35. https://doi.org/10.1016/j.trac.2017.10.015.

Ma, Z. and Michailides, T. J. (2005). Advances in understanding molecular mechanisms of fungicide resistance and molecular detection of resistant genotypes in phytopathogenic fungi. *Crop Protection* 24(10), 853–863. https://doi.org/10.1016/j.cropro.2005.01.011.

Mair, W. J., Thomas, G. J., Dodhia, K., Hills, A. L., Jayasena, K. W., Ellwood, S. R., Oliver, R. P. and Lopez-Ruiz, F. J. (2020). Parallel evolution of multiple mechanisms for demethylase inhibitor fungicide resistance in the barley pathogen *Pyrenophora teres* f. sp. *maculata*. *Fungal Genetics and Biology: FG and B* 145, 103475-103482, 103475. https://doi.org/10.1016/j.fgb.2020.103475.

McCarthy, A. (2010). Third generation DNA sequencing: Pacific Biosciences' single molecule real time technology. *Chemistry and Biology* 17(7), 675–676. https://doi.org/10.1016/j.chembiol.2010.07.004.

McCartney, C., Mercer, P. C., Cooke, L. R. and Fraaije, B. A. (2007). Effects of a strobilurin-based spray programme on disease control, green leaf area, yield and development of fungicide-resistance in *Mycosphaerella graminicola* in Northern Ireland. *Crop Protection* 26(8), 1272–1280. https://doi.org/10.1016/j.cropro.2006.10.027.

Mohd-Assaad, N., McDonald, B. A. and Croll, D. (2016). Multilocus resistance evolution to azole fungicides in fungal plant pathogen populations. *Molecular Ecology* 25(24), 6124–6142. https://doi.org/10.1111/mec.13916.

Noakes, M. T., Brinkerhoff, H., Laszlo, A. H., Derrington, I. M., Langford, K. W., Mount, J. W., Bowman, J. L., Baker, K. S., Doering, K. M., Tickman, B. I. and Gundlach, J. H. (2019). Increasing the accuracy of nanopore DNA sequencing using a time-varying cross membrane voltage. *Nature Biotechnology* 37(6), 651–656. https://doi.org/10.1038/s41587-019-0096-0.

Notomi, T., Okayama, H., Masubuchi, H., Yonekawa, T., Watanabe, K., Amino, N. and Hase, T. (2000). Loop-mediated isothermal amplification of DNA. *Nucleic Acids Research* 28(12), E63. https://doi.org/10.1093/nar/28.12.e63.

Omrane, S., Sghyer, H., Audeon, C., Lanen, C., Duplaix, C., Walker, A. S. and Fillinger, S. (2015). Fungicide efflux and the MgMFS1 transporter contribute to the multidrug resistance phenotype in *Zymoseptoria tritici* field isolates. *Environmental Microbiology* 17(8), 2805–2823. https://doi.org/10.1111/1462-2920.12781.

Pan, L., Li, J., Zhang, W. N. and Dong, L. (2015). Detection of the I1781L mutation in fenoxaprop-p-ethyl-resistant American sloughgrass (*Beckmannia syzigachne* Steud.), based on the loop-mediated isothermal amplification method. *Pest Management Science* 71(1), 123–130. https://doi.org/10.1002/ps.3777.

Pereira, D., McDonald, B. A. and Croll, D. (2020). The genetic architecture of emerging fungicide resistance in populations of a global wheat pathogen. *Genome Biology and Evolution* 12(12), 2231–2244. https://doi.org/10.1093/gbe/evaa203.

Ren, W., Wang, Z., Zhu, M., Zhang, Y., Lian, S., Li, B., Dong, X. and Liu, N. (2022). Detection of cytb point mutation that confers high-level resistance to pyraclostrobin in Glomerella cingulata using LAMP method. *Plant Disease* 107(4), 1166–1171. https://doi.org/10.1094/PDIS-08-22-1992-RE.

Rockman, M. V. (2012). The QTN program and the alleles that matter for evolution: All that's gold does not glitter. *Evolution; International Journal of Organic Evolution* 66(1), 1–17. https://doi.org/10.1111/j.1558-5646.2011.01486.x.

Russell, P. E. (2002). *Sensitivity Baselines in Fungicide Resistance Research and Management*. (FRAC Monograph No. 3). Fungicide Resistance Action Committee, Brussels, Belgium.

R4P network (2021). Monitoring systems for resistance to plant protection products across the world: Between redundancy and complementarity. *Pest Management Science* 77(6), 2697–2709. https://doi.org/10.1002/ps.6275.

Schena, L., Nigro, F., Ippolito, A. and Gallitelli, D. (2004). Real-time quantitative PCR: A new technology to detect and study phytopathogenic and antagonistic fungi. *European Journal of Plant Pathology* 110(9), 893–908. https://doi.org/10.1007/s10658-004-4842-9.

Sereika, M., Kirkegaard, R. H., Karst, S. M., Michaelsen, T. Y., Sørensen, E. A., Wollenberg, R. D. and Albertsen, M. (2022). Oxford nanopore R10.4 long-read sequencing enables the generation of near-finished bacterial genomes from pure cultures and metagenomes without short-read or reference polishing. *Nature Methods* 19(7), 823–826. https://doi.org/10.1038/s41592-022-01539-7.

Sergeeva, V., Nair, N. G., Verdanega, J. R., Shen, C., Barchia, I. and Spooner-Hart, R. (2002). First report of anilinopyrimidine-resistant phenotypes in *Botrytis cinerea* on grapevines in Australia. *Australasian Plant Pathology* 31(3), 299–300. https://doi.org/10.1071/AP02028.

Shao, Y., Zhu, S., Jin, C. and Chen, F. (2011). Development of multiplex loop-mediated isothermal amplification-RFLP (mLAMP-RFLP) to detect *Salmonella* spp. and *Shigella* spp. in milk. *International Journal of Food Microbiology* 148(2), 75–79. https://doi.org/10.1016/j.ijfoodmicro.2011.05.004.

Shrestha, S., Neubauer, J., Spanner, R., Natwick, M., Rios, J., Metz, N., Secor, G. A. and Bolton, M. D. (2020). Rapid detection of *Cercospora beticola* in sugar beet and mutations associated with fungicide resistance using LAMP or probe-based qPCR. *Plant Disease* 104(6), 1654–1661. https://doi.org/10.1094/PDIS-09-19-2023-RE.

Sierotzki, H. and Gisi, U. (2002). Molecular diagnostics for fungicide resistance in plant pathogens. In: Voss, G. and Ramos, G. (Eds). *Chemistry of Crop Protection*, Wiley, pp. 71–88.

Sierotzki, H., Mehl, A. and Stammler, G. (2019). Molecular detection methods for fungicide resistance. In: Stevenson, K. L., McGrath, M. T. and Wyenandt, C. A. (Eds). *Fungicide Resistance in North America* (2nd edn.). The American Phytopathological Society, St. Paul, MN, pp. 305–334.

Spanner, R., Taliadoros, D., Richards, J., Rivera-Varas, V., Neubauer, J., Natwick, M., Hamilton, O., Vaghefi, N., Pethybridge, S., Secor, G. A., Friesen, T. L., Stukenbrock, E. H. and Bolton, M. D. (2021). Genome-wide association and selective sweep studies reveal the complex genetic architecture of dmi fungicide resistance in *Cercospora beticola*. *Genome Biology and Evolution* 13(9), evab209. https://doi.org/10.1093/gbe/evab209.

Stammler, G. and Speakman, J. (2006). Microtiter method to test the sensitivity of *Botrytis cinerea* to Boscalid. *Journal of Phytopathology* 154(7–8), 508–510. https://doi.org/10.1111/j.1439-0434.2006.01139.x.

Stenson, P. D., Mort, M., Ball, E. V., Evans, K., Hayden, M., Heywood, S., Hussain, M., Phillips, A. D. and Cooper, D. N. (2017). The human gene mutation database: Towards a comprehensive repository of inherited mutation data for medical research, genetic diagnosis and next-generation sequencing studies. *Human Genetics* 136(6), 665-677. https://doi.org/10.1007/s00439-017-1779-6.

Suleman, E., Mtshali, M. S. and Lane, E. (2016). Investigation of false positives associated with loop-mediated isothermal amplification assays for detection of *Toxoplasma gondii* in archived tissue samples of captive felids. *Journal of Veterinary Diagnostic Investigation* 28(5), 536–542. https://doi.org/10.1177/1040638716659864.

Talas, F., Kalih, R., Miedaner, T. and McDonald, B. A. (2016). Genome-wide association study identifies novel candidate genes for aggressiveness, deoxynivalenol production, and azole sensitivity in natural field populations of *Fusarium graminearum*. *Molecular Plant-Microbe Interactions* 29(5), 417–430. https://doi.org/10.1094/MPMI-09-15-0218-R.

Toffolatti, S. L., Russo, G., Bezza, D., Bianco, P. A., Massi, F., Marcianò, D. and Maddalena, G. (2020). Characterization of fungicide sensitivity profiles of *Botrytis cinerea* populations sampled in Lombardy (Northern Italy) and implications for resistance management. *Pest Management Science* 76(6), 2198–2207. https://doi.org/10.1002/ps.5757.

Tucker, M. A., Lopez-Ruiz, F., Cools, H. J., Mullins, J. G., Jayasena, K. and Oliver, R. P. (2020). Analysis of mutations in West Australian populations of *Blumeria graminis* f. sp. *hordei* CYP51 conferring resistance to DMI fungicides. *Pest Management Science* 76(4), 1265–1272. https://doi.org/10.1002/ps.5636.

van den Bosch, F., Paveley, N., Shaw, M., Hobbelen, P. and Oliver, R. (2011). The dose rate debate: Does the risk of fungicide resistance increase or decrease with dose? *Plant Pathology* 60(4), 597–606. https://doi.org/10.1111/j.1365-3059.2011.02439.x.

Vielba-Fernandez, A., de Vicente, A., Perez-Garcia, A. and Fernandez-Ortuño, D. (2019). Monitoring methyl benzimidazole carbamate-resistant isolates of the cucurbit powdery mildew pathogen, Podosphaera xanthii, using loop-mediated isothermal amplification. *Plant Disease* 103(7), 1515–1524. https://doi.org/10.1094/PDIS-12-18-2256-RE.

Vrana, K. E. (1996). Advancing technologies in gene amplification. *Trends in Biotechnology* 14(11), 413–415. https://doi.org/10.1016/0167-7799(96)30022-X.

Wang, Y., Zhao, Y., Bollas, A., Wang, Y. and Au, K. F. (2021). Nanopore sequencing technology, bioinformatics and applications. *Nature Biotechnology* 39(11), 1348–1365. https://doi.org/10.1038/s41587-021-01108-x.

Walker, A. S., Micoud, A., Rémuson, F., Grosman, J., Gredt, M. and Leroux, P. (2013). French vineyards provide information that opens ways for effective resistance management of *Botrytis cinerea* (grey mould). *Pest Management Science* 69(6), 667–678. https://doi.org/10.1002/ps.3506.

Wang, L., Tu, H., Hou, H., Zhou, Z., Yuan, H., Luo, C. and Gu, Q. (2022). Occurrence and detection of carbendazim resistance in *Botryosphaeria dothidea* from apple orchards in China. *Plant Disease* 106(1), 207–214. https://doi.org/10.1094/PDIS-06-20-1204-RE.

Weber, R. W. S. and Hahn, M. (2011). A rapid and simple method for determining fungicide resistance in Botrytis. *Journal of Plant Diseases and Protection* 118(1), 17–25. Available at: http://www.jstor.org/stable/43229155.

Weirather, J. L., de Cesare, M., Wang, Y., Piazza, P., Sebastiano, V., Wang, X. J., Buck, D. and Au, K. F. (2017). Comprehensive comparison of Pacific Biosciences and Oxford nanopore Technologies and their applications to transcriptome analysis. *F1000Research* 6, 100. https://doi.org/10.12688/f1000research.10571.2.

Wu, J. Y., Hu, X. R. and Zhang, C. Q. (2019). Molecular detection of QoI resistance in *Colletotrichum gloeosporioides* causing strawberry anthracnose based on loop-mediated isothermal amplification assay. *Plant Disease* 103(6), 1319–1325. https://doi.org/10.1094/PDIS-09-18-1593-RE.

Yan, L., Zhou, J., Zheng, Y., Gamson, A. S., Roembke, B. T., Nakayama, S. and Sintim, H. O. (2014). Isothermal amplified detection of DNA and RNA. *Molecular Biosystems* 10(5), 970–1003.

Yin, D., Chen, X., Hamada, M. S., Yu, M., Yin, Y. and Ma, Z. (2015). Multiple resistance to QoIs and other classes of fungicides in *Botrytis cinerea* populations from strawberry in Zhejiang Province, China. *European Journal of Plant Pathology* 141(1), 169–177. https://doi.org/10.1007/s10658-014-0535-1.

Yuan, H., Hou, H., Zhou, Z., Tu, H. and Wang, L. (2022). Rapid detection of the E198A mutation of carbendazim-resistant isolates in *Colletotrichum gloeosporioides* by loop-mediated isothermal amplification. *Horticultural Plant Journal* 8, 289–296. https://doi.org/10.1016/j.hpj.2022.01.008.

Zhang, X., Lowe, S. B. and Gooding, J. J. (2014). Brief review of monitoring methods for loop-mediated isothermal amplification (LAMP). *Biosensors and Bioelectronics* 61, 491–499. https://doi.org/10.1016/j.bios.2014.05.039.

Zhu, J., Zhang, L., Li, H., Gao, Y., Mu, W. and Liu, F. (2020). Development of a LAMP method for detecting the N75S mutant in SDHI-resistant *Corynespora cassiicola*. *Analytical Biochemistry* 597, 113687. https://doi.org/10.1016/j.ab.2020.113687.

Zulak, K. G., Farfan-Caceres, L., Knight, N. L. and Lopez-Ruiz, F. J. (2023). Exploiting long read sequencing to detect azole fungicide resistance mutations in Pyrenophora teres species using unique molecular identifiers. *bioRxiv*. https://doi.org/10.1101/2023.05.05.539008.

Chapter 4

Crop disease control efficacy and selection for resistance: two sides of the same coin?

Frank van den Bosch, ADAS High Mowthorpe, UK; Stephen Parnell, The University of Warwick Wellesbourne, UK; and Neil Paveley, ADAS High Mowthorpe, UK

1 Introduction

Fungicides are used to control plant disease, so any resistance management strategy of practical relevance needs to provide effective disease control. It is easy to find ineffective fungicide programmes that minimise selection, but such programmes are of no practical use if they compromise effective disease control.

Fungicides should be used 'according to need', i.e. should be used as much as needed to provide effective control of the pathogen populations but not more (Jørgensen et al., 2017). The question then becomes whether some fungicide treatment programmes that provide effective disease control exert less selection for fungicide resistance than other treatment programmes that result in the same level of control. In general, a more intense treatment programme with higher dosages and/or more applications will exert a higher selection pressure on the pathogen to evolve strains less sensitive to the fungicides used.

Are the efficacy of disease control and selection for resistance two sides of the same coin? In this chapter, we explore what is known about the relationship

http://dx.doi.org/10.19103/AS.2022.0116.04

between disease control efficacy and selection for fungicide resistance. We will explore the effects of treatment programme organisation on selection and control. Additionally, we will explore the consequences for control efficacy and selection for resistance when non-fungicide-related disease control measures are introduced.

2 Control versus selection

2.1 The theory

When a fungicide treatment programme is in place, pathogen strains resistant or less sensitive to the fungicide will have a higher population/epidemic growth rate than strains sensitive to the fungicides (Fig. 1a). The fraction of the pathogen population consisting of resistant or less-sensitive strains will thus increase (selection for resistance, Fig. 1b). The larger the difference between the growth rate of the resistant and the sensitive strain, r_R and r_S, respectively, the faster the fraction resistant will increase (the larger the selection rate). We can thus measure selection rate, *s*, as the difference between the growth rate of the resistant and the sensitive strain.

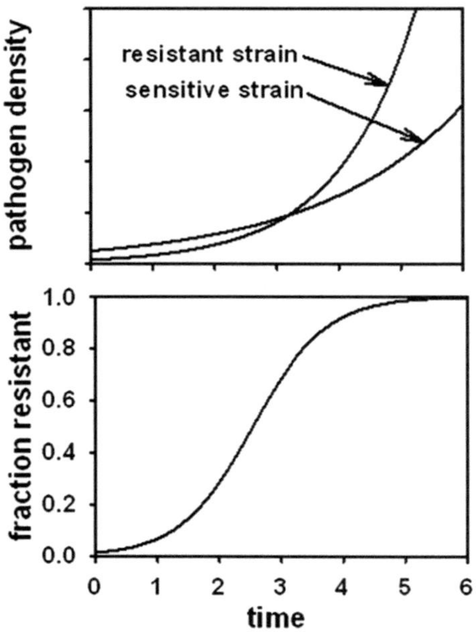

Figure 1 Top panel: The dynamics of the density of the fungicide-sensitive and the fungicide-resistant sub-populations through time when fungicides are applied. Bottom panel: The fraction of the pathogen population consisting of the resistant strain.

$$s = r_R - r_S \tag{1}$$

The amount of selection taking place during a crop growing season is also determined by the length of time the fungicide is active in the crop, *T*. We can thus write for selection per cropping season.

$$sT = (r_R - r_S)T \tag{2}$$

van den Bosch et al. (2014) have shown that this simple equation has a surprisingly strong predictive power. A comparison of predictions made using equation (2) with published field and lab experiments on the effect of fungicide treatment programme on selection showed a close correspondence (van den Bosch et al., 2014).

To quantify the relationship between selection for resistance and disease control efficacy, we need a measure of disease control using the same quantities as equation (2). The growth rate of the total pathogen population, $\overline{r}$, averaged over the two strains is given by

$$\overline{r} = \theta_R r_R + \theta_S r_S \tag{3}$$

where the frequency of the resistant strain in the pathogen population is given by θ_R, and that of the sensitive strain by θ_S, hence $\theta_R + \theta_S = 1$. This growth rate operates during the period the fungicide is present in the crop. In the rest of the growing season, where the fungicide is not active, the pathogen population grows with rate r_0, the growth rate of both strains when no fungicide is present. With an epidemic season of length G, the total growth rate of the pathogen population during one season is given by

$$\text{Seasonal growth rate} = r_0(G - T) + (\theta_R r_R + \theta_S r_S)T \tag{4}$$

or

$$\text{Seasonal growth rate} = r_0 G - \left(r_0 - (\theta_R r_R + \theta_S r_S)\right)T \tag{5}$$

The first part of this equation, r_0G, measures the growth of the pathogen population during a crop growing season when no fungicide is applied. The second part of the equation measures the reduction of the growth due to the fungicide application programme and thus measures disease control efficacy, *eff*. We thus have

$$eff = \left(r_0 - (\theta_R r_R + \theta_S r_S)\right)T \tag{6}$$

Using equations (2) and (6), we can now compare the effect of changes in fungicide application programmes on both selection for fungicide resistance and efficacy of disease control.

2.2 Predictions and a test using published data:

Following the system of van den Bosch et al. (2014), we use equations (2) and (6) to predict the effect of dose, split dose, alternation and mixtures on selection for resistance and efficacy of disease control. These predictions are then confronted with published evidence, using the same set of publications as considered by van den Bosch et al. (2014). Table 1 summarises this and will be referred to throughout the remainder of the chapter.

2.2.1 Total dose in the treatment programmes

Total dose in a treatment programme can be increased by (i) increasing the number of applications and (ii) increasing the dose per application.

- Increasing the number of applications increases the exposure time T, and it is clear from equations (2) and (6) that both selection and control efficacy increase (Fig. 2a).
- An increase in dose per application will decrease r_S and, in the case of partial resistance, r_R. Generally, dose response curves are such that the distance between the curve of the resistant strain and the curve for the sensitive strain increases for increasing dosages. Equations (2) and (6) thus show that increasing the dose per fungicide application increases both selection and control efficacy (Fig. 2a).

Of the 25 experiments summarised by van den Bosch et al. (2014), only 8 experiments published data on both the selection for fungicide resistance and the disease control efficacy (Table 1A). Seven of the eight experiments show,

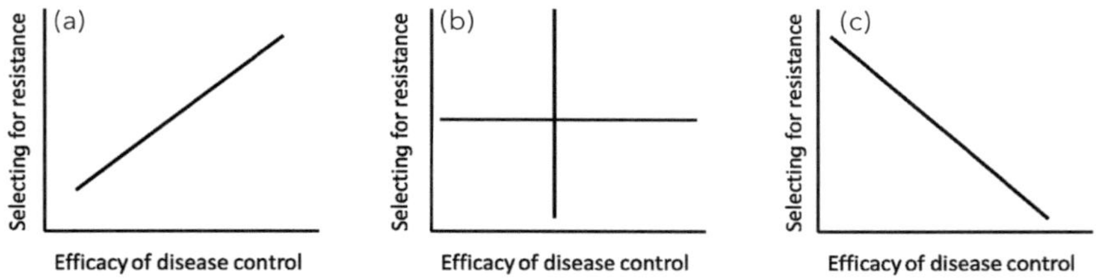

Figure 2 Hypothetical relationships between the efficacy of disease control and the selection for fungicide resistance. On the x-axis is the efficacy of disease control, on the y-axis the rate of selection for resistance.

as the theory predicts, that increasing control efficacy goes hand-in-hand with increasing selection for resistance, as shown in Fig. 2a.

2.2.2 Splitting fungicide dosages

van den Bosch et al. (2014) show that equation (2) predicts that splitting dose increases the selection for fungicide resistance, despite two opposing effects: a decrease in selection resulting from reduced dose per treatment and an increase in selection resulting from more applications increasing exposure time. A similar reasoning shows that equation (6) predicts that splitting fungicide dose increases disease control efficacy. Similar to the effect of total dose in the treatment programme, splitting dose increases both the efficacy of disease control and the rate of selection for resistance, Fig. 2a.

Only two papers describe a split dose experiment and report both the efficacy of disease control and the selection for resistance (Table 1B). One of the studies shows that both efficacy and selection increase when dosages are split. The other paper reports an increase in selection but the effect of dose splitting on efficacy is less clear. For some treatments efficacy increases, for others efficacy stays the same or decreases.

2.2.3 Alternations

Two different types of alternations are relevant here (Fig. 3). We consider the selection for resistance to fungicide A. Fungicide B, which has another mode of action (MOA) than fungicide A, is used as an alternation partner for which resistance is not considered. When the treatment programme consists of applications with fungicide A, Fig. 3, (i) an application with fungicide B is introduced in between two applications of fungicide A, or (ii) half the treatments with fungicide A are substituted by treatments with fungicide B.

In case (i), it is clear from equation (2) that selection for resistance to fungicide A does not change as none of the parameters driving the selection changes. Selection is only exerted when the fungicide is present, and we assume here for simplicity that in the time windows where fungicide B is active, no fungicide A is present. Since in total more fungicide is used, the efficacy of the treatment programme increases. See Fig. 2b for this case.

In case (ii), it is obvious from equation (2) that selection for resistance to fungicide A decreases as T decreases. When the efficacy of fungicide B is similar to that of fungicide A, we expect the efficacy to remain the same. See Fig. 2b for this case.

The published evidence for the alternations is inconclusive. Table 1C and D do not clearly underpin the findings from the theoretical analysis but also do not invalidate it.

Table 1 Summary of the published evidence on the effect of changes to a fungicide treatment programme on the rate of selection for fungicide resistance and on the efficacy of disease control.

A. Total dose in treatment programme

Pathogen	Crop	Fungicide	Dose range	Control	Selection	Reference
Venturia inaequalis	Apple	Benomyl	Range of dosages	Increase	Increase	Lalancette et al., 1987
Blumeria graminis f. sp. *tritici*	Wheat	Fenpropimorph	0 and full dose	Increase	Increase	Lorenz et al., 1992
Blumeria graminis f. sp. *tritici*	Wheat	Triadimenol	0 and full dose	Increase	Increase	Lorenz et al., 1992
Botrytis cinerea	Strawberry	Procymidone	Range of dosages	Increase	Increase	Hunter et al., 1987
Blumeria graminis f. sp. *hordei*	Barley	Ethirimol	Range of dosages	Variable	Increase	Hunter et al., 1984
M. graminicola	Wheat	Fluquinconazole	Range of dosages	Increase	Increase	Mavroeidi and Shaw, 2006
Rhynchosporium secalis	Barley	Epoxiconazole	0 and full dose	Increase	Increase	Cooke et al., 2004
M. graminicola	Wheat	Azoxistrobin	0 and full dose	Increase	Increase	Gisi et al., 2005

B. Split dose

Pathogen	Crop	Fungicide	Control	Selection	Reference
Rhynchosporium secalis	Barley	Epoxiconazole	Increase	Increase	Cooke et al., 2004
Blumeria graminis f.sp. *hordei*	Barley	Azoxystrobin	Variable, year and time in season	Increase	Fraaije, 2006

C. Alternations where sprays with other MOA are added in between two sprays with the to-be-protected fungicide

Pathogen	Crop	Fungicide to be protected	Alternating partner	Control	Selection	Reference
Botrytis cinerea	Grape	Iprodione	Captan	Increased	Increased	Northover and Matteoni, 1986
Rhynchosporium secalis	Barley	Epoxiconazole	Fenpropimorph, azoxystrobin	Increased	No difference	Cooke et al., 2004

D. Alternations where sprays are replaced by sprays with other fungicide

Pathogen	Crop	Fungicide to be protected	Alternate partner	Control	Selection	Reference
Botrytis cinerea	Grape	Iprodione	Captan	Not sign different	Decreased	Northover and Matteoni, 1986
Botrytis cinerea	Strawberry	Dicarboximide/procymidone	Dichlofluanid,	Slightly increased	Slight decrease	Hunter et al., 1987
Rhynchosporium secalis	Barley	Epoxiconazole	Fenpropimorph, azoxystrobin,	Dependent on year	No difference	Cooke et al., 2004

E. Mixtures

Pathogen	Crop	Fungicide	Mixing partner	Control	Selection	Reference
Venturia inaequalis	Apple	Benomyl	Mancozeb	Increase	Decrease	Lalancette et al., 1987
Blumeria graminis f. sp. *tritici*	Wheat	Triadimenol	Fenpropimorph	Increase	Decrease	Lorenz et al., 1992

(*Continued*)

Table 1 (*Continued*)

Pathogen	Crop	Fungicide	Mixing partner	Control	Selection	Reference
E. Mixtures						
Blumeria graminis f. sp. *tritici*	Wheat	Fenpropimorph	Triadimenol	No effect	No effect	Lorenz et al., 1992
Botrytis cinerea	Strawberry	Procymidone	Dichlofluanid	Increase	Variable	Hunter et al., 1987
Blumeria graminis f. sp. *hordei*	Barley	Ethirimol	Triadimenol	Increases	Decrease	Hunter et al., 1984
Phytophthora infestans	Potato	Oxadixyl	Mancozeb	Increase	Decrease	Cohen and Samoucha, 1990
Phytophthora infestans	Potato	Metalaxyl	Mancozeb	Increase	Decrease	Cohen and Samoucha, 1989
M. graminicola	Wheat	Fluquinconazole	Azoxystrobin	No difference	No difference	Mavroeidi and Shaw, 2006
Rhynchosporium secalis	Barley	Epoxiconazole	Fenpropimorph	Increase	Decrease	Cooke et al., 2004
M. graminicola	Wheat	Azoxistrobin	Epoxiconazole	Increase	Decrease	Gisi et al., 2005
Blumeria graminis f.sp. *hordei*	Barley	Azoxystrobin	Fenpropimorph	Increase	Decrease	Fraaije et al., 2006

F. Mixtures of two at-risk fungicides

Pathogen	Host	Fungicides	Selection resistant to F1	Selection resistance to F2	Reference
Blumeria graminis f. sp. *hordei*	Barley	F1 tridemorph F2 ethirimol	Reduced	Reduced	Brent et al., 1989
Blumeria graminis f. sp. *tritici*	Wheat	F1 fenpropimorph F2 triadimenol	Reduced	Reduced	Lorenz et al., 1992
Zymoseptoria tritici	Wheat	F1 pyraclostrobin F2 epoxiconazole	Reduced	Reduced	Thygesen et al., 2009
Zymoseptoria tritici	Wheat	F1 tebuconazole F2 isopyrazam	Reduce	Reduce	Young et al., 2017

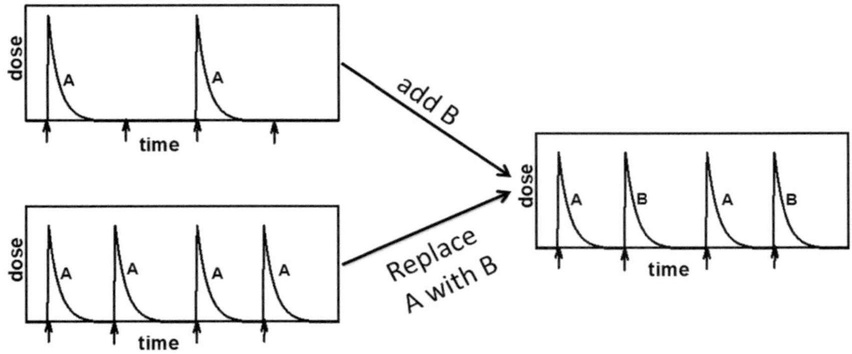

Figure 3 Two possible routes to the introduction of fungicide alternation. When in an existing treatment programme consisting of two applications of fungicide A, a second fungicide, B, is introduced in between the applications of fungicide A. When in a treatment programme of four applications with fungicide A, two of them are replaced by treatments with fungicide B.

2.2.4 Mixtures

Assume we add a fungicide, B, with a different mode of action to the fungicide to which resistance is developing, A. This fungicide B reduces the growth rate of both the sensitive and the resistant strain by a fraction β, as it has a different MOA. Equation (2) then becomes

$$sT = \left(\left(1-\beta\right)r_R - \left(1-\beta\right)r_S\right)T = \left(1-\beta\right)\left(r_R - r_S\right)T \tag{7}$$

and it is clear that adding a mixing partner decreases selection for resistance. The expression for efficacy becomes

$$eff = \left(r_0 - \left(\theta_R\left(1-\beta\right)r_R + \theta_S(1-\beta)r_S\right)\right)T = \left(r_0 - \left(1-\beta\right)\left(\theta_R r_R + \theta_S r_S\right)\right)T \tag{8}$$

From this expression, it is clear that the efficacy of disease control increases when a mixing partner, B, is added to fungicide A. This situation is shown in Fig. 2c. Mixing thus has both the advantage that it increases the efficacy of disease control and it decreases the rate of selection for fungicide resistance. This combination of effects is the most desired one when considering efficacy and selection simultaneously.

From the 49 experiments summarised in van den Bosch et al. (2014), 11 report both selection for resistance and disease control efficacy (Table 1E). Out of these 11, 8 experiments show that in the mixture applications compared to the solo product a decrease in selection for resistance goes paired with an increase in disease control efficacy. This agrees with the theoretical considerations. Two studies report no effect on control nor on selection. The final example reports an increase in efficacy and a variable effect on selection.

2.3 A specific example

A good experimental example of the contrasting effects of fungicide dose and fungicide mixtures on selection for resistance and on efficacy of disease control is given by Lalancette et al. (1987). In a field experiment on apple scab disease caused by the fungal pathogen *Venturia inaequalis*, they assessed the selection for benomyl resistance and the efficacy of disease control for a range of dosages of mixtures of benomyl and mancozeb. The apple trees were artificially inoculated with a mixture of sensitive and resistant *V. inaequalis* strains. Selection was quantified as the fraction resistance in the pathogen population at the end of the experiment. The efficacy of the disease control was quantified as the number of lesions per leaf. Fig. 4 reproduces the results presented in the paper.

The figure clearly shows that increasing the benomyl concentration increases the fraction of the pathogen population resistant to benomyl at the end of the experiment. So increasing dose increases selection. Also increasing benomyl dose decreases the number of lesions per leaf, implying the efficacy of disease control increases with increasing benomyl dose. Increasing efficacy goes paired with increasing selection for resistance. The effect of adding a mixing partner shows exactly the opposite relationship between efficacy of

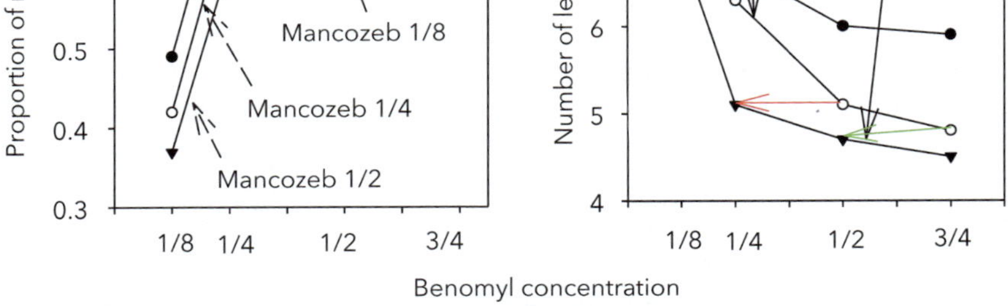

Figure 4 This figure is redrawn from the figure in Lalancette et al. (1987). Proportion of resistant lesions (panel A) and number of lesions per leaf (panel B) as a function of the dose of the benomyl treatment. Lines are for three values of the mancozeb dose. Coloured arrows are data points that keep equal efficacy of control but differ in the rate of selection for resistance.

control and selection. Increasing the dose of the mixing partner mancozeb decreases the rate of selection for resistance and increases the level of disease control.

2.4 Conclusion

We conclude that the theory, built around equations (2) and (6), agrees generally with observations of experiments reporting the efficacy of disease control and the selection for fungicide resistance. Adjusting fungicide application programmes aiming at reducing the selection for fungicide resistance will usually reduce the efficacy of disease control (Fig. 2a). Increasing disease control efficacy by adjusting the fungicide treatment programme normally increases the rate of selection for fungicide resistance (Fig. 2a). Since fungicides are used to keep effective control of the pathogen population, changes in the fungicide resistance programme that reduce the selection for resistance are of little practical relevance when they compromise effective disease control.

The holy grail of fungicide resistance management is to find changes to treatment programmes that are either neutral in their effect on efficacy, or even increase efficacy, and at the same time reduce the selection for resistance, Fig. 2c. The only change to a fungicide treatment programme that reduces the selection for fungicide resistance and increases the level of disease control is to add another fungicide with a different mode of action in mixture with the applied fungicides, Fig. 2c.

3 Reducing selection while maintaining control: fungicide mixtures

The previous section showed that changing a fungicide application programme by adding a mixing partner with a different MOA to a fungicide at risk of developing resistance will reduce the rate of selection for resistance and increase the efficacy of disease control. If the original fungicide treatment programme provided effective control of the target pathogen, then the mixture developed by adding a fungicide provides control over and above what is needed. This can be avoided by reducing to a certain extent the dose of the at-risk fungicide in the mixture. If done to the correct extent, this will bring the efficacy back to the level of the original application programme that provided the control needed. We explore this in the example of using chlorothalonil as a mixing partner for azoxystrobin to control *Zymoseptoria tritici* in wheat.

Both active substances are 'old chemistry', but serve as a useful example to demonstrate principles. Azoxystrobin was used as a case where a high level of resistance against a single-site acting fungicide occurred by a single mutation.

Chlorothalonil was used as an example of an effective multi-site acting fungicide, against which no resistance occurred over decades of widespread use.

Z. tritici developed virtually absolute resistance to the QoI fungicides. We thus restrict the analysis to absolute resistance. The same findings also hold for cases where partial resistance develops but the notational complexity increases. The fungicide affects the rate of increase of the sensitive strain, and we denote by $\alpha_Q(D_Q)$ the fractional reduction in the rate of increase of the sensitive strain when experiencing dose D_Q of the fungicide. We thus have

$$rS = \alpha S\,(D) \tag{9}$$

$$rR = r_0 \tag{10}$$

Chlorothalonil is a multi-site fungicide and we assume no resistance is developing to chlorothalonil during the time scale under consideration. The fractional reduction in the rate of increase of both the QoI-sensitive and the QoI-resistant strain is given by $\alpha_M(D_M)$, where D_M is the dose of the mixing partner. For the mixture of QoI and mixing partner, we then have

$$r_S = r_0 \alpha_M \left(D_M\right) \alpha_Q \left(D_Q\right) \tag{11}$$

$$r_R = r_0 \alpha_M \left(D_M\right) \tag{12}$$

We use the commonly used dose response curves,

$$\alpha_Q\left(D_Q\right) = 1 - RD_Q + RD_Q\, e^{-k_Q D_Q} \tag{13}$$

$$\alpha_M\left(D_M\right) = 1 - RD_M + RD_M e^{-k_Q D_Q} \tag{14}$$

Data from the *Agriculture and Horticulture Development Board (*AHDB) fungicide performance trials were used to parameterise the dose response curves of chlorothalonil and azoxystrobin (https://ahdb.org.uk/fungicide-performance-guide). The parameter values of the system are given in Table 2.

Figure 5 shows the dose response curves of azoxystrobin and chlorothalonil. These are then used in Fig. 6 to draw contour lines of equal efficacy and contour lines of equal selection rate as function of the dose of the fungicides in the mixture. The figure shows, as is to be expected on the basis of the discussion above, that increasing the dose of the QoI (follow the red arrow) increases the rate of selection for resistance to the QoI and it increases the efficacy of the mixture. If we increase the dose of the mixing partner (follow the green arrow), the selection for resistance to the QoI is reduced and the level of control increases. But there is also a possibility to keep the efficacy of

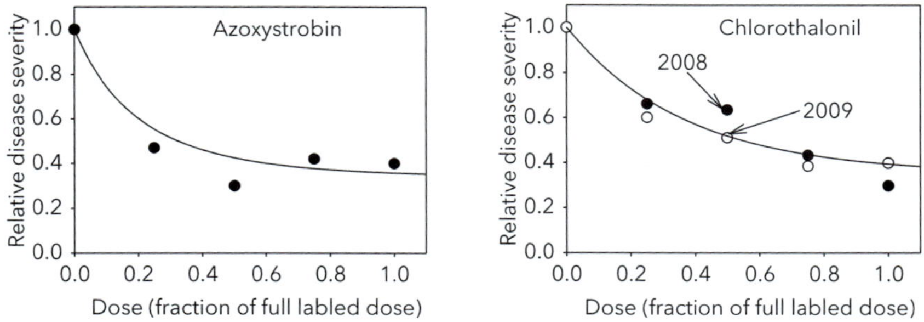

Figure 5 Dose response curves for treatments of Zymoseptoria tritici in wheat with the QoI fungicide azoxystrobin and the multi-site chlorothalonil.

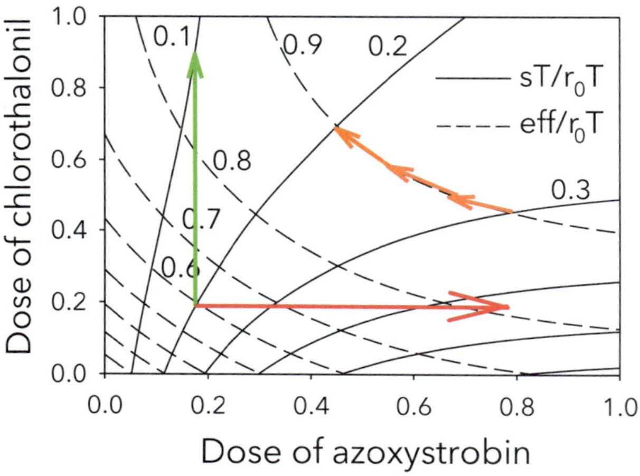

Figure 6 Contour lines of equal disease control efficacy and contour lines of equal rate of selection for resistance as function of the dosages of azoxystrobin and chlorothalonil in the mixture. The red line shows a situation where the dose of the QoI azoxystrobin is increased. The green line is a situation where the dose of chlorothalonil is increased. The orange line shows a case where the dosages of azoxystrobin and chlorothalonil are adjusted such that the level of disease control remains the same.

disease control constant while reducing the selection for resistance by moving along the efficacy contours (follow the orange arrow). We see that this involves increasing the dose of the mixing partner as well as reducing the dose of the QoI fungicide. In this way, control efficacy remains unchanged and selection for resistance decreases both due to adding a mixing partner and due to decreasing the use of the at-risk/QoI fungicide.

The published evidence usually compares full-rate solo with half-rate mixtures. But these mixtures turn out to have a different efficacy from the solo treatments. However, the published evidence on adding a mixing partner clearly shows that this reduces selection and increases control. The published evidence on the effect of dose clearly shows that increasing dose increases selection and increases control. This tells us that the combination of reducing dose of the at-risk fungicide and adding a mixing partner should reduce selection while effective disease control is retained.

The only experimental example from which some of the data form the right comparison is the specific example from Lalancette et al. (1987) described above. Figure 4 has two arrows in panel A and panel B. Following the red arrow from right to left, the benomyl concentration decreases and the mancozeb concentration increases. In both the start and the endpoint, there are 5.1 lesions per leaf on average. So this is a case where disease control remains constant when decreasing the dose of the at-risk fungicide and increasing the dose of the mixing partner. In panel A, the red arrow connects the same data points and it is clear that selection is reduced. The same holds for the green arrow, the level of disease control from right to left changes from 4.8 to 4.7, also most the same. Connecting the same data point in panel A shows that selection decreases for this change where the dose of the at-risk fungicide is decreased and the dose of the mixing partner is increased.

4 Two at-risk fungicides

So far, we considered mixtures of a fungicide to which resistance develops (at-risk fungicide) with a fungicide to which no resistance develops (low-risk fungicide) during the timescale of interest. Would mixing an at-risk fungicide with another at-risk fungicide, with a different mode of action, be a sensible way of developing fungicide mixtures? This is a relevant question because the number of multi-site/low-risk fungicides is small (some highly active multi-site products are no longer approved for use in Europe) and they generally have a lower efficacy than the single-site/at-risk fungicides.

There are two opposite forces that need to be considered. First, as we have seen in this chapter, mixing a fungicide A, to which resistance develops, with another fungicide B of another mode of action, reduces the rate of selection for resistance to A. *Vice versa*, the presence of fungicide A in the mixture also reduces the rate of development of resistance to fungicide B. However, adding fungicide B implies exposing also fungicide B to selection for fungicide resistance. Moreover, there are many examples of pathogens that evolve strains that are less sensitive or resistant to more than one fungicide MOA. If such a strain resistant to both A and B emerges, the mixture will quickly lose its ability to provide effective disease control. There is thus a question whether mixing of two at-risk fungicides will yield a product that maintains effective disease control longer than when the two fungicides would be applied in solo treatments either in different fields or sequentially.

Even though mixtures of two at-risk fungicides is an important topic, very little research has considered these mixtures.

Four papers report experiments on the selection for resistance in mixtures of at-risk fungicides. Brent et al. (1989), Lorenz et al. (1992), Thygesen et al. (2009) and Young et al. (2017) estimated selection rates for fungicide resistance in solo fungicides and in mixtures of two at-risk fungicides. All studies show that in the mixtures the selection for resistance was reduced for both at-risk fungicides (Table 1F). Although the number of studies is small, they show a consistent picture of reduced selection for resistance to both at-risk components of the mixtures. This is however only part of the answer. What the experimental studies did not assess is (i) the development of strains resistant to both fungicide MOAs and (ii) the net effect of the mixture on the durability of disease control.

i. There is a range of examples where a pathogen has acquired a mutation that confers resistance to a fungicide A as well as a mutation that confers resistance to a fungicide B with another MOA. The evolutionary dynamics of the emergence and selection of strains resistant to two MOA is, however, not well studied. One of the reasons for this is that, until recently, the molecular techniques to quantify the fraction of the pathogen population carrying two or more mutations were very labour intensive. The area of emergence of pathogen strains carrying resistance mutations to two of more MOAs needs urgent attention.
ii. Assessing the net effect of mixtures of at-risk fungicides on the durability of disease control requires the definition of effective disease control and the assessment of the time till effective disease control is lost. To this end, the authors have introduced the concept of effective life of a fungicide or fungicide mixtures. The effective life of a fungicide is the time from introduction of the fungicide onto the market till the moment effective disease control is lost. We (Hobbelen et al., 2013) have applied this definition to the example of *Z. tritici* controlled by two hypothetical fungicides (the parameters of the dose response curves were in the range found for QoI and azole fungicides). This model study clearly shows that for a broad range of parameter values mixtures give the same, or a longer effective life than any of the alternations studied. Use of the two fungicides as solo products in different fields generally gave the shortest effective life.

The model study of Hobbelen et al. (2013) clearly shows the potential use of mixtures of two at-risk fungicides with different MOAs. This is however only one model study. Moreover, no experimental evidence exists on the net effect of mixing two at-risk fungicides. Given the small number of low-risk/multi-site fungicides, and given the continuing pressure on their availability, it is of high importance to study the consequences of mixing two at-risk fungicides on effective life.

We note here that the evolution of strains resistant to more than one fungicide is a key concern in the use of mixtures. However, alternations suffer from the same problem with the selection pressures applied making the emergence of strains resistant to all components of the alternation likely. Even the use of fungicide mosaics, where each of the MOAs is used in different fields, may lead to selection for strains resistant to all fungicides in the system when spore exchange between fields is present.

5 Generalisation to integrated pest management measures

As discussed in the previous section, the selection rate of resistance to a fungicide is measured as

$$sT = \left(r_R - r_S\right)T \tag{15}$$

and the efficacy of disease control is measured by

$$eff = \left(r_0 - \left(\theta_R r_R + \theta_S r_S\right)\right)T \tag{16}$$

Assume we introduce a disease control measure that reduces the growth rate of the fungicide-sensitive and the fungicide-resistant/less-sensitive strain equally. Then denoting by β the fractional reduction of the growth rate of the pathogen strains, we find

$$sT = \left(\left(1-\beta\right)r_R - \left(1-\beta\right)r_S\right)T = \left(1-\beta\right)\left(r_R - r_S\right)T \tag{17}$$

$$eff = \left(r_0 - \left(1-\beta\right)\left(\theta_R r_R + \theta_S r_S\right)\right)T \tag{18}$$

We have used this reasoning to show that adding a mixing component to an at-risk fungicide reduces the rate of selection for resistance and will increase the effectiveness of disease control. However, the abstract equations (15) and (16) do not specify in any way that it should be a fungicide that reduces the pathogen population growth rate. It can, according to these equations, be any agricultural control measure that causes the reduction in growth rate. Reducing

Table 2 Parameters, their interpretation and numerical value of the model equations (11)–(14)

Parameter	Interpretation	Value	Dimension	Source
r_0	Epidemic growth rate in the absence of fungicide treatments	0.11	Day −1	Hobbelen et al., 2011
T_Q	Duration of QoI activity after application	10	Day	van den Bosch et al., 2017
T_M	Duration of mixing partner activity after application	15	Day	van den Bosch et al., 2017
RD_Q	Asymptote of QoI dose response curve	0.65	1	See Fig. 5
RD_M	Asymptote of mixing partner dose response curve	0.65	1	See Fig. 5
k_Q	Shape parameter of the QoI dose response curve	3.2	1	See Fig. 5
k_M	Shape parameter of the mixing partner dose response curve	2.8	1	See Fig. 5

the rate of increase of both the sensitive and the resistant strain *by any means* will reduce the rate of fungicide resistance development of the to-be-protected at-risk fungicide. If this theoretical assessment holds through in experimental practice, it opens a range of possible methods to reduce the selection for fungicide resistance.

> *Hypothesis:* Any disease control measure that reduces the epidemic growth rate (of both the fungicide-sensitive and the fungicide-resistant strains) will reduce the rate of build-up of fungicide resistance.

For example, compare the selection for fungicide resistance on a susceptible and on a partially resistant cultivar. On the susceptible cultivar, both the rate of increase of the sensitive and the resistant strain are large. Introducing a partially resistant cultivar will decrease the rate of increase of both sensitive and resistant strain. This thus implies that introducing partially resistant cultivars into the system will reduce the rate of build-up of fungicide resistance as well as increase the efficacy of disease control. Milgroom and Fry (1988) did observe this effect in their simulation model for the build-up of resistance to metalaxyl in potato blight. This observation remained unexplained at that moment.

The authors are working on the effect of partial resistance of wheat to the pathogen *Z. tritici* on the build-up of fungicide resistance. We have used a model to assess the effect of partial resistance on the effective life of a fungicide. The first results, shown in Fig. 7, show that on a more resistant cultivar the effective life of a fungicide is larger, as the governing principle predicts. Further model explorations have shown that introduction of cultivars with a higher level of resistance against the pathogen will reduce the rate of fungicide resistance selection *without any change in the fungicide application programme*. This work in progress will soon be submitted for publication.

We have only been able to find two papers where a susceptible and a partially resistant cultivar were used to assess selection for fungicide resistance. In the paper by McCartney et al. (2007), the difference in the level of resistance of the two cultivars turned out to be too small to reach any conclusions about the effect of cultivar resistance on selection for fungicide resistance/insensitivity. The paper by Kristoffersen et al. (2020) studies mixtures of resistant and susceptible cultivars. One comparison they make is the effect of a mixture of three susceptible and a mixture of three partially resistant cultivars on the selection for fungicide resistance. This paper is the first, and so far only, paper estimating selection for fungicide resistance on a susceptible and a resistant cultivar. The data suggest that selection for fungicide resistance is smaller on the mixture of resistant cultivars.

We conclude that there is an urgent need for research into the effect of cultivar resistance on the selection for fungicide resistance as this could constitute a very effective way of fungicide resistance management.

There is a range of other integrated pest management (IPM) measures that can be taken to reduce the rate of epidemic growth, including:

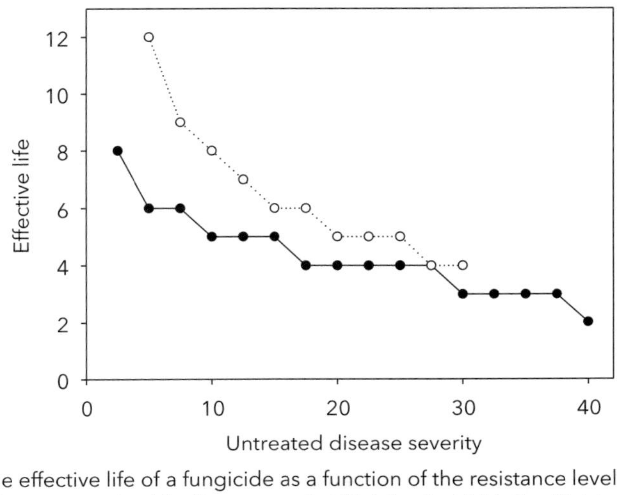

Figure 7 The effective life of a fungicide as a function of the resistance level of a cultivar. The model is parameterised for *Zymoseptoria tritici*; the fungicide is a QoI-type fungicide. This is an example output of our preliminary model. Closed circles are applications at two full dosages per season, and open circles are two applications at half dose rate per season.

- Genetic IPM: Cultivar resistance.
- Cultural IPM: Sowing date, crop rotation, intercropping, different tillage methods.
- Physical IPM: Use of barriers, traps.
- Forecasting IPM: Use a weather-based Decision Support System to guide the need for fungicide treatment.

Our hypothesis predicts that all these individually or in combination will reduce the rate of selection for fungicide resistance. We are not aware of any experimental studies of the effect of IPM methods on the development and selection for fungicide resistance. We conclude that such studies are urgently needed to add to the machinery available for resistance management.

6 Mutual protection

So far, we have considered the development of fungicide resistance. In many crop pathosystems, effective disease control relies on the use of fungicides and cultivar genetic resistance against the pathogen. Just like fungicides, genetic cultivar resistance puts a selection pressure on the pathogen to evolve strains that are less prone to the control. Many cases are reported of the erosion of genetic cultivar resistance due to the development of more virulent pathogen strains. Control of plant pathogens is therefore maintained by two interactions: firstly, between the introduction of new host resistance genes and pathogen populations breaking this resistance; secondly between the introduction of new fungicide MOA and pathogen populations becoming insensitive to them.

Using the same reasoning as for selection for fungicide resistance that led to equation (2), we can argue that the rate of selection of virulent strains breaking cultivar resistance is quantified by the difference between the growth rate of the virulent strain, rV, and that of the avirulent strain, rA. When growing a cultivar with genetic (partial) resistance over a growing season of length T, we can quantify the total selection for virulence as

$$S_V T = \left(r_V - r_A\right) T \tag{19}$$

Following the same logic as for fungicide resistance management, we can argue that a disease control method that reduces the epidemic growth rate of both the virulent and the avirulent strain will reduce the rate of selection for virulence. If the fractional reduction of the epidemic growth rate of the chosen control measure is γ, then (19) becomes

$$S_V T = \left(r_V - r_A\right) T \tag{20}$$

One of the methods that can be applied to reduce epidemic growth and that is indiscriminate of the virulence state of the pathogen strains (unless virulence and fungicide resistance are linked in clonal pathogens) is the use of fungicides.

The upshot of this is that fungicides and cultivar resistance can be used to mutually protect each other from the erosion of disease control efficacy because new strains evolve less prone to the control. We can thus formulate a set of three hypotheses:

H1: The deployment of crop resistance can reduce the selection for fungicide insensitivity.

H2: The deployment of fungicides can reduce selection for virulence.

H3: Integrated control using crop resistance genes and fungicides can greatly increase the durability of effective crop protection.

There is surprisingly little evidence and mechanistic understanding of the combined deployment of fungicides and crop resistance genes. The evolution of fungicide insensitivity and the evolution of virulence are virtually always studied in isolation. We have discussed the little evidence that exists for Hypothesis 1 in the previous sections. We know of only one manuscript that studies a model relating to Hypothesis 2. Carolan et al. (2017) develop a model for the dynamics of potato blight caused by the oomycete pathogen *Phytophthora infestans* and study the effect partial resistance to this pathogen can have on the selection for fungicide resistance. They show that the time from introduction of the cultivar until the moment the frequency of the virulence in the pathogen population has reached 95% significantly increases with the level of partial resistance. This conclusion holds for cultivars with 1, 2 and 3 partial resistance conferring genes. Though a clear example confirming Hypothesis 2, it is the only example existing to date. Moreover, no experimental evidence is available. We have not been able to find any studies testing Hypothesis 3.

We thus conclude that mutual protection of fungicides and cultivar resistance could be an important and effective way of developing sustainable disease control methods, but there is virtually no experimental nor modelling evidence to test the hypotheses underlying mutual protection. Work in this area is urgently needed to be able to improve the durability of crop protection methods.

7 Conclusion

Generally, an intensification of a fungicide treatment programme to increase disease control efficacy is associated with an increase in selection for fungicide-resistant pathogen strains. We have shown that there is one key exception to this, mixtures. Mixing fungicides with different MOAs can both decrease the selection for resistance and increase control efficacy. This dual benefit from mixing is most clear when an effective multi-site is added as a mixture partner to a single-site acting fungicide. The availability of effective multi-site fungicides is becoming restricted by pesticide regulation. Finding two single-site acting fungicides of similar efficacy, to provide mutual protection, is difficult for some important pathogens. Attention is shifting, therefore, towards IPM methods of resistance management.

We have shown that the mixture effect also holds, in principle, for adding another disease control method (such as cultivar resistance) to the control programme. A range of IPM methods is available that will reduce the rate of epidemic growth, thereby reducing the selection for resistance and enabling a reduction in the intensity of fungicide treatment that is needed.

Though the logic is clear and grounded in population genetic theory, there is very little experimental and modelling evidence to support these ideas. Research in these areas is urgently needed to provide a basis for development of long-term durable disease control.

8 Where to look for further information

The review paper by van den Bosch et al. (2014) gives a good introduction to the topic.

The Fungicide Resistance Action Group (FRAG-UK). The book by Ishii and Hollomon gives a further overview of the topic.

Groups of expertise include:

- The Fungicide Resistance Action Group (FRAG): FRAG produces guidance on pesticide resistance issues. Hosted by AHDB, this information can be used to help protect crops and the long-term efficacy of fungicides. https://ahdb.org.uk/knowledge-library/the-fungicide-resistance-action-group-frag-uk.
- Fungicide Resistance Action Committee (FRAC). FRAC works to prolong the effectiveness of fungicides liable to encounter resistance problems and to limit crop losses should resistance occur. https://www.frac.info/.
- Australian Fungicide Resistance Extension Network (AFREN). Supported by the GRDC, AFREN brings together regional plant pathologists, fungicide resistance experts and communications and extension specialists to develop and deliver fungicide resistance resources to growers and advisers across the country. https://afren.com.au/.

9 References

Brent, K. J., Carter, G. A., Hollomon, D. W., Hunter, T., Locke, T. and Proven, M. (1989). Factors affecting build-up of fungicide resistance in powdery mildew in spring barley. Netherl. *Journal of Plant Pathology* 95 (Suppl.1): 31–41.

Cohen, Y. and Samoucha, Y. (1989). Selection for metalaxyl resistance in potato crops infected with Phytophthora infestans: effects of fungicides and initial frequency of resistant sporangia. *Plant Pathology* 38: 382–390.

Cohen, Y. and Samoucha, Y. (1990). Competition between oxadixyl-sensitive and -resistant field isolates of *Phytophthora infestans* on fungicide-treated potato crops. *Crop Protection* 9(1): 15–20.

Cooke, L. R., Locke, T., Lockley, K. D., Phillips, A. N., Sadiq, M. D. S., Colla, R., Black, L., Taggart, P. J. and Mercer, P. C. (2004). The effect of fungicide programmes based on epoxiconazole on the control and DMI sensitivity of *Rhynchosporium secalis* in winter barley. *Crop Protection* 23: 393–406.

Fraaije, B. A., Burnett, F. J., Clark, W. S. and Lucas, J. A. (2006). Development and field testing of fungicide anti-resistance strategies, with particular reference to strobilurin QoI group of fungicides. Project Report No. 392. AHDB.

Gisi, U., Pavic, L., Stanger, C., Hugelshofer, U. and Sierotzki, H. (2005). Dynamics of Mycosphaerella graminicola populations in response to selection by different fungicides. In:. Modern fungicides and antifungal compounds IV 14th International Reinhardsbrunn Symposium, Friedrichroda, Thuringia, Germany, April 25-29, 2004: 89-101. https://eurekamag.com/research/004/418/004418069.php.

Hobbelen, P. H., Paveley, N. D. and van den Bosch, F. (2011). Delaying evolution of fungicide insensitivity by mixing fungicides at a low and high risk of resistance development. *Phytopathology* 101(10): 1224–1233.

Hobbelen, P. H. F., Paveley, N. D., Oliver, R. P. and van den Bosch, F. (2013). The value of alternation or mixtures of fungicides for delaying the selection of resistance against two modes of action in populations of Mycosphaerella graminicola on winter wheat. *Phytopathology* 101: 690–707.

Hunter, T., Brent, K. J. and Carter, G. A. (1984). Effects of fungicide regimes on sensitivity and control of barley mildew. *British Crop Protection Conference, Pests & Diseases* 95: 471–476.

Hunter, T., Brent, K. J., Carter, G. A. and Hutcheon, J. A. (1987). Effects of fungicide spray regimes on incidence of dicarboximide resistance in grey mould (*Botrytis cinerea*) on strawberry plants. *Annals of Applied Biology* 110(3): 515–525.

Ishii, H. and Hollomon, D. W. (editors) (2015). Fungicide Resistance in Plant Pathogens: Principles and a Guide to Practical Management 1st ed. 2015. *Springer verlag*. 502 pp.

Jørgensen, L. N., van den Bosch, F., Oliver, R. P., Heick, T. M. and Paveley, N. D. (2017). Targeting fungicide inputs according to need. *Annual Review of Phytopathology* 55: 181–203.

Kristoffersen, R., Heick, T. M., Müller, G. M., Eriksen, L. B., Nielsen, G. C. and Jørgensen, L. N. (2020). The potential of cultivar mixtures to reduce fungicide input and mitigate fungicide resistance development. *Agronomy for Sustainable Development* 40(5): 36. https://doi.org/10.1007/s13593-020-00639-y.

Lalancette, N., Hickey, K. D. and Cole Jr., H. (1987). Effects of mixtures of benomyl and mancozeb on build-up of benomyl-resistant *Venturia inaequalis. Phytopathology* 77(1): 86–91.

Lorenz, G., Saur, R., Schelberger, K., Forster, B., Kung, R. and Zobrist, P. (1992). Long-term monitoring results of wheat powdery mildew sensitivity towards fenpropimorph and strategies to avoid the development of resistance. In: *Brighton Crop Protection*

Conference: Pests and Diseases (vols. 1, 2, and 3). Farnham: British Crop Protection Council, 171–176.

Mavroeidi, V. I. and Shaw, M. W. (2006). Effects of fungicide dose and mixtures on selection for triazole resistance in *Mycosphaerella graminicola* under field conditions. *Plant Pathology* 55(6): 715–725.

McCartney, C., Mercer, P. C., Cooke, L. R. and Fraaije, B. A. (2007). Effects of a strobilurin-based spray programme on disease control, green leaf area, yield and development of fungicide-resistance in Mycosphaerella graminicola in Northern Ireland. *Crop Protection* 26(8): 1272–1280.

Milgroom, M. G. and Fry, W. E. (1988). A simulation analysis of the epidemiological principles for fungicide resistance management in pathogen populations. *Phytopathology* 78(5): 565–570.

Northover, J. and Matteoni, J. A. (1986). Resistance of *Botrytis cinerea* to benomyl and iprodione in vineyards and greenhouses after exposure to the fungicides alone or mixed with captan. Plant Disease 70: 398–402.

Thygesen, K., Jørgensen, L. N., Jensen, K. S. and Munk, L. (2009). Spatial and temporal impact of fungicide spray strategies on fungicide sensitivity of Mycosphaerella graminicola in winter wheat. *European Journal of Plant Pathology* 123(4): 435–447.

van den Bosch, F., Oliver, R., van den Berg, F. and Paveley, N. (2014). Governing principles can guide the development of fungicide resistance management tactics. *Annual Review of Phytopathology* 52: 175–195.

van den Bosch, F., Paveley, N., van den Berg, F., Hobbelen, P. and Oliver, R. (2014). Mixtures as a fungicide resistance management tactic. *Phytopathology* 104(12): 1264–1273.

van den Bosch, F., Lopez, F., Oliver, R., Paveley, N., Helps, J. and van den Berg, F. (2017). Cost benefit analysis of fungicide applications: to increase or decrease fungicide dose when resistance is developing. *Plant Pathology* 67: 549–560.

Young, C., van den Bosch, F., Fraajie, B., Burnett, F., Knight, S., Gosling, P. and Paveley, N. (2017). Strategies to delay the development of fungicide resistance: alternation, mixtures or restricting the number of treatments? *Aspects of Applied Biology* 134: 37-43

Chapter 5

Fungicide resistance risk assessment

Michael Grimmer, ADAS Boxworth, UK

1 Introduction

New fungicide active substances are at risk of losing efficacy due to the development of resistance in target pathogen populations. In some cases, resistance has been reported within a few years of introduction of a new fungicide (e.g. Georgopoulos, 1982); in other cases, it has taken decades for resistance to be identified (e.g. Noble et al., 1966). Efficacy has in some instances been almost entirely lost at once (e.g. Lucas, 2003); in other instances, the loss in efficacy has been incremental (e.g. Kendall et al., 1993). Fungicide resistance risk can be considered to encompass both the speed and the extent of resistance development.

Maintaining the efficacy of fungicide active substances has important implications for the grower, the crop protection industry and the environment. The grower benefits from having a range of effective disease control options. The crop protection industry benefits from a greater return on investment from fungicide products when their effective life is prolonged. In cases where efficacy is gradually eroded, a higher dose of a fungicide needs to be applied to a crop to achieve the same level of disease control. There is therefore broad interest in maintaining the efficacy of fungicide active substances. To achieve this, it is important to be able to understand the resistance risk and introduce appropriate anti-resistance strategies to slow or prevent resistance where there is perceived to be a higher risk. This chapter focuses on the assessment of fungicide resistance risk. The deployment of effective anti-resistance strategies is covered in Chapter 6, 'Good practice in minimising the development of fungicide resistance in crop pathogens', by Neil Paveley and Frank van den Bosch.

http://dx.doi.org/10.19103/AS.2022.0116.05

In the European Union, an assessment of resistance risk is required by regulatory authorities for the approval of each fungicide product (Anon, 2015). The risk assessments are used to guide the extent of anti-resistance strategies (termed 'risk modifiers') such as constraining the number of applications or requiring the mixture or alternation of a fungicide product with other effective fungicides. Resistance risk is also taken into account in the implementation of Substitution and Comparative Assessment (Anon, 2009). Within this process, plant protection products that fail to meet certain hazard criteria can have their approval revoked when a safer product for the same use becomes available and can be substituted for it. A greater diversity of modes of action (MoA) should remain available if the risk of resistance is high; a paucity of available MoA provides flexibility to regulatory authorities to maintain product approvals under these circumstances.

This chapter will review the development and scientific rationale of the widely used risk matrix approach to fungicide resistance risk assessment. This type of approach relies on previous experience of resistance with a particular fungicide mode of action (MoA), pathogen and agronomic system to assess the level of risk. A more recent trait-based approach that doesn't rely on prior knowledge will also be reviewed. Finally, this chapter will consider future trends in fungicide resistance risk assessment.

2 Historical development: the risk matrix approach

2.1 Background

For fungicide approval applications in the European Union, resistance risk is assessed by following guidance in European and Mediterranean Plant Protection Organisation (EPPO) Standard PP 1/213 (4) 'Resistance risk analysis' (Anon, 2015). Under this process, applicants are first required to consider the 'inherent risk' of resistance, described as being due to the interaction between certain characteristics of the target pathogen and the plant protection product and cannot be changed by the pattern of use. Applicants must then consider the 'agronomic risk' which is described as influenced by the particular characteristics of the crop, the geographic area in which the product is applied and the use pattern. The three main risk components described - pathogen risk, fungicide risk and agronomic risk - are discussed in more detail in subsequent sections. If the resistance risk of unrestricted use of the fungicide product (termed the 'unmodified risk') is considered unacceptable as it could lead to the rapid development of resistance, then the applicant is required to consider the risk if anti-resistance strategies are used (termed the 'modified risk'). If either the unmodified risk or the modified risk is considered acceptable, then the application can proceed towards product registration.

EPPO Standard PP 1/213 (4) advocates visualizing the 'inherent risk' of resistance on a risk matrix, whereby the pathogen risk and the fungicide risk are represented by two different axes. One of the first such resistance risk assessment schemes estimated comparative risk based on previous experience of resistance (Brent and Hollomon, 2007a). This scheme was later refined (Brent and Hollomon, 2007b) to include numerical values for 'basic fungicide risk' and 'basic disease risk'. The risk values for basic fungicide risk and basic disease risk were multiplied together to determine the combined risk. Kuck and Russell (2006) further refined the resistance risk assessment scheme by adding agronomic risk as a third factor in the matrix (Fig. 1). Agronomic risk was described as having two components: (i) an environment that was more or less

Fungicide Risk	**Risk**	**Combined Risk**			**Agronomic Risk**
Benzimidazoles	High = 3	3	6	9	High = 1
Dicarboximides		1.5	3	4.5	Medium = 0.5
Phenylamides		0.75	1.5	2.25	Low = 0.25
QoI fungicides					
Carboxanilides	Med = 2	2	4	6	High = 1
SBI fungicides		1	2	3	Medium = 0.5
Anilinopyrimidines		0.5	1	1.5	Low = 0.25
Phenylpyrroles					
Multi-site fungicides	Low = 0.5	0.5	1	1.5	High = 1
MBI-R inhibitors		0.25	0.5	0.75	Medium = 0.5
SAR inducers		0.125	0.25	0.3	Low = 0.25
		Low =1	Medium = 2	High = 3	**Pathogen Risk**
		Seed borne & soil borne pathogens Rusts	Eyespot *Rhynchosporium secalis* *Zymoseptoria tritici*	*Blumeria graminis* spp. *Botrytis cinerea* *Plasmopara viticola* *Magnaporthe grisea* *Venturia* spp.	

Figure 1 A risk matrix for fungicide resistance risk assessment (Kuck and Russell, 2006). The combined risk is calculated by multiplying together the risk values for fungicide risk, pathogen risk and agronomic risk.

conducive to the development and spread of disease and (ii) a cultural effect influenced by growers, including factors such as cropping intensity and choice of resistant or susceptible cultivars. Although certain regions of Europe were highlighted as representing a high risk for specific pathogens, the method for fitting examples into the low, medium or high agronomic risk categories was not made explicit.

2.2 Pathogen risk

The EPPO Standard on resistance risk assessment includes a table giving examples of pathogens considered to present a high risk of resistance development (see Table 1). It is stated that these pathogens have shown themselves to have the capacity to become resistant to particular fungicides in a short time. Whilst quantitative evidence regarding the speed and extent of resistance development for these pathogens relative to others is not made explicit here, such tables draw upon decades of industry experience, peer-review papers and industry publications. Many of these pathogens cause diseases of major arable and horticultural crops that have become difficult for growers to control because of the development of resistance to various fungicide MoA. In addition to listing perceived high-risk pathogens, the EPPO Standard states that the risk of resistance has generally been shown to be significantly higher for fungi possessing a combination of some of the following characteristics:

- Short life cycle and/or the ability to infect during a high proportion of the crop-growing season, resulting in many life cycles during the growing season;
- High epidemic growth rates (resulting, for example, from prolific spore production or a short latent period arising from the inherent characteristics of the pathogen and/or the environment including climate);

Table 1 Examples of plant pathogens considered to present a high risk of resistance development to fungicides, from EPPO Standard PP 1/213 (4)

Pathogen	Crop
Plasmopara viticola	Grapevine
Blumeria graminis spp.	Wheat and barley
Sphaerotheca spp.	Various
Mycosphaerella fijiensis	Banana
Magnaporthe grisea	Rice
Gibberella fujikuroi	Rice
Botrytis cinerea	Various, especially grapevine
Venturia spp.	Apple and pear
Ramularia collo-cygni	Barley

- Fungi which are 'specialist' pathogens (defined as pathogens which are only able to infect one or a few crop host species); and
- High genetic variability, e.g. seen as ability to easily generate stable insensitive mutant variants in the laboratory.
- Furthermore, it is stated that the speed of geographical spread of insensitive strains, and the impact of resistance, will be greater for pathogens which;
- can achieve widespread dispersal of spores or other propagules in space and time;
- have high epidemic growth rates; and
- infect host crop species which are extensively grown with short crop rotations, monocropping or continuous cropping.

Whilst there appears to be a good mechanistic rationale for linking some of these characteristics to resistance risk, evidence demonstrating a significant association appears to be lacking. It is not clear which of these traits are particularly associated with the listed high-risk pathogens (Table 1).

2.3 Fungicide risk

The EPPO Standard states that risk factors relating to the fungicide are very difficult to define for a new compound but more easily accepted for an established compound. This underlines a key limitation to the traditional risk matrix approach to fungicide resistance risk assessment. When fungicide products containing an active substance belonging to a new fungicide MoA are put forward for approval, there is no practical experience of their use to guide an assessment of resistance risk. It takes many years of practical use, perhaps a decade or longer, for resistance to develop in a sufficient number of pathogens to allow the speed of resistance development to be compared in broad terms with the existing fungicide MoA. The EPPO Standard applies a precautionary principle under these circumstances. It states that for new compounds, unless data are presented to prove the risk is low, a significant risk of potential resistance development should be assumed.

For a well-established fungicide active substance for which there are many years of practical experience against a wide range of pathogens, the degree of risk can be assessed with relative confidence. Where resistance has developed relatively quickly in different pathogens and to such an extent that efficacy has been severely affected, then the fungicide risk can be considered high. Where an active substance has been used against a wide range of pathogens over many years and resistance has occurred relatively slowly or to a limited extent, then the fungicide risk can be considered low.

For a new fungicide active substance within an 'old' fungicide MoA, the picture may be a little more complicated. Usually if a new fungicide has a similar structure and/or MoA to existing fungicides against which resistance has developed, then cross-resistance is found (Brent and Hollomon, 2007b). This means that, in effect, resistance to the new fungicide active substance is already present in the target pathogen population. However, cross-resistance patterns within a fungicide MoA can be complicated, and it is possible there may be no cross-resistance or even negative cross-resistance, in which pathogen isolates sensitive to one active substance are insensitive to another and vice versa. Applicants are encouraged within the EPPO Standard to conduct bioassay tests to explore the degree of cross-resistance with other active substances in the same fungicide MoA group. Whilst cross-resistance is stated to form part of the inherent risk, negative cross-resistance is also referred to as a potential anti-resistance strategy ('risk modifier'). However, caution is advisable here as further evolution of the target pathogen population could alter the cross-resistance patterns that existed at the time of testing. It seems likely that – all else being equal – a pathogen strain with resistance to more fungicides within a MoA has a greater selective advantage than a strain carrying resistance to fewer.

In addition to consideration of past experience with the MoA and cross-resistance, applicants are required to consider characteristics of the fungicide active substance that 'may favour' the development of resistance: persistent activity, monogenic resistance, ease of metabolism or a single-site MoA. However, evidence of a general link between the characteristic and the speed or extent of resistance development is in some cases lacking or unclear. Whilst it seems reasonable to assume that resistance will progress more quickly if determined to be monogenic through *in vitro* testing, there is potential for fitness costs to be expressed *in vivo* during practical use that may complicate this pattern (e.g. Scalliet et al., 2012). There is however a clear rationale for determining that multi-site fungicides carry a lower resistance risk than single-site fungicides as there is good evidence that resistance tends to develop more slowly with multi-site fungicides (Brent and Hollomon, 2007b).

2.4 Agronomic risk

The EPPO Standard states that the risk of resistance inherent in the plant protection product and the pathogen can be increased by certain conditions of use. Furthermore, it states that this agronomic risk affects selection pressure on the development of resistance and is influenced by the particular characteristics of the crop, the geographic area in which the product is applied and the use pattern. It is stated that the factors influencing the agronomic risk may include:

- Widely grown crop with short rotations;
- Monocropping or continuous cropping;

- Fungicide application techniques;
- Other cultural practices (e.g. fertilizers and cultivation);
- Need for high numbers of applications or long exposure to obtain control, because of the features of the crop environment;
- Use of cultivars susceptible to the pathogen(s);
- Geographical isolation of populations preventing the re-entry of sensitive forms;
- Environmental conditions favouring more frequent generations or higher population densities of the pathogen, e.g. in protected crops; a greater risk of resistance has been demonstrated where fungicides are used on protected crops (defined as a crop grown in a glasshouse or polytunnel) than on outdoor crops. This is also known for insecticides;
- Exclusive reliance on a single active substance; and
- Lack of diversity of available control measures.

At first glance, there appears to be some speculation involved in linking this long list of agronomic factors to the speed or extent of resistance development; no scientific evidence for such a link is provided. However, progress in our understanding can be made by grouping these factors together according to the processes which they influence. Firstly, there are those non-chemical factors that can influence disease severity, such as rotations, use of susceptible or resistant cultivars, growing a crop in a protected environment and cultural practices. Where such measures are used to constrain a disease epidemic, this reduces the need for chemical control and reduces the rate of selection for resistant pathogen strains. Secondly, there are those factors that relate to chemical control, such as the application technique, number of applications and availability of diverse active substances. Repeated application of a single active substance would likely result in a high rate of selection and be considered poor resistance management. The impact of non-chemical disease control methods and chemical control strategies on the rate of resistance development is explored in Chapter 6 of this book. Thirdly, there is a factor relating to geographical isolation of pathogen populations preventing the re-entry of sensitive forms, such as may occur in protected cropping. There is more on this topic in subsequent sections.

2.5 Predictive value of the matrix approach

Grimmer et al. (2014) provided the first empirical evaluation of the risk matrix approach, through examination of 67 unique cases of fungicide resistance that had occurred in Europe. The risk variable to be predicted by the risk matrix was the time from introduction of a fungicide MoA to control a particular pathogen to the 'first detection of resistance' in that pathogen (termed 'FDR time'). Each of the resistance cases was assigned to a pathogen risk and fungicide risk category according to the Kuck and Russell (2006) risk matrix. Since not all the pathogens and fungicides in the 67 cases were represented in the risk matrix, only 40 of them could be included in the analysis. This in itself represents a limitation of the risk matrix approach. Cases were also assigned to a medium or high agronomic risk category by Phil Russell, one of the authors of the Kuck and Russell (2006) scheme.

The analysis identified a statistically significant difference in FDR time between multi-site fungicides in the low-risk category and single-site fungicides in the medium- and high-risk categories. However, there was no significant difference in FDR time between the single-site fungicides in the medium-risk category and those in the high-risk category. It was considered that this represented a severe limitation to the risk matrix, given that future fungicide product approval requests would likely be predominately for single-site fungicides due to perceived regulatory difficulties in Europe of registering a multi-site active substance. The Kuck and Russell risk matrix was found in this analysis to have no significant predictive value for fungicide risk for single-site active substances.

Pathogen risk was found to have some predictive value, with a significantly longer FDR time for pathogens in the low-risk category compared with those in the medium- and high-risk categories. However, there was no significant difference between the medium- and high-risk categories. FDR time was significantly shorter for the high agronomic risk category than for the medium risk category. However, there was some confounding between agronomic and fungicide risks, as the five cases involving multi-site fungicides were in the medium agronomic risk category. This perhaps highlights a fundamental difficulty in developing and using a fungicide resistance risk matrix such as the Kuck and Russell scheme. If we consider, for example, the rapid breakdown in efficacy of methyl-benzimidazole carbamate (MBC) and demethylation inhibitor (DMI) fungicides against cucumber powdery mildew in Norway (Magnus, 1987), there are a number of potential factors that could have influenced the speed of resistance development. Did the resistance occur rapidly due to the active substances, or the MoA, or the evolutionary potential of the pathogen, or the particular crop and the way it is grown? Furthermore, particular MoA may be used against particular types of pathogen, which occur on particular types of crop. There is strong potential for confounding when subjectively assigning fungicides, pathogens and agronomic systems to risk categories. An objective method of determining risk would, in principle, be a more rigorous approach.

3 Beyond the matrix: a trait-based approach

Grimmer et al. (2015) developed a new, objective approach to fungicide resistance risk assessment that sought to overcome the main drawbacks of the risk matrix approach and to deepen scientific understanding of resistance development across pathosystems. The first drawback was the need for long experience of resistance with a particular fungicide MoA or pathogen before its risk level could be determined with any degree of confidence. The new methodology addressed this by attempting to associate risk not with particular fungicide MoA or pathogens, but with their underlying traits. Therefore, the MoA and pathogen traits could be used to determine the risk level, without prior experience of resistance occurring with them. The second drawback of the matrix approach is that it was found to have low predictive value for single-site fungicide MoA. Grimmer et al. (2014) found that the combined risk values (pathogen, fungicide and agronomic risk) of the matrix approach had good predictive value (R^2 = 0.73) across the whole data set for the time taken for resistance to first be detected (FDR time). However, once the small number of low-risk, multi-site-acting fungicide cases were removed from the analysis, the predictive value fell considerably (R^2 = 0.26) for single-site-acting fungicides. The objective, trait-based approach offered the potential to increase the predictive value by using mathematical modelling to identify the key traits of the system that were driving FDR time and weight them according to their importance.

The trait-based approach built on the fundamental principles of the risk matrix approach by starting with the hypothesis that the rate of fungicide resistance development depends on certain traits of the pathogen, fungicide and the agronomic system within which the fungicide is deployed. A set of candidate traits of fungicides, pathogens and agronomic systems were identified for which there was a rationale that they could influence the rate of resistance evolution, and hence the FDR time. The authors then obtained the data for these traits for each of the 61 European resistance cases from primary, peer-reviewed literature (preferably) or from books and published reports. In certain cases where data could not be found in the literature, expert judgement of the authors and colleagues was used. The traits were described as either variables (consisting of numeric values with a continuous distribution) or factors (consisting of discrete categories of numerical or descriptive information). All the trait variables and factors were individually tested for a significant association with FDR time, as a preliminary test of which traits might have predictive value. Following confirmation of this, the key traits explaining the variation in FDR time were identified through stepwise forward regression.

The traits that came out in the model as explaining the variation in FDR time were pathogen latent periods per year, fungicide molecular complexity, number of crop host species (1-9 or 10+) and agronomic system (outdoor or protected crop). The model combining these traits explained 61% of the variation in FDR time which represented a strong improvement over the risk matrix approach (26% of the variation in FDR time explained). Validation showed that this predictive power was highly unlikely to have occurred by random chance. Interestingly, the findings supported the prevailing, general view of fungicide resistance risk assessment which is that the pathogen, fungicide and agronomic system all are important components of the risk. The trait-based risk assessment approach has been adopted into EPPO Standard PP 1/213 (4) 'Resistance risk analysis' (Anon, 2015) as an alternative to the risk matrix approach.

The findings from the trait-based approach offer new insights into our understanding of why resistance occurs relatively quickly in some cases but relatively slowly in others. Pathogen latent periods per year (the epidemic period per year divided by the pathogen latent period) explained a large proportion of the variation in FDR time. This variable was considered likely to be highly correlated with pathogen generations per year (which was not included in the analysis owing to insufficient information). The rate of evolution will be the product of the selective advantage of the resistant strain per generation and the number of generations per year. Pathogen species with more crop hosts had longer FDR times, which was explained by the refugia effect whereby pathogens can reside in a range of hosts where no selection (fungicide use) is imposed. It was suggested that the effect of the agronomic system (protected or arable cropping) could operate through protected cropping reducing the refugia effect or alternatively by allowing a larger number of generations per year, owing to higher temperatures in protected environments shortening the pathogen latent period. Regarding fungicide molecular complexity, it was stated that the probability that a molecule binds to another molecule decreases as molecular complexity increases, with low-complexity molecules tending to be 'promiscuous' and high-complexity molecules being more specific. Therefore, for high-complexity fungicide molecules, any given amino acid variant in the target protein (arising through DNA mutation) would thus be more likely to result in reduced fungicide binding, compared with low-complexity molecules.

4 Conclusions and future trends

4.1 Strengthening trait-based risk assessment

The current trait-based risk assessment was published by Grimmer et al. in 2015. It is known that further cases of fungicide resistance have occurred and been documented in Europe since then. These would likely include cases of resistance in various pathogen species to succinate dehydrogenase inhibitor fungicides, providing added depth to the existing data set. Repeating the mathematical modelling work with a larger, richer data set would strengthen the trait-based risk assessment and may provide new insights into the key components governing fungicide resistance risk. This work should be a research priority.

4.2 Simplifying trait-based risk assessment

The current method of trait-based risk assessment requires applicants to engage with mathematical modelling outputs and perform calculations to determine the resistance risk. This could be simplified by instead providing look-up tables or by converting the trait-based assessment into a risk matrix. For example, the important, quantitative trait 'pathogen latent periods per year' could be divided into low-, medium- and high-risk categories. Where the number of latent periods per year is already known for a particular pathogen, it could be assigned to one of the risk categories. In effect, it should be feasible to develop a risk matrix similar to the one in Fig. 1 but with the categories determined by trait-based methodology and therefore offering improved predictive value.

4.3 Developing our understanding of the key traits

The findings of the trait-based approach offered new insights into the key processes driving fungicide resistance development. Further investigation into these is warranted, to improve our understanding and to explore the potential to leverage these processes as a means of slowing pathogen evolution. The potential role of refugia in resistance development could be investigated through *in vitro* and *in vivo* studies, potentially combined with mathematical modelling to explain the impact of refugia on resistance selection. The relationship between fungicide molecular complexity and resistance development could also be explored through experimental studies.

5 Where to look for further information

Chapter 6 of this book entitled 'Good practice in minimising the development of fungicide resistance in crop pathogens' by Neil Paveley and Frank van den Bosch provides sources of further information that are all of relevance for this chapter also.

6 References

Anon. (2009). Concerning the placing of plant protection products on the market and repealing Council Directives 17 79/117/EEC and 91/414/EEC. Regulation (EC). In: Official Journal of the European Union 18 1107/2009 of the European Parliament and of the Council of: L 309 24.11.2009 p. 1.

Anon. (2015). EPPO PP1/213 (4) Resistance risk analysis. *Bulletin OEPP/EPPO Bulletin* 45: 371-387. Available at: https://pp1.eppo.int/standards/PP1-213-4.

Brent, K. J. and Hollomon, D. W. (2007a). *Fungicide Resistance in Crop Pathogens: How Can It Be Managed?* FRAC Monograph No. 1 (second, revised edition), Fungicide Resistance Action Committee, Brussels, Belgium, 57 pp.

Brent, K. J. and Hollomon, D. W. (2007b). Fungicide Resistance: the assessment of risk. In: FRAC Monograph No. 2 (second, revised edition), Fungicide Resistance Action Committee, Brussels, Belgium, 53 pp.

Georgopoulos, S. G. (1982). Case study 1. *Cercospora beticola* of sugar beets. In: *Fungicide Resistance in Crop Protection*, Dekker, J., Georgopoulos, S. G. and Pudoc (Eds), Wageningen, Netherlands, pp. 187–194.

Grimmer, M. K., van den Bosch, F., Powers, S. J. and Paveley, N. D. (2014). Evaluation of a matrix to calculate fungicide resistance risk. *Pest Management Science* 70(6): 1008–1016.

Grimmer, M. K., van den Bosch, F., Powers, S. J. and Paveley, N. D. (2015). Fungicide resistance risk assessment based on traits associated with the rate of pathogen evolution. *Pest Management Science* 71(2): 207–215.

Kendall, S. J., Hollomon, D. W., Cooke, L. R. and Jones, D. R. (1993). Changes in sensitivity to DMI fungicides in *Rhynchosporium secalis*. *Crop Protection* 12(5): 357-362.

Kuck, K.-H. and Russell, P. E. (2006). FRAC: combined resistance risk assessment. *Aspects of Applied Biology* 78: 3–10.

Lucas, J. (2003). Resistance to QoI fungicides: Implications for cereal disease management in Europe. *Pesticide Outlook* 14(6): 268–270.

Magnus, H. A. (1987). Fungicide resistance. Present situation and fungicide strategy in Norway. *Vaextskyddsrapporter, Jordbruk* 48: 36–40.

Noble, M., Maggarvie, Q. D., Hams, A. F. and Leaf, E. L. (1966). Resistance to mercury of *Pyrenophora avenae* in Scottish seed oats. *Plant Pathology* 15(1): 23–28.

Scalliet, G., Bowler, J., Luksch, T., Kirchhofer-Allan, L., Steinhauer, D., Ward, K., Niklaus, M., Verras, A., Csukai, M., Daina, A. and Fonné-Pfister, R. (2012). Mutagenesis and functional studies with succinate dehydrogenase inhibitors in the wheat pathogen *Mycosphaerella graminicola*. *PLoS ONE* 7(4): e35429.

Chapter 6

Good practice in minimising the development of fungicide resistance in crop pathogens

Neil Paveley and Frank van den Bosch, ADAS High Mowthorpe, UK

1 Introduction

Good practice requires research to provide reliable evidence on resistance management and the translation of new evidence into changes in practical guidance for farmers and advisers. But research and guidance have opposing requirements:

- Research requires constant questioning of current theories, adversarial arguments between those with opposing evidence or opposing interpretations of the evidence, and openness to uncertainty and change.
- Guidance requires consistent messaging over long periods from many stakeholders, stated with sufficient confidence to be convincing.

Unhelpful outcomes can arise. The continuous flow of new research evidence could lead to constantly changing guidance and, without consensus, stakeholders may deliver differing messages. Or the opposite may occur

http://dx.doi.org/10.19103/AS.2022.0116.06

whereby groups of stakeholders delivering consistent messages over long periods may become collectively invested in the messages and become reluctant to change guidance when change is justified. The former confuses farmers and advisers. The latter leads to guidance that is out of line with current evidence.

A balance needs to be maintained. This chapter aims to inform that balance by considering the limitations and constraints affecting research and the translation of evidence into guidance. Areas where guidance appears out of line with evidence are highlighted.

2 Resistance management guidance: historical development

To start, we consider the end of the research and translation process – the guidance presented to farmers and advisers.

In 1986, responding to early cases of fungicide resistance in cereal pathogens, ADAS (then the UK government's agricultural development and advisory service) published a booklet 'Use of fungicides and insecticides on cereals' (Jenkins, 1986). The guidance on resistance management stated:

1. Reduce disease severity by non-chemical means.
2. Avoid the frequent application to a crop of either the same fungicide or different fungicides from the same mode of action. Wherever possible, a fungicide from a different mode of action should be used for the subsequent application.
3. Make full use of multi-site fungicides which are less prone to fungicide resistance problems.
4. Make full use of any appropriate formulated fungicide mixtures or label-recommended tank mixture. To minimise the risk of resistance, both fungicides in the mixture should be effective against the disease or diseases concerned and should be from different groups.
5. Avoid the application of fungicides sprays at times when they are not cost effective or when they are likely to be only marginally cost effective.
6. On winter cereals, minimise the use of seed treatments containing active ingredients with the same mode of action as fungicides used for control of foliar diseases.

In 1994, the Fungicide Resistance Action Committee (FRAC; a specialist technical group of fungicide manufacturers) commissioned a broad review of progress worldwide in dealing with fungicide resistance, and of the outstanding difficulties that needed to be overcome across a range of global crops. This was published as FRAC Monograph No 1 (Brent, 1995) 'Fungicide resistance in crop pathogens: How can it be managed?', and subsequently revised in a second edition (Brent and Hollomon, 2007).

The resistance management strategies advocated were:

1. Do not use the product exclusively. Apply it as a mixture with one or more fungicides of a different type, or as one component in a rotation or alternation of different fungicide treatments.
2. Restrict the number of treatments applied per season and apply only when strictly necessary.
3. Maintain manufacturers' recommended dose.
4. Avoid eradicant use.
5. Integrated disease management.
6. Chemical diversity.

These strategies and wording have remained the headline FRAC guidance for the intervening decades (https://www.frac.info/fungicide-resistance-management/background) and form the basis of the detailed resistance management guidance published by FRAC for individual crop species and MoA (FRAC, 2021). Items 3 and 4 were the notable additions compared with the earlier ADAS guidance. The point about maintaining the manufacturers' recommended dose has been amended within some of the individual MoA guidance, to recognise the practical reality that many farmers routinely adjust dose. Various forms of wording are used in the guidance for different MoA, within the broad theme of avoiding strongly reduced doses and multiple low dose applications.

The current UK Fungicide Resistance Action Group guidance for cereals (FRAG-UK, 2021) gives the following general guidance:

1. Fungicide input is only one aspect of crop management and other integrated control measures should always be used, such as disease-free seed lots, good hygiene through disposal of crop debris and control of volunteers which may harbour disease.
2. Always aim to select varieties exhibiting a high degree of resistance to diseases known to be relevant in your area, in addition to the main agronomic factors you desire.
3. Avoid growing large areas of any one variety, particularly in areas of high disease risk where the variety is known to be susceptible.
4. Only use fungicides in situations where the risk or presence of disease warrants treatment.
5. Use a dose that will give effective disease control and which is appropriate for the variety and disease pressure.
6. Make full use of effective fungicides with different modes of action in mixtures or as alternate sprays.

7. Ensure that mixing partners are used at doses that give similar efficacy and persistence.
8. Monitor crops regularly for disease and treat before the infection becomes well established.
9. Avoid repeated applications of the same product or mode of action and never exceed the maximum recommended dose or number of applications.

The 1980s guidance was exceptionally prescient, given the scarcity of evidence at the time, and it is striking that key themes have remained largely unchanged for over 35 years:

- Use Integrated Pest Management.
- Use alternation.
- Use mixtures.
- Treat according to need.

During that period, there have been huge advances in research, particularly in the development of high-throughput resistance genotyping and phenotyping, and in mathematical modelling. These advances have revolutionised our understanding of resistance and accelerated the accumulation of evidence on resistance management. Despite, or perhaps because of, these advances, there remain many debates about practical interpretation of the evidence into guidance. We consider the progress and limitations of research and its translation.

3 Phases of resistance evolution

Before considering how to manage a process, it is helpful to define what the process is and to understand its mechanisms. The development of fungicide resistance progresses through three key evolutionary phases:

- The 'emergence phase', during which the resistant strain has to arise through mutation and invasion.
- The 'selection phase', the resistant strain is present in the pathogen population and the fraction of the pathogen population carrying the resistance increases due to the selection pressure caused by the fungicide.
- During the final 'adjustment phase' the intensity of treatment or choice of fungicide or mixture may need to be changed to maintain effective control over a pathogen population where resistance has developed to intermediate levels.

These phases are described further in Table 1. Key points to note are: Firstly, that the emergence phase is completed well before detection, when resistance is often referred to, loosely, as 'emerging'. Secondly, the emergence phase has a strong random and probabilistic (stochastic) element and is about the number of resistant individuals, whereas the selection phase is substantially deterministic and about the proportion of resistant individuals in a population. These differences have important consequences for resistance management, as we discuss later.

In this chapter we focus on managing the selection phase, because that phase has become well understood through experimentation and modelling, and managing selection can substantially delay loss of effective control. In contrast, the emergence phase is much less well understood (see later section on the emergence phase) and the adjustment phase is often too late for interventions to provide a strong benefit – it is adaptation rather than mitigation.

Table 1 Definitions of words or concepts

Guidance	For brevity 'guidance' is used here to include both recommendations and conditions of PPP use. The former are implemented voluntarily and the latter are legally binding as defined on the product label.
Mode of action (MoA)	For single-site acting fungicides, MoA is taken here to mean acting on a specific target site. The implication being that fungicides of differing MoA are likely to have limited or no cross-resistance. Fungicides within a MoA may have complete or incomplete cross-resistance.
Emergence	When a fungicide mode of action is introduced and the resistance is not present in the genetic background of the population, it has to arise through mutation and invasion. Mutation refers to one or more heritable changes in individual cells conferring some level of resistance. Invasion refers to the increase in frequency – by selection or genetic drift – to a sufficiently high number that the resistant strain is unlikely to die out by random chance. The sequence of mutation and invasion is called emergence and may occur more than once independently (Torriani et al., 2009).
Selection	The process that determines the relative contribution of a genotype to the next generation. When a fungicide is used and a resistant genotype is present in the population, its proportional contribution to the next pathogen generation increases relative to that of the fungicide sensitive strain(s) in the population.
Per capita growth rate	The *per capita* rate of increase (also often called the intrinsic rate of increase) is the net increase in density per cycle of infection – which can be loosely described as the epidemic growth rate. In population genetics, the fitness of a pathogen strain (Crow and Kimura, 1970) is defined as the intrinsic rate of increase, r, of that pathogen strain.
Strategy	The 'what' aspect of resistance management. What do we want to achieve with resistance management?
Tactic	The 'how' aspects of resistance management. How do we achieve our strategy of resistance management?

4 Governing principles of resistance evolution as a foundation of guidance

In 1988, Milgroom and Fry defined the underlying processes and principles of resistance evolution and the possible resistance management strategies which flow from them. In summary: in the presence of fungicide, resistant pathogen strains will have a higher *per capita* growth rate of the population than sensitive strains. The resistant fraction of the pathogen population will therefore increase. The larger the difference in growth rates of the two strains, the faster the fraction resistant will increase. The selection rate is therefore determined by the difference between the growth rate of the resistant and the sensitive strain. This is the fundamental evolutionary mechanism, by which traits are selected in all living things; where a mutation leads to a fitness advantage (quantified through the increased growth rate of the mutant population).

In earlier work, Staub and Sozzi (1983) noted that the amount of selection taking place during a crop growing season is determined partly by the length of time the pathogen is exposed to the fungicide. In van den Bosch et al. (2014), we added exposure time to Milgroom and Fry's principles for determining rate of selection; so the total amount of selection is determined by the rate of selection and the exposure time during which the selection occurs. For further details and analysis, see chapter 4 by van den Bosch et al. in this volume. We then demonstrated that these 'governing principles' have strong predictive power. This was achieved by comparing predictions from the principles against experimental and modelling studies on a wide range of crops, pathogen species and MoA, which have been published in the decades since Milgroom and Fry's 1988 paper.

Three strategies can be derived from the governing principles:

Strategy 1: Selection for fungicide resistance is reduced when the *per capita* rate of increase of both the sensitive and the resistant strain (r_R and r_S respectively) are reduced.

Strategy 2: Selection for fungicide resistance is reduced when the *per capita* rate of increase of the resistant strain, r_R, is reduced relative to that of the sensitive strain, r_S.

Strategy 3: Selection for fungicide resistance is reduced when the timespan over which selection takes place, (exposure time), is reduced.

These guiding strategies underpin fungicide resistance management.

In this chapter we use 'strategy' to refer to one of the three strategies derived from the governing principles and 'tactic' to refer to a particular practice that aims to exploit one of the three strategies to manage resistance (see Table 1 for definitions). Examples of tactics which exploit each strategy include:

- Mixtures of MoA that operate through the mixing partner slowing the selection rate for strains resistant and sensitive to the 'at-risk' partner, thus reducing the difference in their growth rates (strategy 1).
- Reducing the dose applied (in circumstances where effective control can be obtained at a dose below the maximum recommended dose) (strategy 2).
- Alternation that operates by reducing exposure time of a MoA (strategy 3).

The governing principles, and the three strategies derived from them, provide a foundation for guidance because:

- The principles have been shown to apply to such a wide range of crops, pathogens and MoA, that tactics exploiting the strategies are highly likely to be effective - even if little is known about the characteristics of resistance that might occur to a new fungicide.
- The principles identify where there are opposing effects on selection, and therefore where the outcome might be uncertain. For example, dose splitting increases exposure time, but decreases the difference in resistant and sensitive population growth rates.
- If a tactic does not address one of the resistance management strategies identified from the governing principles, then it is questionable whether there is a clear mechanism for it to be effective. A theoretical exception being if fungicide treatment could completely kill all the resistant strain - this is generally impossible to achieve in practice.

5 Resistance management guidance: future development

The evidence base for resistance management guidance has accumulated rapidly in recent years, due to the development of techniques for testing resistance management tactics. In the next section we consider the complementary roles of experiments and modelling to identify good tactics. The following section describes the limitations of research; particularly areas of resistance management which are difficult or impossible to investigate. Subsequent sections outline the numerous barriers to translation of research into reliable guidance, including:

- Evidence from research which is inconsistent or seems contradictory;
- Findings from research which seem counter-intuitive and are therefore difficult to accept;
- Widely held beliefs which are inadequately substantiated or contradicted by evidence;

- Uncertainty about resistance risk, leading to inadequate or overly-restrictive resistance management;
- Trade-offs between resistance management tactics - where implementing one tactic weakens another;
- Limited availability of effective MoA to implement tactics;
- Balancing long-term resistance management and short-term economics of crop production;
- Conflating guidance for achieving effective control with guidance for resistance management; and
- Deciding if changes to guidance are justified when resistance is detected.

These barriers and limitations provide plenty of challenges and opportunities for the next generation of resistance management researchers and practitioners. In the meantime, awareness may improve our collective understanding of what constitutes good guidance.

6 Complementary roles of experiments and modelling to inform resistance management guidance

The development of high-throughput genotyping methods has enabled field, controlled environment and laboratory experiments (Genet et al., 2006; Fisher and Lang, 2016; Gutiérrez-Alonso et al., 2017; Paveley et al., 2019) to compare the effect of resistance management methods on selection for specific target site mutations. Mathematical modelling techniques have enabled similar studies *in silico*. Experimentation and modelling are complementary for two main reasons:

Firstly, experimental studies are usually restricted to measuring resistance selection within one cropping season or a limited number of pathogen generations. Modelling can then be used to estimate the cumulative effect of different levels of selection across seasons and (given data on the resistance phenotype) estimate the number of years until loss of effective control under different resistance management tactics. Early modelling studies typically used the increase in the resistant fraction as the measure of success; often resulting in the tactics which provided poor disease control appearing to be the most successful. The move towards estimating time to loss of effective control provided a more practically useful metric to compare tactics (see Chapter 4 by van den Bosch et al., this volume, for further analysis of the interplay between resistance management and efficacy).

Secondly, experimental studies can usually only test a small proportion of the possible treatments under a limited range of conditions. The question then is whether the experimental findings resulted from the particular conditions of the system or are generalisable. Modelling enables exploration of the outcomes from a wide range of tactics across a multi-dimensional parameter space. If the experimental findings are found to hold true in the model across the practically plausible parameter space, the findings can be implemented through guidance with greater confidence.

One important example of experiments being limited to a particular set of conditions is that the frequency of resistance must be sufficiently high to be quantified by high-throughput genotyping methods. Pyrosequencing, for example, typically has a detection level around 4%. Resistance management experiments in the field are therefore limited to cases where resistance is already established. For resistance management guidance to have maximum effect, it needs to be implemented when a new fungicide is introduced; well before resistance has reached detectable levels. Modelling enables exploration of the effect of the resistance frequency at implementation ('starting frequency') on the effect of resistance management tactics. In general, the lower the starting frequency at which tactics are implemented, the more effective life can be gained (Kabel and Jeffery, 1980; Hobbelen et al., 2011). There is no obvious mechanistic reason why, once in the selection phase (and hence the stochastic effects which affect emergence are no longer important) that the rank order of effectiveness of different tactics should change as the frequency of resistance changes. A consequence of this is that there is no general rationale to change tactics at the point when resistance is detected. We consider this point in more detail in a later section.

There is a limited 'window of opportunity' for field experiments to measure the effect of resistance management tactics on selection. Experiments can only start in a season when the frequency of resistant mutants reaches a level quantifiable by high-throughput assays. Unless mutants are unfit in the absence of treatment, the frequencies of mutations increase over a run of seasons to levels which constrain the ability to discriminate between treatments. There is an optimum pre-treatment frequency which maximises the ability to discriminate treatment effects. Modelling, based on the field experiments, suggests that pre-season frequencies of mutants in the range 10–50% give the best chance of detecting differences between treatments. High starting frequencies lead to running out of 'headroom' to detect differences. Low starting frequencies result in small absolute percentage differences, which are more likely to be below the limits of statistical significance. Within the optimum range, lower starting frequencies are optimal for detecting treatment differences where there are larger resistance shifts and higher frequencies where there are smaller resistance shifts (Young et al., 2021).

Resistance monitoring is subject to similar limitations in the respect that the frequency of resistance needs to be detectable. Nevertheless, it is tempting to look at resistance monitoring observations across seasons and make inferences that resistance developed first in that country/region due to particular agronomic or fungicide use practices prevalent there. There is little in the literature to support the idea that such observations of differences in the development of resistance between different geographic regions have provided any reliable evidence of which resistance management tactics are most effective. Such comparisons are prone to many extraneous variables (Russel, 2009), which may lead to erroneous conclusions. The stochasticity of emergence means that where and when resistance first emerges has a substantial random element. And countries and regions vary in many agronomic and fungicide use practices, making it difficult to identify one practice, or a combination of practices, as being causal. Nevertheless, spatial and temporal monitoring of resistance of key pathogens over many seasons has increased our quantitative understanding of how resistance occurs (vagndorf et al., 2018; Garnault et al., 2020).

6.1 Intractable areas of resistance research

There are some resistance topics which remain difficult to address.

6.1.1 Integrated pest management for resistance management

It is generally accepted that IPM is a key component of resistance management and that resistance management is a key component of IPM (European Directive, 2009/128/EC annex III). But hard evidence of the effect of IPM on the emergence or selection for resistance is limited to a few papers (e.g. Kristoffersen et al., 2020). A key reason for this is that IPM is comprised of many cropping practices; so if the task is to compare resistance evolution with and without implementation of IPM, the question is 'what to compare with what?' This multiplicity of experimental and observational options makes the task daunting, although the principles could be evidenced by a reasonable number of test examples.

In the meantime, there is a strong rationale that any IPM practice that reduces the intensity or duration of fungicide treatment required (i.e. strategies 2 and 3) or slows the *per capita* growth rate of epidemics during the period when fungicide is present (i.e. strategy 1), will slow selection for resistant strains (refer to chapter 4 by van den Bosch et al in this volume for further explanation of mechanisms). This rationale is sufficiently strong to support the continued advocacy of IPM as a resistance management tactic; by inference from the effectiveness of strategies 1, 2 and 3 having a sound mechanistic basis and being well evidenced for other tactics.

6.1.2 Exploiting within mode-of-action diversity

The use of mixtures (strategy 1) and alternation (strategy 3) for fungicides with different modes of action (MoAs) is well established for fungicide resistance management and is based on the general absence of cross-resistance between MoAs. Mixtures of fungicides with the same MoA and complete cross-resistance are of no use for resistance management. However, in some cases (for example, within the azole/imidazole group of demethylation inhibitors (DMIs)) cross-resistance is only partial within a MoA. In some extreme cases cross-resistance can be negative; such as between prochloraz (an imidazole) and metconazole. The potential resistance management benefits of mixing or alternating active substances within a MoA have not been fully investigated.

Experiments to test the effect of within MoA mixtures/alternation on selection are feasible and have demonstrated useful effects (Dooley et al., 2016), but few such studies have been reported. It remains unclear, for example, the extent to which the benefits of partial or negative cross-resistance might prove ephemeral, when put under selection pressure.

Modelling within MoA is complex because it requires mathematical methods to:

- Keep track of multiple pathogen haplotypes (strains with differing haploid genotypes);
- Vary the degree of cross-resistance;
- Represent the joint action of fungicides with varying degrees of cross-resistance; and
- Quantify the success of resistance management strategies so they can be compared.

These methods have been addressed in the groups of the authors (manuscript in preparation).

6.1.3 Landscape mosaics of different mode of action

Mixtures and alternation are advocated in guidance and implemented at the level of individual fields. These tactics are therefore within the control of individual farmers and their advisers. A free market economy, where each farmer has a choice of fungicides, often results in a mosaic of different fungicides applied to different fields. At any point in the cropping season, this mosaic will comprise fields with mixed or solo active substances (the latter, where an alternation tactic is being implemented).

There is a rationale that mosaics should be beneficial. In pathosystems where there is substantial spore exchange between fields, successive generations may be exposed to fungicides of different MoA if spores produced on a field treated with one MoA land on a field treated with a different MoA; thus reducing the exposure time to each (strategy 3).

Although the form of mosaic that occurs is not under any direct control, the extent to which there is landscape diversity of MoA, depends on the extent to which mixtures or alternation are advocated. Taking an extreme example, if there are only two effective MoA to control a particular pathogen, then making those two MoA available only as a mixture (by making them available only in formulated mixture or by a label requirement to tank mix) removes all landscape diversity.

The effects of mosaics are not amenable to testing by direct empirical experimentation in the field, because of the landscape scale at which they operate. Some form of controlled environment study may be feasible, if methods can be devised to create and control spore movement between small fungicide treated areas representing fields. The authors are not aware of any publications using such an approach. Observational comparisons between large geographic regions with differing mosaics could be made in principle, although it may prove impossible to control for extraneous variables which would confound the comparisons. Modelling approaches are feasible, as demonstrated by analogous studies on the effect of deployment of host resistance genes in the landscape on pathogen virulence evolution (Rimbaud et al., 2018). But there is no literature on modelling mosaics for fungicide resistance management.

6.1.4 Management of resistance emergence

A new resistant pathogen strain has to emerge through mutation (conferring some level of insensitivity) and invasion. Invasion refers to the increase to a sufficiently high number that the resistant strain is unlikely to die out by random chance. Note that we use 'emergence' here in its strict sense (van den Bosch et al., 2011), not in the sense of a strain emerging when it is detected or begins to have an impact in the field.

Even if it does not carry a fitness cost, a resistance mutation is likely to occur hundreds or thousands of times, and die out each time, before a mutant individual produces sufficient offspring to invade. Anti-emergence strategies which increase the likelihood of mutations dying out could, therefore, extend the number of years before resistance emerges. By the time monitoring detects a resistant strain, the emergence phase is already over; all we can do at that stage is to try to slow down its subsequent selection.

The importance of the emergence phase has been recognised by biologists working on antibiotic resistance in microbial organisms (Iwasa et al., 2004; Handel et al., 2006). But the opportunity to delay fungicide resistance emergence has been largely neglected as an area of research, with the exception of some modelling and *in vitro* studies (Hobbelen et al., 2014; Mikaberidze et al., 2017; Gutiérrez-Alonso et al., 2017). These suggest, for example, that mixtures may also be a useful anti-emergence tactic. However, there is insufficient evidence to identify effective 'anti-emergence' strategies which should be implemented in practice, through guidance.

The emergence phase is largely a 'black box' currently and is inherently difficult to research, but questions which still need to be addressed include:

- Which fungicide strategies can delay the emergence of fungicide resistance?
- Do the characteristics (traits) of different pathogen: fungicide systems affect which 'anti-emergence' strategies are likely to be effective?
- Do 'anti-emergence' strategies differ from 'anti-selection' strategies?
- If so, would we extend the effective life of products by implementing different strategies at different stages after release of a new fungicide.

6.1.5 Management of mutations with a small effect on sensitivity

The experimental evidence from the literature is usually based on studies where the effect of a known target site mutation on fungicide sensitivity is sufficiently large to be clearly measurable and leads to a selection rate which produces measurable changes in mutation frequency within one season. It is theoretically possible (van den Bosch et al., 2011) that optimal resistance management tactics may be different in cases where resistance occurs through the accumulation of many mutations, each of small effect.

In practice, it is mutations of moderate or large effect on sensitivity that are the greatest resistance concern and the main target of resistance management. Even if there was good evidence that different tactics were more effective against mutations of small effect, it would be risky to implement if they were sub-optimal for management of mutations of greater consequence.

6.2 Contradictory evidence

One consequence in the increased flow of experimental and modelling evidence is that there are many papers that need to be considered for translation into practical guidance and some of the evidence is apparently contradictory. Some of the variation in findings is due to the natural variability of biological systems and error variation associated with sampling. But the governing principles of resistance evolution (van den Bosch et al., 2014) identify particular tactics, or comparisons between tactics, for which there is a logical explanation for studies finding differing effects.

Tactics where reasonably consistent findings might be expected are those where the effects of the tactics act only to increase or decrease selection. For example, increasing the number of applications of fungicide, with each application at a given dose, will increase the exposure time and thus increase selection. Tactics where less consistent effects might be expected are those where there are two opposing effects on selection (through different strategies) and/or where the outcome depends on particular quantitative variables.

In the case of dose splitting (where a given total dose is applied over a larger number of applications) the increase in number of applications increases selection (i.e. the reverse of strategy 3), whereas the decrease in dose per application decreases selection (strategy 2) (e.g. Genet et al., 2006). The effects of dose splitting tend therefore to be relatively small and may be below the limits of statistical significance in field experiments (Paveley et al., 2019; Young et al., 2021).

In the case of comparisons between two resistance management tactics which are both effective, the difference in effectiveness between the two tactics may be small relative to the error variation of experiments. Which tactic is actually best may depend on particular quantitative variables which differ between studies. For example, the question of whether alternation or mixtures are the most effective tactic has been debated for decades. This is another case where the comparison includes two opposing effects. Moving from mixtures increase exposure time compared to alternation (i.e. the reverse of strategy 3, thus increasing selection), but the addition of a mixture partner slows the *per capita* growth rate of mutants which are both sensitive and insensitive to the 'at-risk' fungicide (i.e. strategy 1, thus decreasing selection). There are good theoretical arguments for circumstances under which alternation or mixtures should be the most effective at reducing selection, but experimental proof of these theories has proved elusive. Thus far, the conclusion is that the choice of tactic should be determined by practical considerations.

In the case of spray timing, applying fungicide earlier or later could result in the fungicide having its effect during a period when the *per capita* growth rate of the epidemic is high or low (due, for example, to the weather conditions affecting infection). Applying a fungicide earlier or later could increase or decrease selection, depending on the particular circumstances of the epidemic.

6.3 Counter-intuitive truths

Epidemics and evolution are driven by a set of interacting processes, some of which operate non-linearly. These processes can lead to outcomes that seem contrary to our intuition and common sense.

During the emergence phase, mutations (some of which may lead to reduced fungicide sensitivity) occur at a certain rate per unit of population size. Hence, it seems desirable to keep the total population size small, in order to reduce the number of resistant mutants that arise. But in a small population, mutants face less competition and have a higher probability of surviving and reproducing. Depending how these opposing effects work out, a smaller population size (achieved, for example, by more intensive fungicide use) can lead to more, rather than fewer, mutants surviving (Hobbelen et al., 2014; Mikaberidze et al., 2017).

During the selection phase, a certain proportion of the pathogen population is resistant. It seems desirable to keep the total population size small, in order that the population size of the resistant strain is also minimised. In films and television dramas, fewer 'bad guys' is good. But in resistance management aiming to 'kill the bad guys' leads to poor resistance management decisions, for the reasons described next.

Loss of efficacy of a MoA is determined jointly by the size of the sensitivity shift caused by a mutation and the proportion of the pathogen population carrying the mutation. The key point is that it is the rising *proportion* of mutants that eventually leads to inability to control epidemics. The size of the pathogen population rises and falls within and between seasons, but (in the absence of a substantial fitness penalty; Hawkins and Fraaije, 2018) selection drives the proportion inexorably upwards, until control is lost. Attempts to reduce the number of mutants by more intensive fungicide treatments (more applications or more dose per application) are counter-productive (the reverse of strategies 2 and 3), as the sensitive proportion of the population is controlled to a greater extent than the mutants, thus increasing selection (van den Bosch et al., 2011).

It also seems intuitive that it would reduce resistance development to treat epidemics early, when the population size is small, rather than later when the population size has increased. Early treatment may well be needed to ensure good efficacy, but it is not a resistance management tactic. None of the three strategies apply. The rate of selection is determined by the difference in the *per capita* growth rates of the epidemics between the resistant and sensitive populations. *Per capita* growth rate can be thought of as analogous to the rate of compound interest by which financial investments grow (see Table 1 for definition). Epidemics increasing in a unit time from, for example, 5–25% severity or from 0.5% to 2.5% or from 0.05% to 0.25% severity all have the same *per capita* growth rate, and hence the same increase in selection due to fungicide treatment. This theory is supported by the studies reviewed by van den Bosch et al. (2014) which tested the effect of spray timing on selection and by modelling (van den Berg et al., 2013). The evidence showed that selection could be increased or decreased by changing spray timing. If weather conditions and host resistance expression remain stable through time, then the *per capita* growth rate of the epidemic will tend to remain constant through time, as severity increases. In such consistent conditions, fungicide treatment at any point in time would lead to similar selection, regardless of population size. More typically, changing weather leads to changing *per capita* growth rates. Treating early may coincide with a period or high or low growth rates, despite the population being small – leading to the inconsistent effects seen in the experimental evidence.

If a small population size is obtained by integrated control measures that reduce the *per capita* growth rate of the epidemic, then the growth rate effect will decrease selection and the small population size benefits crop yield.

One aspect of treatment timing which is relevant to resistance management is whether earlier or later treatment minimises the number or dose of treatments needed to obtain effective control (strategy 3) and hence minimises selection. Later treatment may lead to the epidemic becoming more difficult to control, resulting in 'chasing' the disease with repeated applications. But treating prophylactically or very early in the epidemic means that the treatment decision is being made at the time when the future progress of the epidemic is most difficult to predict. If treatment is delayed and the crop monitored, it may become clear that the epidemic is not progressing, making treatment (and the resulting selection) unnecessary. Farmers and advisers should be free to use their best judgement on such efficacy decisions, according to the characteristics of the pathogen and the circumstances of the season. They should not be placed under an erroneous resistance management obligation to treat early.

6.4 Unsubstantiated tactics

The guidance to avoid curative use was introduced in original FRAC Monograph Number 1. Brent (1995) was candid about the lack of evidence and in the second edition of the monograph Brent and Hollomon (2007) state that to the authors' knowledge there is no experimental evidence to support the guidance. Two arguments were put forward to avoid curative use. The first argument was specific to circumstances where a systemic single-site acting fungicide with curative activity is applied in mixture with a multi-site. In eradicant situations, the multi-site would not be effective, leaving the single-site acting alone. This is a well founded argument, although the circumstances in which it applies to an important extent may be limited. At any point in an epidemic when a fungicide might be applied, there is usually some combination of latent infections still in a sufficiently early stage to be controlled by eradicant action, and new infections that occur after treatment. Both components of the mixture would be effective against the latter (so strategy 1 applies to protectant control of new infections after treatment). The balance between latent and new infections will vary depending on circumstances. After a period of conditions which are not conducive to infection, control is likely to be more towards the protectant end of the spectrum. After a period of conditions conducive to infection, control will be towards the eradicant end of the spectrum – particularly if spray application was delayed by adverse weather. The guidance on avoiding curative treatment is often interpreted to mean 'treat early', but it is by no means clear that a higher *proportion* of infections are in the latent stage later in epidemics. As the epidemic develops there will usually be a higher density of latent infections, but also a higher density of infectious lesions (which are beyond control in many pathogen species) and a higher density of new infections that occur after treatment. As presented above, selection is not determined by the size of the epidemic.

So how effective is the multi-site likely to be as a mixture partner to protect the single site against resistance, under typical practical conditions? The mixture effect occurs where the mixture partner (in this case the multi-site) acts to reduce the *per capita* growth rate of strains both resistant and sensitive to the single site, during the period when the single-site is acting. Such an overlap of effect is more likely where the mixture partners share similar physicochemical properties and activity. But in practical terms, if the multi-site is making a useful contribution to control, it is also making a useful contribution to protecting the single site.

The second argument put forward to support avoiding curative use was more general.

Avoidance of eradicant use could possibly delay resistance for another, more widely applicable reason. To wait until a threshold population of the pathogen appears, usually means that many sporulating lesions are exposed to the fungicide. Opportunity for selection could be much greater than if the fungicide had been applied prophylactically to keep populations permanently low.

This argument has not been supported by subsequent evidence, for the reasons described in the preceding section.

6.5 Uncertainty about resistance risk

Most resistance management guidance constrains the flexibility available to farmers in how they can use fungicides, for example by limiting the number of treatments or requiring MoA to be used in mixtures or alternation. These constraints are in addition to regulatory limits on use that are required to protect non-target species. The degree to which these additional constraints are justified depends on the degree of resistance risk: strong guidance is justified by high risk. Decisions about the strength of guidance are, therefore, informed by resistance risk assessment – which is an uncertain art.

When the QoI fungicides were introduced arguments about the risk of resistance were put forward, but the speed at which resistance would develop was not foreseen (Heaney et al., 2000). These arguments had a logical basis, but turned out to be erroneous. The QoIs were highly effective and opinions about solo use were divided at the time. Some argued that such use was acceptable, based on the risk assessment. Others considered solo use reckless.

The QoIs are one case. Grimmer et al. (2014a) reviewed all the unique cases of resistance in Europe (i.e. the first reported cases of resistance for each pathogen by fungicide MoA combination). The data set was used to assess the predictive value of the type of 'resistance matrix', widely used for resistance risk assessment. The resistance matrix tested was shown to have predictive value, but the majority of the predictive value was in the distinction between the resistance risk for multi-site acting (low risk) versus single-site acting fungicides (typically moderate or high risk). The value for assessing risk within single-site acting fungicides was poor. An alternative risk assessment method, based on the traits of the pathogen, fungicide and agronomic system, was reported subsequently (Grimmer et al., 2014b), with improved predictive value. This has been included in the European and Mediterranean Plant Protection Organisation (EPPO) standard on resistance risk assessment (Anon, 2015). Nevertheless, predicting resistance risk still contains considerable uncertainty.

Resistance management is most effective when introduced at the point of introduction of a new MoA. At that point, information to inform resistance risk assessment is limited, so there are judgements to be made about how precautionary resistance management guidance should be.

6.6 Trade-offs between resistance management tactics

Much of the complexity of implementing resistance management arises from 'trade-offs'. Elsewhere in this chapter we consider trade-offs between the long-term aim of resistance management and the short-term need to obtain cost-effective control. Here we consider trade-offs between different resistance management tactics, particularly where the aim is to manage resistance evolving concurrently against more than one MoA. In these circumstances, exploiting one of the underlying strategies can reduce the ability to exploit another strategy. An example of concurrent resistance is provided by *Zymoseptoria tritici* populations which are shifting for sensitivity against both succinate dehydrogenase inhibitor (SDHI) and demethylation inhibitor (DMI) fungicides (Blake et al., 2018; Rehfus et al., 2018). There are many more examples in other pathosystems (Trkulja et al., 2017).

Experimental and modelling studies have shown that three tactics are widely effective across diverse pathosystems (van den Bosch et al., 2014): limiting the number of applications of a MoA per crop season (exploiting strategy 3), alternating MoA (exploiting strategy 3), and mixing MoA (exploiting strategy 1).

Consider a pathosystem where several fungicide applications are required to obtain effective control and a limited number of effective MoA are available. If the number of applications of fungicide A is limited in order to manage resistance against that MoA, then there will be treatments in the programme when fungicide A cannot be used as a mixture partner to protect fungicide B. The tactic of limiting treatments is therefore constraining the tactic of using mixtures. One MoA is, therefore, well protected against resistance, but other MoA are consequently less well protected (if other effective mixtures partners are not available). This situation is likely to become more common in Europe, where regulation is limiting the availability of products and is removing effective multi-site acting fungicides which are useful mixture partners. Similarly, we noted earlier that use of mixtures could reduce the diversity of landscape mosaics of MoA.

Virtually all the experimental and modelling evidence on resistance management strategies comes from studies where selection was measured for strains resistant to one MoA. The few studies where resistance management against more than one MoA was considered (Brent et al., 1989; Lorenz et al., 1992; Thygesen et al., 2009; Hobbelen et al., 2013) do not fully resolve the trade-offs raised above. In some circumstances, the trade-off between limiting the number of treatments and the use of mixtures could be overcome by using alternation rather than mixtures. But mixtures are often the preferred practical approach, for example where a mixture is necessary to achieve a wider spectrum of activity or to ensure that resistance management is implemented by the use of formulated mixtures.

A modelling analysis by van den Berg et al. (2016) suggested that similar effective lives for a MoA could be obtained by limiting the number of treatments or by limiting total dose. Limiting total dose, and allowing growers some flexibility on the number of applications, would resolve a trade-off by allowing effective mixtures to be used throughout the spray programme, even where there are few MoA available.

Because increasing the number of applications, but decreasing the dose per application has two opposing effects on selection, the net effect of 'dose splitting' cannot be readily predicted. But the effect of increased exposure time from increased number of treatments might be expected to outweigh the decrease in selection due to lower dose per application. This theoretical expectation was reflected in the studies reviewed by van den Bosch et al. (2014), where the effect of dose splitting varied between studies, but generally led to small or moderate increases in selection.

The effect of dose splitting also has implications for whether mixture or alternation is likely to be more effective at reducing selection. Consider two options for the simple case of a two-spray programme. The first option is to alternate fungicide A followed by fungicide B (where A and B are different MoA), each at the maximum permitted individual dose. The second option is to apply A at both spray timings, mixed with B. If both fungicides are of similar efficacy and there is no antagonism, then a mixture containing half the maximum individual dose of A and B will provide at least the same efficacy as the alternation programme and will apply the same total dose. Effectively, the mixture programme is dose splitting (half the dose of A is applied twice as often). If dose splitting has little effect on selection, then the mixtures are likely to be more effective at reducing selection than alternation. This is because the increase in exposure time is counteracted by the decreased dose per application, and the effect of mixture partner B reducing the *per capita* growth rates of strains resistant and sensitive to A will reduce selection further. In cases where dose splitting increases selection, depending on the size of the increase in selection, mixtures and alternation may be similarly effective at reducing selection or alternation may be more effective.

The effects of restricting total dose, rather than number of treatments, were explored in experiments on concurrent resistance in *Z. tritici* against SDHIs and DMIs, as reported by Young et al. (2021). The work also compared the effects of mixtures or alternation on selection for mutants on the SDHI target gene *sdh*. Results were as follows: (i) the field experiments showed no significant effect of splitting the SDHI dose (within a fixed total dose) on selection for *sdh* mutants, (ii) modelling predicted that dose splitting would increase selection, but that the increases would often be below the limits of detection from the experiments, (iii) evidence from the field experiments and from modelling show that the benefit of splitting the dose of SDHI, to allow more of the DMI treatments to be protected by an SDHI mixture partner, was small and below the detection limits of the experiments, (iv) the field experiment data show that dose splitting was generally neutral for efficacy of disease control and for the economics of fungicide programmes, (v) in principle, alternation is only likely to be better than mixtures for reducing selection where dose splitting increases selection, and the mixture partner is relatively ineffective. This rationale was not corroborated experimentally and will be explored further by modelling and (vi) alternation and mixtures were similarly effective at reducing selection and for disease control. It is likely that the potential benefit of having fewer unprotected DMI treatments by allowing dose splitting was counteracted by the lower dose per treatment making the SDHI a less effective mixture partner.

6.7 Limited numbers of effective mode of action

A common constraint on the translation of principles into guidance arises when the number of effective MoA against a particular pathogen becomes very limited, due to loss of effective products to resistance or regulation. The concept in medicine of being down to 'the antibiotic of last resort' has become common in crop protection. In cases where the number of available MoA dwindles, the need for resistance management becomes stronger (to avoid further loss of effective chemistry), but the ability to deliver it becomes weaker. Diversity of MoA for mixing, alternation or landscape mosaics reduces. The aim of achieving 'balanced mixtures' where both components contribute similarly to control, becomes more of an aspiration than an achievable goal. This creates difficult decisions for companies bringing formulated mixtures to market and for regulators in countries where they need to determine what constitutes an acceptably balanced mixture.

One scenario is where only one or two MoA remain available to control a key pathogen and the control from those MoA is deteriorating. Then the crop protection industry produces a new molecule 'just in time'. For example, in 2011 the QoIs were no longer effective for control of *Z. tritici*, the activity of azoles was declining and control was becoming increasingly reliant on the multi-site chlorothalonil. In 2012, a new generation of SDHIs were launched in Europe, just in time to manage a difficult septoria season.

Operating a 'just in time' system is a risky game. As we have seen in other pests - such as control of black-grass (*Alopecurus myosuroides*) and cabbage stem flea beetle (*Psylliodes chrysocephala*) – sometimes the next molecule does not arrive in time. At that point, the industry reverts to IPM practices to be able to continue to grow the crop. This is too late. Application of IPM is required when there is good diversity of chemistry available, to slow the loss of activity.

6.8 Balancing resistance management and economics

6.8.1 Short-term versus long-term economics for farmers

Many crop commodities are produced by farmers who make small or unpredictable margins. One season of high crop losses to disease may threaten the viability of a farm, so short-term economics requires effective control, at an affordable cost. In horticultural crops, market requirements for blemish-free produce add to the pressure for a high level of efficacy. Long-term economics depend on managing resistance to ensure effective control in future seasons.

Guidance to maintain the manufacturers' recommended dose is counter-productive for both short- and long-term economics. The recommended dose is set by manufacturers, and approved by regulators, at a level which gives a high level of control in a high proportion of circumstances. The recommended dose is an important reference point and may be required under high disease pressure situations and/or a high degree of uncertainty about future disease severity (te Beest et al., 2013). Under lower disease pressure, or where IPM practices are reducing the need for treatment and constraining the bounds of uncertainty, a lower dose per treatment or fewer treatments may be appropriate to maximise gross margin (Jørgensen et al., 2017) and reduce selection pressure.

There can be trade-offs between short- and long-term economic goals. For example, adding a multi-site mixture partner to a single-site fungicide (or to a mixture of single-site fungicides) adds cost. If the multi-site clearly increases disease control or enables reduction of the dose of the single-site fungicide in the mixture, then there is usually a strong short- and long-term argument for its inclusion. Where the additional benefit of the multi-site has a neutral effect on gross margin, there remains a good long-term argument for inclusion, although the addition is adding to the environmental load of plant protection product (PPP). Where the short-term economics of adding a multi-site are negative, there may be cases where the short-term costs would be outweighed by the long-term benefits. But the long-term benefits are uncertain. If the riskier approach of not including the multi-site is followed, the long-term economic effect depends on whether a new effective fungicide is released before the existing chemistry is lost.

6.8.2 Short-term versus long-term economics for PPP manufacturers

Developing a new fungicide and registering it for use requires a huge investment that must be repaid by sales revenue. There are strong commercial pressures to secure the return on investment at the earliest opportunity by driving sales. Increasing sales through high market share and intensity of use creates the potential for conflicts with resistance management. If a manufacturer has a unique MoA and sees little foreseeable prospect of other companies producing products, then managing resistance makes commercial sense to protect the MoA and maximise sales in the medium and long-term – even if that means tight restrictions on use. Driving high market share will shorten product life.

In other circumstances, the development of a new MoA by one company often results in other companies identifying molecules to block the same target site, with resulting cross-resistance – a legitimate practice which Russell (2005) referred to as 'patent busting'. The competition this creates is mainly for market share. If cross-resistance within a MoA is strong, then one product replacing use of another within the same MoA has little effect on selection, unless the product has higher efficacy (through intrinsic activity or a higher loading of active substance). In such a competitive environment, there is less certainty about sales in the medium- or long-term and more incentive to maximise sales in the short term. The companies have a strong collective interest in managing resistance, but there may be little incentive for tight restrictions on use. These types of interplay are beginning to be a topic for academic study (Lemarié and Marcoul, 2018).

6.9 Conflating resistance management guidance and efficacy guidance

Decisions about fungicide product choice, dose and timing are driven by many factors, including the requirements to obtain good efficacy and to manage resistance . Actions for which there is inadequate evidence of resistance management benefit sometimes appear in resistance management guidance. It is unclear whether this happens because the guidance has not kept up with the evidence or because the guidance seems practically justifiable on efficacy grounds.

For example, there may be many circumstances where it is necessary to treat early, use the maximum recommended dose or avoid curative applications, to obtain effective control. But that does not justify inclusion in resistance management guidance. If a document is providing guidance on efficacy and resistance management, then the aim of each action should be made clear.

6.10 Changing guidance when resistance is detected

Because the frequency of resistant individuals is exceptionally low when the emergence phase is complete (Hobbelen et al., 2014), resistance is generally well into the selection phase when new resistant strains are first detected. Nothing changes in the evolutionary process at detection. All that changes is that we become aware of a process which has been happening for some time and there is now less time for resistance management to have a useful effect. The situation is not like Schrodinger's cat in quantum physics where our observation changes the conditions.

Resistance management is most effective when implemented at introduction. Assuming that was done, then changes to guidance are only justified if new knowledge indicates that the original guidance was wrong. Two scenarios might illustrate cases where change could be justified. First, if the original resistance risk assessment at introduction underestimated the risk, the original restrictions on use may have been too lax. If detection makes it clear that the risk was in fact high, then implementation of more stringent guidance could be justified – although there may be insufficient time for revised guidance to have a useful effect. The second scenario is where detection and characterisation of mutants provides some new information relevant to the guidance. Evidence of a strong fitness penalty (e.g. Schepers et al., 2018) might justify a change to the guidance, but the difficulty remains that the new information relates to currently known strains, whereas the greater threat may be from future strains which have yet to be detected and characterised.

7 Updating resistance management guidance

The following guidance is supported by current evidence:

1. Resistance management is most effective when implemented at the introduction of a new fungicide.
2. The total dose of fungicide applied in a season is a key driver of resistance selection.
3. Integrated Pest Management practices to reduce the need for fungicide treatment and to target treatment according to need, can reduce the total dose of fungicides required and thereby reduce resistance selection.
4. Limiting the number of applications per season of an MoA will reduce selection for resistance, provided higher individual doses are not used to compensate for fewer applications.
5. Mixtures or alternation of fungicides with different MoAs are generally effective resistance management strategies.
6. The choice between mixtures or alternation can be made based on practical considerations. There is no proven rationale to determine, in advance, which will be most effective at reducing selection.
7. Application timing can be determined by when it provides the best targeting of treatment and minimises the total dose of fungicide required during the season.
8. There is a specific exception to the flexibility of timing described above. If a mixture of an eradicant fungicide and a protectant fungicide (such as a multi-site) is used, then application in strongly eradicant situations should be avoided, as the protectant may not provide good protection against resistance.
9. When resistance is first detected, implementation of stronger resistance management guidance may be justified if the original resistance risk assessment underestimated the risk.

8 Future trends in research

Three potentially disruptive advances, with implications for practical resistance management, are already starting to have some impact or show promise.

8.1 'Resurrection' of mode of action

Previously, when the target site of a pathogen changed to confer resistance, the efficacy of a MoA was lost and the search moved to identifying new target sites. Examples include ethirimol, the MBCs and QoIs (see the review by Russell, 2005). The exception to this was the azoles, where a succession of new effective molecules was developed over decades as the performance of earlier chemistry deteriorated. There are recent examples from a wider range of MoA where new molecules are being developed or introduced which are effective against populations containing high frequencies of strains resistant to earlier active substances. Examples include QoI (Matsuzaki and Iwahashi, 2019), azole (Strobel et al., 2019) and SDHI (Sierotzki et al., 2017) fungicides. Molecules within a MoA (or new sub-groups/groups/classes of molecules within a MoA) have been found which bind effectively to mutated target sites. This resurrection of multiple MoA is a potential game changer.

8.2 Gene editing

Disease resistance genes introgressed into improved crop varieties are a powerful component of IPM, which can reduce the intensity of fungicide treatment required. But resistance genes are frequently overcome by pathogen evolution towards virulence (Rimbaud et al., 2018). Plant breeders are therefore in a race to introduce new host resistance genes, but it takes several generations to introgress new genes and then backcross to remove undesirable alleles linked to the desirable genes. Gene editing holds the potential to introduce resistance genes without undesirable alleles, thus speeding up the breeding process - provided technical challenges and societal opposition can be overcome.

8.3 Sensing, diagnostics and decision algorithms

In many climates, the risk of disease epidemics fluctuates substantially within and between seasons, due to varying weather. With the benefit of hindsight it is easy to identify seasons where the level of fungicide input was higher than necessary - leading to selection for resistance from treatments which did not provide an economic benefit. Identifying such seasons in real time, to guide treatment decisions is much more difficult. New sensing and diagnostic technologies, perhaps linked through the Internet of Things and feeding data to decision algorithms, have the potential to improve disease prediction and targeting of treatment according to need, although many barriers remain to implementation (see review by Fenu and Malloci, 2021).

9 Where to look for further information

The Fungicide Resistance Action Groups (FRAG) are national or multi-national groups, usually comprising a combination of research, advisory and PPP manufacturer members. Active groups include NorBaRAG (for the Nordic/Baltic countries) www.norbarag.au.dk and FRAG-UK https://ahdb.org.uk/knowledge-library/the-fungicide-resistance-action-group-frag-uk.

The FRAC is a specialist technical group of CropLife international. FRAC members are manufacturers of PPPs. www.frac.info provides many sources of information, including resistance management guidance for each of the major fungicide groups in major crops, monitoring information, definitions of fungicide groups and educational material.

R4P is a publicly funded network of six teams and laboratories working on resistance in France. www.r4p-inra.fr provides independent information and resources.

The EPPO publishes 'Standards' which define various aspects of resistance management, such as the assessment of resistance risk. The standards are used by manufacturers and regulators of plant protection products and guide specific technical aspects of how PPPs are regulated. A database of resistance cases is maintained by EPPO. www.eppo.int.

10 References

Anon. (2015). Resistance risk analysis. *EPPO* PP1. Available at: https://pp1.eppo.int/standards/PP1-213-4.

Anonymous. (2009). Directive 2009/128/EC (annex III). Official Journal of the European Union. The European Commission. Available at: https://eur-lex.europa.eu.

Blake, J. J., Gosling, P., Fraaije, B. A., Burnett, F. J., Knight, S. M., Kildea, S. and Paveley, N. D. (2018). Changes in field dose-response curves for DMI and QoI fungicides against *Zymoseptoria tritici*. *Pest Management Science* 74(2): 302–313.

Brent, K. J. (1995). *Fungicide Resistance in Crop Pathogens: How Can It Be Managed?* FRAC Monograph 1. GIFAP, Brussels, 48pp.

Brent, K. J., Carter, G. A., Hollomon, D. W., Hunter, T., Locke, T. and Proven, M. (1989). Factors affecting build-up of fungicide resistance in powdery mildew in spring barley. *Netherlands Journal of Plant Pathology* 95(S1): 31–41.

Brent, K. J. and Hollomon, D. W. (2007). *Fungicide Resistance in Crop Pathogens: How Can It Be Managed?* FRAC Monograph 1, (2nd rev. edn.). Available at: www.frac.info.

Crow, J. F. and Kimura, M. (1970). *An Introduction to Population Genetic Theory*. Harper & Row, New York.

Dooley, H., Shaw, M. W., Spink, J. and Kildea, S. (2016). Effect of azole fungicide mixtures, alternations and dose on azole sensitivity in the wheat pathogen *Zymoseptoria tritici*. *Plant Pathology* 65(1): 124–136. Doi: 10.1111/ppa.12395.

Fenu, G. and Malloci, F. M. (2021). Forecasting plant and crop disease: An explorative study on current algorithms. *Big Data and Cognitive Computing* 5(1): 5010002. Doi: 10.3390/bdcc5010002.

Fisher, K. J. and Lang, G. I. (2016). Experimental evolution in fungi: An untapped resource. *Fungal Genetics and Biology: FG and B* 94: 88–94.

FRAC. (2021). *Summary of Fungicide Resistance Management Guidelines by FRAC Working Group and Expert for a Relevant for Cereals and Corn*. Available at: www.frac.info.

Frag, U. K. (2021). *Fungicide Resistance Management in Cereals*.

Garnault, M., Duplaix, C., Leroux, P., Couleaud, G., David, O., Walker, A. S. and Carpentier, F. (2021). Large-scale study validates that regional fungicide applications are major determinants of resistance evolution in the wheat pathogen *Zymoseptoria tritici* in France. *New Phytologist* 229(6): 3508–3521. Doi: 10.1111/nph.17107.

Genet, J. L., Jarworska, G. and Deparis, F. (2006). Effect of dose rate and mixtures of fungicides on selection for QoI resistance in populations of *Plasmopora viticola*. *Pest Management Science* 62(2): 188–194.

Grimmer, M. K., van den Bosch, F., Powers, S. J. and Paveley, N. D. (2014). Evaluation of a matrix to calculate fungicide resistance risk. *Pest Management Science* 70(6): 1008–1016.

Grimmer, M. K., van den Bosch, F., Powers, S. J. and Paveley, N. D. (2015). Fungicide resistance risk assessment based on traits associated with the rate of pathogen evolution. *Pest Management Science* 71(2): 207–215.

Gutiérrez-Alonso, O., Hawkins, N. J., Cools, H. J., Shaw, M. W. and Fraaije, B. A. (2017). Dose-dependent selection drives lineage replacement during the experimental evolution of SDHI fungicide resistance in *Zymoseptoria tritici*. *Evolutionary Applications* 10(10): 1055–1066.

Handel, A., Regoes, R. R. and Antia, R. (2006). The role of compensatory mutations in the emergence of drug resistance. *PLoS Computational Biology* 2(10): e137.

Hawkins, N. J. and Fraaije, B. A. (2018). Fitness penalties in the evolution of fungicide resistance. *Annual Review of Phytopathology* 56: 339–360.

Heaney, S. P., Hall, A. A., Davies, S. A. and Olaya, G. (2000). Resistance in fungicides in the QoI-STAR cross-resistance group: Current perspectives. *Proceedings of the BCPC Conference – Pests and Diseases* 2: 755–762.

Hobbelen, P. H. F., Paveley, N. D., Oliver, R. P. and van den Bosch, F. (2013). The usefulness of fungicide mixtures and alternation for delaying the selection for resistance in populations of *Mycosphaerella graminicola* on winter wheat: A modeling analysis. *Phytopathology* 103(7): 690–707.

Hobbelen, P. H. F., Paveley, N. D. and van den Bosch, F. (2011). Delaying selection for fungicide insensitivity by mixing fungicides at a low and high risk of resistance development: A modelling analysis. *Phytopathology* 101(10): 1224–1233.

Hobbelen, P. H. F., Paveley, N. D. and van den Bosch, F. (2014). The emergence of resistance to fungicides. *PLoS ONE* 9(3): e91910. Doi: 10.1371/journal.pone.0091910.

Iwasa, Y., Michor, F. and Nowak, M. A. (2004). Evolutionary dynamics of invasion and escape. *Journal of Theoretical Biology* 226(2): 205–214.

Jenkins, J. E. E. (1986). The Use of Fungicides and Insecticides in Cereals. ADAS Booklet No. 2257. Published: Ministry of Agriculture Fisheries and Food, Great Britain.

Jørgensen, L. N., van den Bosch, F., Oliver, R. P., Heick, T. M. and Paveley, N. D. (2017). Targeting fungicide inputs according to need. *Annual Review of Phytopathology* 55: 181–203.

Kabel, P. F. and Jeffery, H. (1980). Selection for tolerance in organisms exposed to sprays of biocide mixtures: A theoretical model. *Phytopathology* 70(1): 8–12.

Kristoffersen, R., Heick, T. M., Müller, G. M., Eriksen, L. B., Nielsen, G. C. and Jørgensen, L. N. (2020). The potential of cultivar mixtures to reduce fungicide input and mitigate fungicide resistance development. *Agronomy for Sustainable Development*. Springer Verlag/EDP Sciences/INRA.

Lemarié, S. and Marcoul, P. (2018). Coordination and information sharing about pest resistance. *Journal of Environmental Economics and Management* 87: 135–149.

Lorenz, G., Saur, R., Schelberger, K., Forster, B., Kung, R. and Zobrist, P. (1992). *Long-Term Monitoring Results of Wheat Powdery Mildew Sensitivity Towards Fenpropimorph and Strategies to Avoid the Development of Resistance*. British Crop Protection Council, Farnham Brighton Crop Protection Conference: Pests and Diseases (Vols. 1), 2, and 3, 171–176.

Matsuzaki, Y. and Iwahashi, F. (2019). Biological profile of Pavecto – A new QoI -fungicide: Discovery of tetrazolinone chemical class and in vitro/in vivo analysis of activity against QoI resistant strains. In: Deising, H. B., Fraaije, B., Mehl, A., Oerke, E. C., Sierotzki, H. and Stammler, G. (Eds). *Modern Fungicides and Antifungal Compounds* (Vol. IX). Deutsche Phytomedizinische Gesellschaft, Braunschweig, pp. 265–270.

Mikaberidze, A., Paveley, N., Bonhoeffer, S. and van den Bosch, F. (2017). Emergence of resistance to fungicides: The role of fungicide dose. *Phytopathology* 107(5): 545–560.

Milgroom, M. G. and Fry, W. E. (1988). A simulation analysis of the epidemiological principles for fungicide resistance management in pathogen populations. *Phytopathology* 78(5): 565–570.

Milgroom, M. G. and Fry, W. E. (2020). A simulation analysis of the epidemiological principles for fungicide resistance management in pathogen populations. *Phytopathology* 78: 565–570.

Paveley, N., Young, C., Fraaije, B., Kildea, S., Burnett, F., Clark, W., Gosling, P., Boor, T., Corkley, I. and van den Bosch, F. (2019). Managing resistance evolving concurrently against two modes of action. In: Deising, H. B., Fraaije, B., Mehl, A., Oerke, E. C., Sierotzki, H. and Stammler, G. (Eds). *Modern Fungicides and Antifungal Compounds* (Vol. IX). Deutsche Phytomedizinische Gesellschaft, Braunschweig, pp. 141–146.

Rehfus, A., Strobel, D., Bryson, R. and Stammler, G. (2018). Mutations in sdh genes in field isolates of *Zymoseptoria tritici* and impact on the sensitivity to various succinate dehydrogenase inhibitors. *Plant Pathology* 67(1): 175–180.

Rimbaud, L., Papaix, J., Rey, J. F., Barrett, L. G. and Thrall, P. H. (2018). Assessing the durability and efficiency of landscape-based strategies to deploy plant resistance to pathogens. *PLoS Computational Biology* 14(4): e1006067.

Russell, P. E. (2005). A century of fungicide evolution. *Journal of Agricultural Science* 143(1): 11–25.

Russel, P. E. (2009). Fungicide dose rates and resistance development in *Mycosphaerella graminicola*: A spurious correlation or reasonable hypothesis? *Outlooks on Pest Management* 20(5): 194–196.

Schepers, H. T. A. M., Kessel, G. J. T., Lucca, F., Forch, M. G., van den Bosch, G. B. M., Topper, C. G. and Evenhuis, A. (2018). Reduced efficacy of fluazinam against *Phytophthora infestans* in the Netherlands. *European Journal of Plant Pathology* 151(4): 947–960. Doi: 10.1007/s10658-018-1430-y.

Sierotzki, H., Haas, H. U., Oostendorp, M., Stierli, D. and Nuninger, C. (2017). Adepidyn TM fungicide: A new broad spectrum foliar fungicide for multiple crops. In: Deising, H. B., Fraaije, B., Mehl, A., Oerke, E. C., Sierotzki, H. and Stammler, G. (Eds.), *Modern Fungicides and Antifungal Compounds* (Vol. VIII). Deutsche Phytomedi¬zinische Gesellschaft, Braunschweig, pp. 77–84.0

Staub, T. and Sozzi, D. 1983. Recent practical experiences with fungicide resistance. In *Proceedings of the 10th International Congress Plant Protection*, pp. 591–598. Alton, UK: BCPC Publ.

Strobel, D., Bryson, R., Semar, M., Stammler, G., Kienle, M. and Smith, J. (2019). Mefentrifluconazole (Revysol) – The first isopropanol azole. In: Deising, H. B., Fraaije, B., Mehl, A., Oerke, E. C., Sierotzki, H. and Stammler, G. (Eds.), *Modern Fungicides and Antifungal Compounds* (Vol. IX). Deutsche Phytomedizinische Gesellschaft, Braunschweig, pp. 259–264.

te Beest, D. E., Paveley, N. D., Shaw, M. W. and van den Bosch, F. (2013). Accounting for the economic risk caused by variation in disease severity, in fungicide dose decisions: Exemplified for *Mycosphaerella graminicola* on winter wheat. *Phytopathology* 103(7): 666–672.

Thygesen, K., Jorgensen, L. N., Jensen, K. S. and Munk, L. (2009). Spatial and temporal impact of fungicide spray strategies on fungicide sensitivity of *Mycosphaerella graminicola* in winter wheat. *European Journal of Plant Pathology* 123(4): 435–447.

Torriani, S. F. F., Brunner, P. C., McDonald, B. A. and Sierotzki, H. (2009). QoI resistance emerged independently at least 4 times in European populations of *Mycosphaerella graminicola*. *Pest Management Science* 65(2): 155–162.

Trkulja, N. R., Milosavljevic, A. G., Mitrovic, M. S., Jović, J. B., Toševski, I. T., Khan, M. F. R. and Secor, G. A. (2017). Molecular and experimental evidence of multi-resistance of *Cercospora beticola* field populations to MBC, DMI and QoI fungicides. *European Journal of Plant Pathology* 149(4): 895–910.

Vagndorf, N., Heick, T. M., Justesen, A. F., Andersen, J. R., Jahoor, A., Jørgensen, L. N. and Orabi, J. (2018). Population structure and frequency differences of CYP51 mutations in *Zymoseptoria tritici* populations in the Nordic and Baltic regions. *European Journal of Plant Pathology* 152(2): 327–341.

van den Berg, F., Paveley, N. D. and van den Bosch, F. (2016). Dose and number of applications that maximize fungicide effective life exemplified by *Zymoseptoria tritici* on wheat – A model analysis. *Plant Pathology* 65(8): 1380–1389.

van den Berg, F., van den Bosch, F. and Paveley, N. D. (2013). Optimal fungicide application timings for disease control are also an effective anti-resistance strategy: A case study for *Mycosphaerella graminicola* on wheat. *Phytopathology* 103(12): 1209–1219.

van den Bosch, F., Oliver, R., van den Berg, F. and Paveley, N. (2014). Governing principles can guide fungicide-resistance management tactics. *Annual Review of Phytopathology* 52: 175–195.

van den Bosch, F., Paveley, N., Shaw, M. W., Hobbelen, P. and Oliver, R. (2011). The dose rate debate: Does the risk of fungicide resistance increase or decrease with dose? *Plant Pathology* 60(4): 597–606.

Young, C., Boor, T., Corkley, I., Fraaije, B., Clark, W. S., Havis, N., Kildea, S. and Paveley, N. (2021). *Managing Resistance Evolving Concurrently against Two or More Modes of Action to Extend the Effective Life of New Fungicides*. Agriculture and Horticulture Development Board, UK. Project Report No. 637.

Chapter 7

Fungicide resistance: Evolutionary questions and practical implications

Nichola Hawkins, NIAB, UK

1 Introduction

1.1 Selection of resistance

The rise of resistance, whether against clinical drugs or agricultural pesticides, is a clear example of evolution in action, under anthropogenic selection, at rapid speeds. Unlike the deliberate artificial selection of desirable traits in domesticated crops and livestock, crop protection measures exert a self-defeating selection: that is, in controlling a pest or pathogen population, they are also exerting a selection pressure in favour of control failure. As a result, resistance management is often a balancing act between achieving the required level of pest or disease control now, and conserving the effectiveness of the control measure for future use.

Fungicide resistance is selected when the average per capita growth rate of resistant genotypes is higher than that of sensitive genotypes (van den Bosch et al., 2014). This will occur when a fungicide is applied, killing (or preventing growth of) sensitive strains, if other strains are better able to survive and reproduce in the presence of the fungicide, i.e. more resistant (unless that advantage is outweighed by other impacts on fitness: see Section 6).

http://dx.doi.org/10.19103/AS.2022.0116.07

Resistant genotypes may already be present at low levels in the population, or ongoing mutation in the pathogen population may, by chance, produce a resistant variant (Hawkins et al., 2019; Section 3). Therefore resistance risk depends on the genetic variability of the pathogen population (McDonald and Linde, 2002; Section 2), as well as the probability of available genetic variation conferring resistance for a given fungicide (Grimmer et al., 2015); Section 4). In assessing resistance risk, these are referred to as pathogen risk factors and fungicide risk factors, respectively (Brent and Hollomon, 2007b). Resistance risk in practice will also depend on agronomic risk factors: that is, fungicide use and other crop management practices, which determine the strength of selection and may also affect pathogen population size. Therefore, a given pathogen and fungicide will have an inherent level of resistance risk, but resistance management involves changing agronomic factors to reduce the realised level of risk (Brent and Hollomon, 2007a).

1.2 Resistance management

Practical measures used to slow the evolution of resistance include mixing or alternating fungicides with different modes of action (Elderfield et al., 2018), limiting repeat treatments with the same fungicide, using optimal dose rates (van den Bosch et al., 2011) and timing of treatments (van den Berg et al., 2016), and combining fungicide use with resistant varieties, cultural control and other measures as part of an integrated pest management strategy (Corkley et al., 2022).

In some cases, such as limiting the number of repeated treatments, there is a trade-off between reducing selection for resistance (fewer treatments), and achieving higher levels of disease control for the current crop (more treatments). In other cases, such as dose rates and timing, or whether to use different fungicides in mixtures or alternations, even deciding which approach is best purely for resistance management is less clear-cut. Therefore, to establish the optimal approach, these trade-offs must be calculated, including quantifying the strength of selection for resistance under the different management scenarios. Approaches to evaluating different management strategies include modelling (van den Bosch et al., 2014), monitoring pathogen populations in different regions for sensitivity shifts (Jorgensen et al., 2020), and field trials with different fungicide programmes selecting from the natural population (Dooley et al., 2016) or from a defined mixed-genotype inoculum as has been demonstrated for selection by different host varieties (Zhan and McDonald, 2013).

Optimised resistance management strategies can slow the selection of resistance, but are even more effective if they are implemented early, in time to delay the initial emergence of resistance too (Hobbelen et al., 2014). Therefore, resistance risk should be considered even before a new product is released, and specific management guidelines put in place ahead of its first use.

If possible, resistance risk should be considered even earlier, during product development: so rather than developing a product and then assessing it as high risk, efforts could be focused on targets and leads that could result in a lower-risk product. Whilst the ideal situation would be an 'evolution-proof' measure, this is unlikely to be possible in practice, but it is still useful to understand the evolutionary factors that result in higher or lower risk and so select targets for which resistance is likely to emerge more rarely and be selected more slowly (Bell and Maclean, 2018). Resistance risk and management should be considered throughout the crop protection product life cycle, to maximise durability of control (Fig. 1); and at each stage, practical decisions depend on the underlying evolutionary processes driving the emergence and selection of resistance.

Improved understanding of the evolutionary origins and predictability of resistance against different fungicide types would enable more accurate resistance risk assessments (Hawkins and Fraaije, 2021). This then informs resistance management guidelines, as the level of mitigation needed is tailored to the fungicide level (EPPO, 2015), and the precise management methods can be optimised according to the resistance levels, cross-resistance patterns and fitness penalties associated with the expected resistance mechanisms. Resistance monitoring programmes can also be designed, and diagnostic tests developed, based on predicted resistance mechanisms (Malandrakis et al., 2011; Miao et al., 2016), but also considering the limits to evolutionary predictability and the need to include some monitoring methods that would also pick up any sensitivity shifts due to alternative, unexpected mechanisms (Hawkins and Fraaije, 2021).

For the first site-specific fungicides, the methyl benzimidazoles (MBCs), the first resistant isolates were found in 1969 (Schroeder and Provvidenti, 1969) but mutagenesis and molecular studies into the mode of action and resistance mechanisms did not follow until years later (Davidse and Flach, 1977; Koenraadt et al., 1992). Likewise, for the azole fungicides, resistance had been reported in multiple pathogens in the field before the mode of action and resistance mechanisms had been elucidated (Kelly et al., 1993). For the Quinone-outside inhibitor (QoI) fungicides, the mode of action of related compounds had been characterised before any agricultural fungicides in that class were released (Di Rago et al., 1989). From 1994, resistance risk assessments have been required for product registration in the European Union (Russell, 2004). For the newer succinate dehydrogenase (SDHI) and Quinone inside Inhibitor (QiI) fungicides, resistance risk assessments including mutagenesis studies of potential mechanisms were carried out prior to registration (Scalliet et al., 2012; Fouché et al., 2022). This has enabled a more proactive approach to

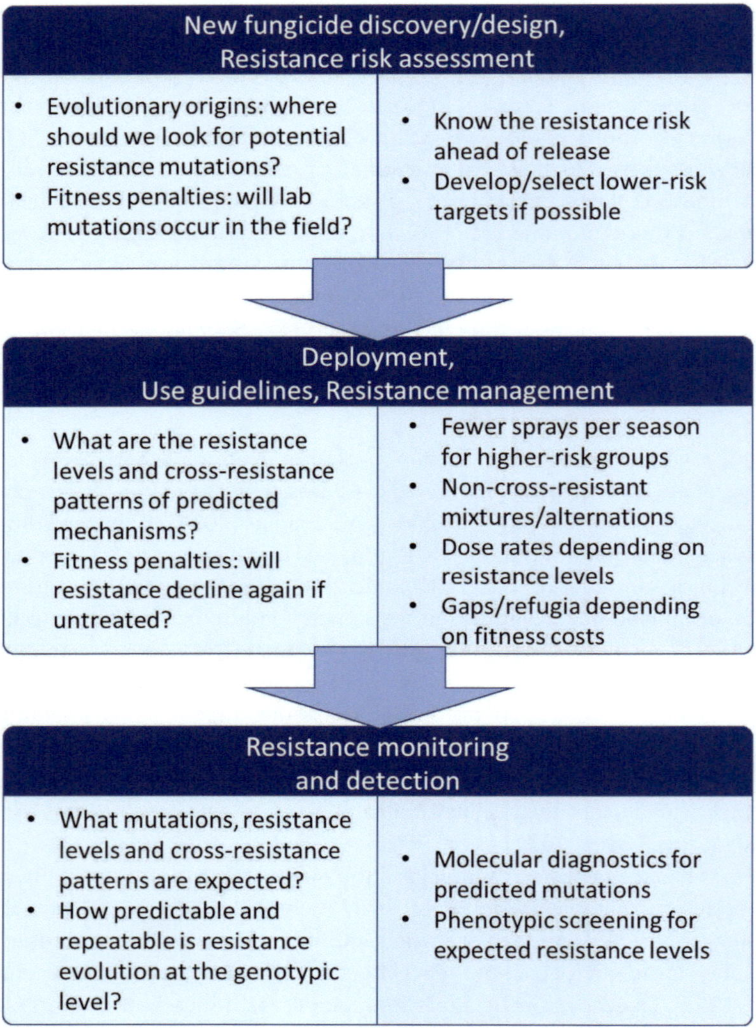

Figure 1 Evolutionary questions and practical implications at each stage of the crop protection product life cycle.

resistance management, with higher-risk actives approved only in mixtures, and for a limited number of sprays per season. This more pre-emptive approach to resistance should also be followed for future crop protection measures, including non-chemical approaches. The better we can understand the risk and drivers of resistance evolution, the more accurately we can predict and mitigate against it.

2 Evolutionary origins

2.1 Standing variation or de novo mutation

Resistant genotypes may be selected form standing variation: that is, pre-existing genetic variation within the pathogen population happened to include some variants with lower fungicide sensitivity already present at low frequency, and then once the fungicide was introduced, fungicide selection favoured the resistant genotypes and they increased in frequency in the population. Alternatively, resistant genotypes may arise through *de novo* mutation: that is, a fungicide is applied to an initially sensitive population, but ongoing mutation within that population continues to produce new genetic variants, and if by chance those variants include fungicide-resistant genotypes then those will be selected by fungicide use (Hawkins et al., 2019).

In those cases where evolutionary origins have been specifically investigated or can otherwise be inferred, point mutations and promoter inserts conferring fungicide resistance appear to have arisen *de novo*. Mitochondrial haplotypes of QoI-resistant *Plasmopara viticola* (Chen et al., 2007) and *Zymoseptoria tritici* (Torriani et al., 2009) indicated multiple parallel *de novo* mutations emerging over the pathogen's range within a short time, while a *CYP51* promoter insert associated with azole resistance in *Uncinula necator* was linked to synonymous mutations in a single genetic background (Frenkel et al., 2015), indicating a single recent *de novo* origin. Azole resistance in *Z. tritici* has shifted over time due to the stepwise accumulation of sequential *CYP51* mutations, indicating *de novo* origins of each new mutation in the previously selected background (Cools et al., 2011).

Resistance due to paralogue number variation has been selected from standing variation, as inferred from gene phylogenies (Hawkins et al., 2014) or the presence of resistant genotypes in fungicide-naïve populations (Steinhauer et al., 2019), but so far these seem to be isolated cases. This is in contrast to the higher prevalence of selection from standing variation for cases of herbicide resistance in weeds, a more even mix of standing variants and *de novo* origins for insecticide resistance (Hawkins et al., 2019), and the importance of horizontal gene transfer of plasmid-based antibiotic resistance in bacteria (Perry and Wright, 2013).

2.2 Impact on resistance risk and management

Standing variation may lead to a higher risk of more rapid evolution of resistance than *de novo* mutation for two main reasons. First, the variant is already present, whereas for *de novo* mutations, evolution may be delayed until a relevant mutation occurs. Second, a *de novo* mutation will initially occur within a single individual, whereas a standing variant will be present as a low but non-zero proportion of the population (Hermisson and Pennings, 2005). This means *de novo* variants are more vulnerable to loss by genetic drift, and any increase in frequency under selection is from a lower starting point. However, these factors are less applicable for pathogens with extremely large effective population sizes and high mutation rates, so these pathogens are at higher resistance risk (McDonald and Linde , 2002).

Resistance management strategies are designed to reduce the selection of resistance. However, for *de novo* mutations, strategies should also account for the initial emergence of resistance. During this emergence phase, reducing the total pathogen population, by using fungicide mixtures or combining with other control measures, may be especially effective as a resistance management strategy, reducing the population size in which mutations can occur, and reducing the growth rate of resistant as well as sensitive genotypes while resistant genotypes are still at sufficiently low frequency to vulnerable to loss by genetic drift (Hobbelen et al., 2014). Furthermore, when resistance has been detected in some regions but not others, biosecurity measures to limit the spread of resistant genotypes may be justified if the mutation has arisen *de novo* in a pathogen with a relatively low mutation rate; whereas for variants already present at lower frequencies throughout the population, the focus should be on resistance management to slow the parallel selection of those same variants in other growing regions. However, in pathogens with high mutation rates, even relatively complex resistant genotypes can emerge *de novo* independently in different regions (McDonald et al., 2019).

2.3 Implications for risk assessments

The main methods used in resistance risk assessments are baseline screening, and mutagenesis studies. In baseline screening, sensitivity testing is carried out on a large number of isolates collected from a fungicide-naïve population (EPPO, 2015). A bimodal or 'disruptive' distribution (Russell, 2004), with a small but distinct grouping of more resistant variants, indicates higher resistance risk as those resistant variants could increase in frequency under pesticide selection. This can give an indication of the risk of resistance evolving by selection from standing variation, but not the risk of new mutations arising. The reliability of baseline screens for predicting the evolution of resistance from standing variation will depend on the initial frequency of resistant variants and the number of individuals screened: a small sample means rarer variants, or those found only in some populations or regions, may be missed (Russell, 2004). It will also depend on how many different resistant variants are present: if there are multiple resistant variants, this adds uncertainty as to which will be selected, but also raises the risk that even higher levels of resistance may emerge through recombination of resistance alleles at different loci (Burke et al., 2014).

In mutagenesis studies, isolates are exposed to fungicide selection, often *in vitro* but sometimes *in planta* by passaging in a glasshouse, and often with UV treatment but sometimes with chemical mutagens or with spontaneous mutations only. Generating mutants with monogenic resistance, especially if a single mutation confers a high level of resistance without detectable fitness penalties, indicates high resistance risk (EPPO, 2015). This can give an indication of the resistance risk from *de novo* mutations, but not from standing variation. The reliability of mutagenesis studies for predicting the evolution of resistance by *de novo* mutation will depend on the population size and mutation rate. under experimental conditions, and the underlying evolutionary repeatability of the resistance mechanisms involved: for example, if there are multiple possible mutations, with finely balanced trade-offs between resistance levels and fitness penalties, it is more difficult to predict which lab mutations will then be selected in the field; this is discussed further in Section 6.

Baseline screening has been used in risk assessments for herbicides (Espeby et al., 2011; Ulber et al., 2013), for which resistance is often selected from standing variation (Delye et al., 2013; Baucom and Mauricio, 2010). For fungicides, *de novo* mutation appears to be the more common route to resistance, and mutagenesis has been used in risk assessments (Scalliet et al., 2012; Fouché et al., 2022), but baselines still need to be established for comparison against later populations to determine whether a sensitivity shift has occurred (Russell, 2004).

3 Adaptive potential and pathogen risks

3.1 Mutational supply

The overall risk of fungicide resistance for a given pathogen against a given fungicide is a combination of the inherent pathogen-associated risk (pathogen risk), the inherent fungicide-associated risk (fungicide risk), and the agronomic risk which varies according to management (Brent and Hollomon, 2007b) (Fig. 2). Therefore, if pathogen risk is higher, stricter control of agronomic risk is needed, especially for fungicides that are also high risk. Risk levels for pathogen groups have primarily been assigned based on the history of resistance evolution against previous fungicide classes, which is problematic for newly-emerging pathogens (Grimmer et al., 2014). Understanding the general risk factors for rapidly evolving pathogens allows more predictive models to be developed (Grimmer et al., 2015). These risk factors include traits that influence the level of variation in a pathogen population, and traits affecting the selection and spread of resistant variants.

Natural selection acts on the heritable variation present in a population. In the case of fungicide resistance, the most relevant variants are target-site single-nucleotide polymorphisms (SNP), and target site or efflux pump

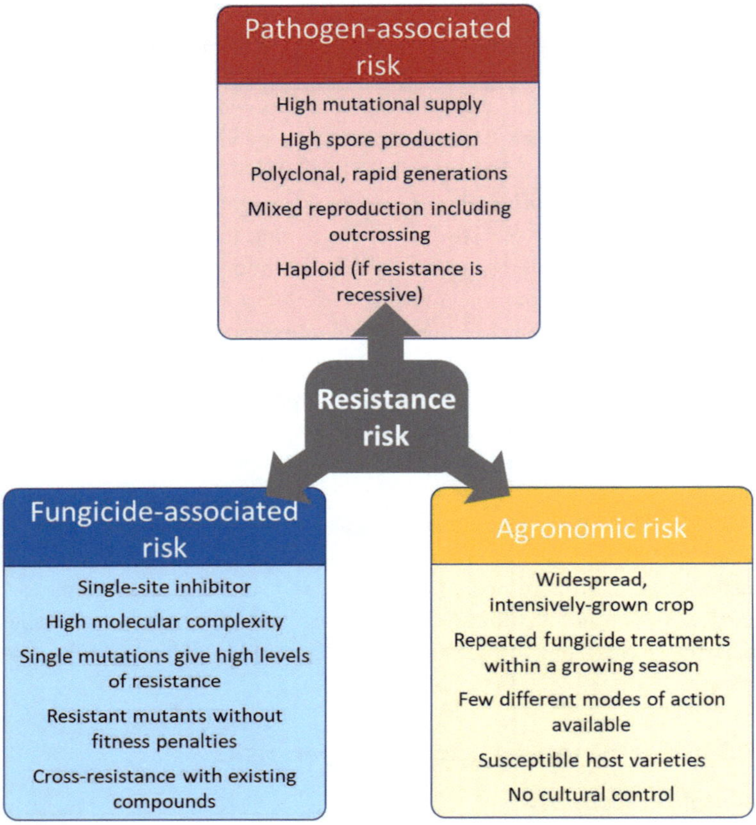

Figure 2 Factors affecting fungicide resistance risk, which can be divided into pathogen-related, fungicide-related and agronomic risk factors.

promoter changes. The resistance risk for a given pathogen will depend partly on the mutational supply of relevant variants, which is a product of mutation rate, population size and reproductive rate. The relevant mutational supply for a given fungicide will also depend on the number of different possible mutations that would actually confer resistance: this is an aspect of fungicide risk, discussed further in Section 4. Some fungicides may increase mutational supply, for example by stress-related transposon activation (Schnabel and Chen, 2013), but this effect may depend on genetic background (Gambhir et al., 2021), and differences from untreated lines measured in lab conditions have not so far been reported in the field (Dowling et al., 2016).

For *Z. tritici*, it has been estimated that in an average untreated infected wheat crop, three pycnidial generations into a growing season, over two trillion pyncidiospores per hectare are produced, and the mutation rate has been measured as 3×10^{-10} substitutions per site per cell division. It has therefore been calculated that if substitutions at any of ten possible nucleotide positions would confer a specific selective advantage, for example conferring fungicide resistance or cultivar-specific virulence, over 28 million spores per hectare per growing season would have one or more of those adaptive mutations (McDonald and Linde, 2002). Therefore, in this system, the evolution of resistance is unlikely to be mutation-limited, and the effective lifespan of a new crop protection measure is not likely to benefit from a prolonged lag period before relevant mutations occur. *Z. tritici* is generally considered to be a medium-risk pathogen (Brent and Hollomon, 2007b), although multi-fungicide-resistant Septoria leaf blotch has become a major problem in some wheat-growing countries (Kildea et al., 2019). Comparable estimates of mutational supply have not yet been calculated for other, higher- or lower-risk pathogens.

The high mutation rate in *Z. tritici* is reflected in the multiple parallel origins of QoI resistance in *Z. tritici* (Torriani et al., 2009). In the oomycete *Plasmopara viticola*, QoI resistance has also emerged at least twice (Chen et al., 2007). In *Pyrenophora teres* f. sp. *maculata*, both target-site SNPs and promoter inserts of a long terminal repeat (LTR) transposon had emerged in parallel multiple times across pathogen populations (Mair et al., 2020). Similarly, in *Pyrenopeziza brassicae*, *CYP51* overexpression is associated with various promoter inserts of different lengths in different field isolates, indicating multiple parallel origins (Carter et al., 2014). However, in *Uncinula necator*, a *CYP51* promoter insert was found in a single haplotypic background, consistent with a selective sweep following a single emergence of target-site overexpression (Frenkel et al., 2015).

3.2 Pathogen life-history traits

In addition to high mutation rates and large population sizes, a mixed reproductive strategy including outcrossing as well as clonal reproduction is associated with high adaptive potential and therefore high risk of rapidly overcoming control measures (McDonald and Linde, 2002). Whilst clonal reproduction can produce mutations, recombination can lead to higher levels of resistance by combining resistant alleles at multiple loci, such as target-site mutations and over-expressed efflux pumps, or even combining multiple mutations within a target-site-encoding gene in the case of intragenic recombination (Brunner et al., 2008). Furthermore, recombination can also facilitate the evolution of multiple resistance to different modes of action, if resistance mutations in different target-site-encoding genes were initially present in different isolates; it can also produce novel combinations of fungicide resistance and virulence, or genotypes with resistance in other favourable genetic backgrounds and escaping linkage drag (McDonald and Linde, 2002). Reproductive mode can also influence dispersal (see Section 3.4): for example, where conidia are splash-dispersed but ascospores are airborne, sexual reproduction will lead to longer-distance spread (Fraaije et al., 2005).

A shorter generation time results in higher resistance risk, not only because more reproductive cycles allow more opportunities for mutation (McDonald and Linde, 2002), but also because a faster epidemic growth rate results in more rapid selection of advantageous genotypes (van den Bosch et al., 2014). Monocyclic pathogens, with a single generation per year, are generally lower risk than polycyclic pathogens with multiple generations within a growing season. Therefore, mixed reproductive strategies, with rapid clonal generations during each growing season as well as periodic sexual cycles to provide new recombinant genotypes, result in the highest resistance risk (Grimmer et al., 2015).

Population structure and patterns of gene flow will determine whether resistance is likely to emerge in different regions through parallel localised evolution, or the spread of resistant genotypes after a single emergence event (Croll and McDonald, 2017). Dispersal patterns will depend on the method of spread, such as airborne spread following prevailing winds (Brunner et al., 2008), or human-mediated spread in produce, soil or seeds (Thomas et al., 2014). Only human-mediated spread can be modified by phytosanitary measures, but if genotypes are spreading by other means there is still a need for international data-sharing. Gene flow will also determine the spatial scale at which resistance management must be applied. For a soilborne pathogen, management within a single field will benefit the individual farmer, whereas for airborne pathogens, an individual farmer cannot prevent the ingress of resistant genotypes (Estep et al., 2015), regional or wider co-operation would be required. This, unfortunately, can lead to a 'tragedy of the commons' whereby growers will not follow best stewardship practice if they believe their neighbours are not (Corkley et al., 2022).

Haploid pathogens are at higher risk than diploid species, especially for fungicides where resistance mutations are recessive or semidominant and so the full selective advantage is only expressed in homozygotes, such as resistance against zoxamide or carboxylic acid amides in diploid oomycetes (Gisi et al., 2007; Young et al., 2001). Rust fungi are diploid and have been considered as low resistance risk, but this may be partly due to a lower number of effective fungicide classes in the first place (and therefore fewer cases of resistance that could evolve), and to a cytochrome *b* intron that lowers resistance risk specifically for the QoIs (Oliver, 2014). Resistance against both azole and SDHI fungicides has now been reported in yellow rust of wheat (Cook et al., 2021). Resistance against multiple fungicides has evolved in the soybean rust *Phakopsora pachyrhizi*, partly because of the high agronomic risk of an intensively grown and treated crop (Dalla Lana et al., 2018). High pathogen risk in practice results from combination of intrinsic pathogen-related risk factors such as mutation rate and life-history traits, and agronomic risk in the host crop, such as intensive cultivation and treatment; susceptibility and lack of diversity of crop varieties; and lack of alternative control measures (Chong et al., 2021). These factors can even confound apparent fungicide risk for fungicide classes with a narrow spectrum of use: for example, the dicarboximides rapidly selected for resistance in the high-risk pathogen *Botrytis cinerea* on heavily treated crops such as grapevine (Hahn, 2014).

4 Trait complexity and fungicide risk

4.1 Single-site and multi-site inhibitors

The inherent fungicide-related risk of resistance depends on the evolvability and relative selective advantage of resistant genotypes. Whilst mutational supply is an important component of pathogen risk, the likelihood of any such mutations conferring resistance, and the resulting level of resistance, is a key component of fungicide risk (Brent and Hollomon, 2007b; McDonald and Linde, 2002). Small, simple genetic changes are more probable: an SNP within a single gene is the simplest possible change, and so fungicides for which one SNP is sufficient to confer resistance will be at high risk, whereas resistance requiring rarer or multiple genetic changes would take longer to evolve.

The clearest difference in fungicide risk levels is between single-site and multi-site inhibitors (Grimmer et al., 2014). Resistance against single-site fungicides can generally evolve through mutations in a single gene, and often through a single mutation. In contrast, target-site resistance against a multi-site inhibitor would require resistant variants of all essential targets to be present for a resistant variant to be selected, although non-target site mechanisms are also possible. Consequently, multi-site inhibitors are the lowest risk fungicides in terms of resistance, with most multi-site fungicide classes having either no reported resistance, or field/glasshouse resistance reported in a single high-risk pathogen, *Botrytis cinerea* (FRAC, 2020). The morpholines inhibit two different enzymes in the sterol biosynthetic pathway, and reduced sensitivity in field isolates has been reported only in cereal powdery mildews, *Erysiphe graminis* (FRAC, 2020).

However, this broad activity has led to concerns about off-target effects on humans or the environment. Consequently, those fungicides which have the low resistance risk of being lost from the toolbox due to resistance are instead increasingly being lost through regulatory changes (AHDB, 2020). In evolutionary terms, the high specificity of single-site fungicides is a double-edge sword: a narrower taxonomic range of organisms are affected, limiting undesirable off-target effects, but this also means the evolutionary distance for the intended target to escape inhibition is smaller, so the risk of resistance is higher. An ideal control measure would escape this trade-off, for example by finding compounds with robust binding to targets that are highly conserved in pathogens but very different or absent in other organisms.

Fungicides with inherently low resistance risk are valuable as mixing partners to protect higher-risk fungicides (Elderfield et al., 2018): in mixtures where both fungicides are at risk of resistance, a finer balance is needed between the level of disease control contributed by each (Mikaberidze et al., 2014). Therefore, the loss of lower-risk multi-site fungicides leaves the remaining modes of action at even higher risk. Where all available fungicides are at risk of resistance developing by very small, simple genetic changes, evolutionary complexity should be added to the disease control programme in other ways: for example, by combining fungicides with host resistance or cultivar mixtures (Kristoffersen et al., 2020).

4.2 Single or multiple target-site mutations

There are smaller, but nonetheless important, differences within the broad grouping of 'single-site' fungicide classes. The highest risk groups are those where a single mutation is sufficient to confer a high level of resistance, such as MBCs or QoIs. Resistance evolution also tends to be more repeatable for these fungicides, as discussed in Section 6. For other single-site fungicides such as azoles, pathogens can still evolve resistance through mutation of a single target-site-encoding gene, but individual mutations confer lower levels of resistance. Higher levels of resistance can still evolve if multiple mutations accumulate, but this takes longer, because multiple mutational steps are needed, and because the incremental increases in resistance at each step tend to confer a smaller advantage. Therefore these fungicide groups are considered medium risk (Brent and Hollomon, 2007b).

For the QoI fungicides, a single target-site substitution, G143A in cytochrome *b*, is sufficient to confer a high level of resistance. G143A rapidly emerged in many pathogen species, sometimes after as little as two years of QoI use (Heaney et al., 2000), and by 2020 it had been reported in 35 different pathogens (Hawkins and Fraaije, 2021). Alternative substitutions F129L and G137R/S confer moderate to high levels of resistance, also occurring as single mutations (Fernandez-Ortuno et al., 2008). For the MBCs, a wider range of β-tubulin mutations have been reported, including several different substitutions at codon 198 or F220Y, but again, even highly resistant isolates have a single mutation (Hawkins and Fraaije, 2016).

For SDHI fungicides, the mutations that have evolved are far more variable between pathogen species (Hawkins and Fraaije, 2021). Single mutants show a mixture of moderate to high levels of resistance, and field isolates reported so far are predominantly single mutants, with a small number of double mutants. For example, in *Pyrenophora teres*, a range of single mutations have evolved, including sdhB-H277Y, sdhC-G79R, and sdhD-D145G (Rehfus et al., 2016). In *Zymoseptoria tritici*, sdhC-H152R confers higher levels of resistance, whilst other mutations such as sdhC-T79N, sdhC-N86S or sdhB-T268I have intermediate resistance (Rehfus et al., 2018), but double mutants such as sdhC-N86S + sdhB-N225T or sdhC-N86S + sdhC-L85P are more resistant than their component single mutants (FRAC SDHI Working Group, 2022). In *Corynespora cassiicola,* double mutants have been reported with various combinations of one *sdhB* and one *sdhD* mutation: for those *sdhB* mutations that confer only low resistance in single mutants, the double mutants were more resistant (Sun et al., 2022). Overall, SDHIs are considered medium-to-high risk (Fungicide Resistance Action Committee, 2021): highly resistant genotypes have not yet reached high frequencies in many pathogens, but intermediate resistance has emerged relatively rapidly, and more resistant genotypes could increase in future.

For azole fungicides, single mutations mostly confer low to moderate levels of resistance, especially against newer, more active azoles, but multiple mutations combine to give further decreases in sensitivity (Cools and Fraaije, 2008; Huf et al., 2018). Multiple mutants are more common than SDHIs, and can have greater numbers of mutations. Of 18 plant pathogen species with reported CYP51 mutations associated with reduced azole sensitivity in field isolates, the most common mutation, Y137F, has been reported in 11, but was only the sole reported mutation in four (Hawkins and Fraaije, 2021). In *Ramularia collo-cygni*, up to three mutations per haplotype have been reported so far (Rehfus et al., 2019); in *Pseudocercospora fijiensis*, four-mutation haplotypes are common but up to seven have been found (Chong et al., 2021); and in *Z. tritici*, haplotypes with six or seven *CYP51* mutations are common and up to nine have been reported (Huf et al., 2018). However, even these multiple mutants are not fully resistant, and newer azoles, at robust doses and protectant timing, are still a valuable component of control programmes. Therefore, azoles are considered medium risk fungicides: they have a low risk of sudden, total control failure due to resistance, but a high risk of incremental loss of sensitivity.

Where multiple changes are needed for high levels of resistance, the evolutionary accessibility of resistance will depend on the selective advantage of intermediate steps. If a selective advantage only applies once multiple mutations are present, the complexity of the multi-mutation change would reduce the probability of resistance emerging. However, if each mutation in turn confers a stepwise increase in resistance, then a more complex resistant genotype can evolve incrementally. The impact of adaptive landscapes on the evolution of resistance is discussed further in Section 6.

These differences between single-site fungicide classes are harder to predict than the overall difference between single-site and multi-site fungicides

(Grimmer et al., 2014), but further analysis of traits associated with higher resistance risk, measured as a shorter time from the introduction of the fungicide class to the first detected resistance, identified fungicide molecular weight as a key risk indicator (Grimmer et al., 2015). This was considered to be a proxy measure of structural complexity, associated with highly specific fungicide-target binding which could be severely disrupted by a small change in the target site. However, the predictive power of these general traits is limited, and so relative risk must still be assessed individually for each fungicide class, largely based on empirical mutagenesis studies and any existing field resistance. Other factors affecting resistance risk, which cannot yet be predicted based on general fungicide traits, include functional constraints on the target site: if potential mutations that would disrupt fungicide binding would also affect substrate binding, they may be selected more slowly if at all due to associated fitness penalties, as discussed in Section 5.

4.3 Overexpression and non-target-site resistance

The most common non-target-site resistance mechanism in plant pathogens is efflux, in which ATP-binding cassette (ABC) or major facilitator superfamily (MFS) transporter proteins remove fungicide molecules from the fungal cells. Overexpression of the transporter-encoding gene results in reduced sensitivity (Del Sorbo et al., 2000). Overexpression of the target-site-encoding gene can also confer resistance, by shifting the stoichiometric ratio of inhibitor:target. For fungicides, *CYP51* overexpression has been detected in azole-resistant isolates of several species (Schnabel and Jones, 2001; Nikou et al., 2009, Hamamoto et al., 2000). Another non-target-site resistance mechanism common for other drug and pesticide classes is metabolic resistance due to detoxifying enzymes (Hawkins et al., 2019), but there have been a very limited number of suspected cases in fungicide resistance, e.g. Barak and Edgington (1984).

Overexpression, whether of the target site or an efflux transporter, is most commonly caused by inserts in the promoter region of these genes. The exact nature of the inserts varies: sometimes they comprise tandem repeats derived from the promoter itself; others are derived from transposable elements. In *Penicillium digitatum*, resistant isolates have five copies of a 126 base pair section of the *CYP51* promoter that is present as a single copy in sensitive isolates (Hamamoto et al., 2000), whereas the *CYP51* promoter insert in azole-resistant isolates of *Blumeriella jaapii* has been identified as a long interspersed nuclear element retrotransposon (Ma et al., 2006), and the *MgMFS1* promoter insert leading to enhanced efflux in *Z. tritici* appears to be derived from an LTR retrotransposon (Omrane et al., 2015). In some cases, multiple different inserts have been found in different isolates of the same species (Carter et al., 2014; Diaz-Trujillo et al., 2016). This indicates multiple origins of overexpression in the pathogen population, suggesting that these are relatively common and probable genetic change in those cases; however, this will be gene and species dependent, with some genomic regions predisposed to transposon activity. Other mechanisms for promoter inserts include transcription factor mutations (Kretschmer et al., 2009), or copy number variants (Stalder et al., 2023).

In *Botrytis cinerea*, the MDR1 phenotype is associated with overexpression of the ABC transporter AtrB due to amino acid substitutions in the transcription factor mrr1, whereas the MDR2 phenotype is associated with overexpression of the MFS transporter *mfsM2* due to a retrotransposon-derived insertion within the *mfsM2* promoter. Isolates with both the *mrr1* mutation and *mfsM2* promoter inserts have higher levels of resistance, referred to as the MDR3 phenotype (Kretschmer et al., 2009). In this case, the more complex MDR3 resistance, with changes in two separate genes, is likely to have evolved stepwise, as each mutation individually confers partial resistance.

All currently known cases of fungicide resistance due to efflux involve overexpression of pre-existing transporters for which the fungicide happened to be a substrate. Gain of function mutations, change of substrate or the evolution of whole transporter or metabolic genes *de novo* would be far less likely to evolve rapidly following the introduction of a fungicide. For drug and pesticide classes where metabolic resistance is more common, it is often selected from standing genetic variation (Hawkins et al., 2019) or acquired by horizontal gene transfer (Perry and Wright, 2013), and pre-existing ability to metabolise a range of substrates is considered a risk factor for resistance (EPPO, 2015).

There have been two published cases of paralogue number variation conferring fungicide sensitivity differences. A *CYP51A* paralogue is found in addition to the usual *CYP51B* in *Rhynchosporium commune* isolates with reduced azole sensitivity (Hawkins et al., 2014), and an additional sdhC paralogue, named alt-SDHC, was identified in *Z. tritici* isolates with reduced sensitivity to some SDHIs (Steinhauer et al., 2019; Yamashita and Fraaije, 2018). In both cases, it was found that the additional paralogue had been selected from standing genetic variation pre-dating fungicide selection, rather than a new paralogue emerging *de novo*. However, the azole resistance mechanism in *R. commune CYP51A* is target-site overexpression (Hawkins et al., 2014), which has evolved in many other pathogens through promoter insertions (Becher and Wirsel, 2012), and the reduced SDHI sensitivity of the additional *Z. tritici* sdhC paralogue is largely due to a single amino acid difference which has also occurred as a *de novo* mutation within the standard *sdhC* orthologue in lab mutants (Steinhauer et al., 2019; Scalliet et al., 2012). Therefore, whilst the presence of a functionally diverged paralogue is a more complex trait and less likely to evolve *de novo* in populations where it is not present as standing variation, equivalent resistance mechanisms have evolved *de novo* from existing paralogues, representing a simpler evolutionary path to a similar phenotypic destination.

Mutagenesis studies are more likely to generate SNPs than promoter inserts, resulting in a bias towards target site mutations (Hawkins and Fraaije, 2021) and potential failure to consider overexpression-based mechanisms. Target-site mutations are often seen as the major resistance concern for fungicides as they are often associated with higher resistance factors than other mechanisms, but potential non-target-site resistance can still add to the overall risk. This should be considered especially for multi-site fungicides where target-site resistance is less likely (Barak and Edgington, 1984); where existing non-target-site resistance to current fungicides confers cross-resistance to new classes (Fouché et al., 2022); or where target-site resistance and efflux could combine to result in higher resistance factors than target-site resistance alone (Leroux and Walker, 2011).

5 Fitness penalties

5.1 Costs and trade-offs

Fungicide-resistant genotypes confer a selective advantage under fungicide selection, but in some cases, they also have negative effects on the pathogen's metabolism, growth or reproduction (Lalève et al., 2014). Pleiotropic fitness costs may result from target-site mutations that reduce fungicide binding but also impair the target protein's original function, for example by affecting substrate binding or protein stability (Hawkins and Fraaije, 2018). Fungicides are effective because their targets are essential for the fungus, for example in respiration or membrane biosynthesis; therefore, it is not beneficial to evolve a target site variant that evades fungicide binding if it can no longer fulfil its original cellular function anyway. This imposes functional constraints that limit which target-site mutations can evolve: at one extreme, lethal mutations will not evolve at all, but where the impact on protein function is less severe, there will be a trade-off between fungicide selection and other selective pressures. The outcome of a fitness trade-off may depend on the strength of fungicide selection, for example the dose rate (Gutiérrez-Alonso et al., 2017); or on environmental conditions, for example if fitness costs are temperature-dependent (Ma et al., 2003).

For resistance mechanisms involving overexpression, fitness penalties are likely to result from allocation costs. The pathogen must devote more resources to producing the protein product in question, and in the case of efflux transporters, active efflux is energy-dependent (through ATP or transmembrane proton-motive force), and so there are energy costs too. Reduced fitness has been measured in some field isolates with efflux-mediated fungicide resistance (Samaras et al., 2020; Chen et al., 2016), although the functional basis of these costs has not yet been confirmed.

Evolutionary trade-offs can also result from different patterns of cross-resistance between fungicides (Cools et al., 2013). Target site mutations confer cross-resistance to other fungicides with the same mode of action, but in some cases, there is only partial cross-resistance, or even negative cross-resistance, between different compounds in the same class. The G143A substitution confers cross-resistance to all QoI fungicides (Heaney et al., 2000) except the new Subgroup A which was developed for its effectiveness against QoI-resistant strains (Suemoto et al., 2019), whereas QiIs bind to a different site on cytochrome *b* and so are not affected by target-site cross-resistance (Fouché et al., 2022).

For the SDHIs, some mutations affect the whole class, whereas for other mutations, the stretched heterocycle amide SDHIs (SHA-SDHIs) such as fluopyram (Förster et al., 2022) form a distinct cross-resistance group from the pyrazole-4-carboxamides such as bixafen and fluxapyroxad (Fungicide Resistance Action Committee, 2021). For example, in *Alternaria alternata*, field isolates with sdh mutations sdhB-H277L, sdhC-H134R and sdhC-G79R were more resistant to pyrazole-4-carboxamides than SHA-SDHIs. In contrast, *Z. tritici* isolates with the sdh point mutations sdhB-T268I, sdhC-T79N, sdhC-N86S and sdhC-H152R showed cross-resistance to all tested SDHIs (Rehfus et al., 2018), while the presence of alt-SDHC is associated with reduced sensitivity to SHA-SDHIs and not pyrazole-4-carboxamides (Steinhauer et al., 2019). Incomplete cross-resistance to different azoles is also seen with many *CYP51* mutations (Becher and Wirsel, 2012).

The first reported case of negative cross-resistance in fungicides was between the benzimidazoles and N-phenyl-carbamates: E198A and other codon 198 substitutions confer resistance against benzimidazoles, but loss of intrinsic resistance to N-phenylcarbamates (Koenraadt and Jones, 1993). Negative cross-resistance has also been reported for some *CYP51* variants in *Z. tritici*: the substitution V136A reduced sensitivity to prochloraz but not tebuconazole, whereas I381V reduced sensitivity to tebuconazole but not prochloraz (Cools et al., 2010).

5.2 Impact on risk levels and management

When assessing the resistance risk of new fungicides, fitness penalties could reduce the risk of a given mutation evolving in the field (EPPO, 2015). For example, in *Zymoseptoria tritici*, the sdhC-H152R substitution confers high levels of SDHI resistance, but laboratory selection experiments show evidence of fitness penalties (Gutiérrez-Alonso et al., 2017). This mutation has been detected in the field, but so far at relatively low levels, indicating slower selection than other sdh mutations (FRAC SDHI Working Group, 2022). However, caution is needed to avoid over-interpreting apparent penalties as a guarantee of low resistance risk, as fitness penalties detected in lab mutants, under controlled or *in vitro* conditions, may not apply in the field.

Resistance could evolve in combination with compensatory mutations to overcome fitness penalties: for example, the *Z. tritici CYP51* mutation I381V occurs in combination with mutations at codons 459–461 (Cools et al., 2010), and experimental evolution of fludioxonil-resistant *Aspergillus nidulans* with resistance-associated fitness penalties resulted in the evolution of resistant strains with increased fitness (Schoustra et al., 2006). Resistance could also evolve through alternative resistant mutations without penalties, such as the evolution of QoI resistance through F129L or G137S in cytochrome *b* genes with an intron that would make G143A lethal (Standish et al., 2019; Klosowski et al., 2016). Where mutagenesis studies produce multiple possible mutations, fitness penalties could give an indication of which of those mutations are more likely to evolve in the field (Hawkins and Fraaije, 2016).

Fitness penalties can reduce the inherent risk for the fungicide-pathogen combination in question, but the realised risk level in practice will still depend on agronomic risk and management. Whether a resistant genotype comes with fitness penalties determines whether resistance is static or reversed in the absence of selection (Allen et al., 2017). If the fitness cost is not lethal and fungicide selection is strong, the resistant genotype will still be selected during fungicide treatment and so resistance management is still needed, but certain resistance management measures will be more effective: treatment gaps or alternations, or spatial mosaics or refugia, will be far more effective where fitness penalties mean that resistance frequency will actually decrease and not just stay at a similar level whenever the fungicide is not used.

Quantifying fitness costs would allow such strategies to be optimised. In theory, the length of treatment gaps could be calculated to return to pre-treatment levels after each treatment and hold back resistance indefinitely, but in practice the time required is likely to be prohibitively long for a relatively small fitness cost compared to a very large selective advantage of resistance when treated (Mikaberidze and McDonald, 2015). Fitness costs can also result in dose-dependent selection of resistance (Gutiérrez-Alonso et al., 2017), so the optimal dose rate can be calculated accordingly. In general, higher fungicide doses exert stronger selection of highly resistant mutants: in the absence of fitness costs, increasing dose rates from low to high would result in a gradual increase from weaker to stronger positive selection for resistance, but in the presence of fitness costs, there will be tipping point from negative to positive selection. Keeping dose rates below that point could avoid selection for the highly resistant mutation, but this will only be an option in practice if that dose provides sufficient disease control; otherwise reducing the dose will delay rather than prevent resistance. Models also indicate that mixtures with a lower-risk mixing partner can be more effective as a resistance management strategy where resistance against the higher-risk fungicide also carries a fitness cost: the mixing partner reduces overall epidemic growth rate, and the fitness cost reduces the selective advantage of resistant over sensitive genotypes (Mikaberidze and McDonald, 2015).

Incomplete or negative cross-resistance could also be exploited for resistance management. Mixtures or cycling strategies could be designed around fungicides with negative cross-resistance, but this carries the risk of selecting alternative, positively cross-resistant mutations: mixing benzimidazoles, to which E198A β-tubulin mutants are resistant, and N-phenyl-carbamates, to which E198A mutants are sensitive and wild-type strains are resistant, resulted in the selection of F200Y mutants, which are resistant to both (Katan et al., 1989). Even mixing or alternating different modes of action risks selection of non-target-site mechanisms such as efflux, leading to trade-offs in resistance management decisions as mixtures or alternations to limit the selection of specific resistance may increase selection for generalised resistance mechanisms.

6 Predictability of resistance

6.1 Fitness landscapes

Fitness landscapes are a model of evolution in which topology represents fitness and the horizontal plane represents the genetic or phenotypic space through which an evolving lineage can move. Therefore, when natural selection drives a lineage to increase in fitness under given environmental conditions, the lineage will 'climb an adaptive peak'. The simplest landscape would have a single peak, and so any lineage starting at any point on the landscape can be expected to climb that peak and ultimately reach the same adaptive end-point. In contrast, a more 'rugged' landscape has multiple peaks, and so natural selection may drive lineages from some starting points to climb a local rather than the global peak. In such cases, the optimal state may not be evolutionarily accessible without crossing a 'fitness valley', and natural selection lacks the foresight to accept a reduction in immediate fitness in order to reach a better state in the longer term.

The fitness landscape was initially envisaged as a metaphor (Wright, 1932), but more recently there has been more interest in reconstructing empirical fitness landscapes, with the horizontal plane representing phenotypes in morphometric space (Martin and Wainwright, 2013) or a genotypes in a mutational network (Khan et al., 2011). Fitness landscapes have also been reconstructed for some cases of antibiotic resistance: for example, the accumulation of β-lactamase mutations resulting in increased antibiotic detoxification activity (Weinreich et al., 2006). A simple example from fungicide resistance is shown in Fig. 3: under QoI fungicide selection,

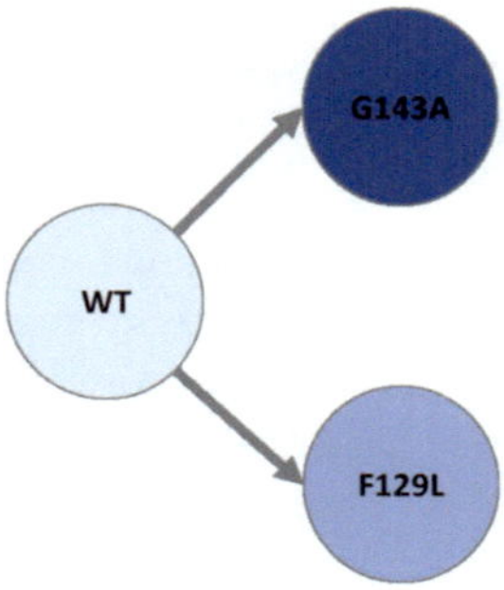

Figure 3 A simple fungicide resistance fitness landscape, showing the relative fitness (darker shade = higher fitness) of different cytochrome b genotypes under QoI fungicide selection.

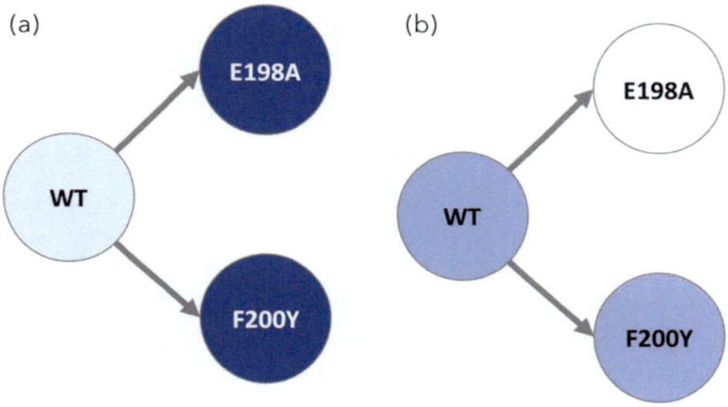

Figure 4 Fitness landscapes (darker shade = higher fitness) for β-tubulin genotypes, under selection by (a) benzimidazole and (b) N-phenyl-carbamate fungicides.

the G143A mutation confers a high level of fungicide resistance (with no detectable fitness penalty) and therefore higher fitness, whereas F129L confers a lower level of resistance and therefore an intermediate fitness advantage.

The topology of a fitness landscape can depend on the selective environment: for example, a lineage may be adapting to infect a specific host, or to grow at certain temperatures; similarly, resistance generally becomes selectively advantageous only when a fungicide is used. Therefore, the predictability of evolutionary pathways may be affected if fluctuating selection creates a changeable fitness landscape (Trubenová et al., 2019), and the accuracy of predictions will depend on understanding how the fitness landscape changes under different selective pressures. Figure 4 shows an example of a fitness landscape that changes due to negative cross-resistance between two different fungicides.

6.2 Epistasis

Fitness landscapes also become more complex where the fitness effect of a given mutation varies according to the genetic background; that is, it has epistatic interactions, with other mutations in the same gene or with other variable loci in the genome. In the absence of epistasis, the effects of multiple mutations are additive; that is, the combined impact is equal to the sum of its parts (or in the case of resistance factors, the product of its parts, i.e. additive on a log scale). Magnitude epistasis means the combined effect differs in size from the sum of its parts; either greater (positive epistasis, synergistic interaction), or smaller (negative epistasis, diminishing returns) than expected based on the effect of each mutation separately. Sign epistasis means a mutation can be either deleterious or beneficial depending on the genetic background, and reciprocal sign epistasis means this applies to both interacting genes: i.e. each is deleterious alone but the combination is beneficial, or each is beneficial but the combination is deleterious (Fig. 5).

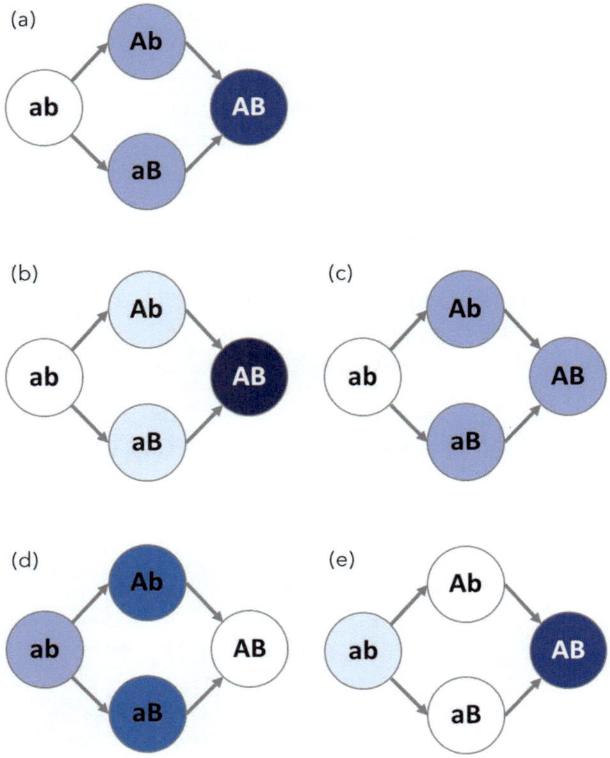

Figure 5 Fitness landscapes (darker shade = higher fitness) for two genes, A and B, with different epistatic interaction: (a) additive interaction, no epistasis; (b) positive magnitude epistasis; (c) negative magnitude epistasis; (d) and (e) sign epistasis.

Epistatic interactions between mutations produce more rugged fitness landscapes, resulting in less predictable evolutionary trajectories, since the outcome depends on the genetic background, the selective history of a population and order in which mutations emerge, and potentially on chance events such as rarer genetic changes to cross a fitness valley. For example, in the case of sign epistasis where a double mutant has a fitter advantage but each single mutation is deleterious, the double mutant would only emerge if both mutations arise simultaneously; or if selective conditions temporarily change to favour a single mutant; or if relaxed selection or genetic drift allow a maladaptive single mutant to survive for long enough for the second mutation to occur in that background or for both single mutants to recombine.

A simple case of sign epistasis in fungicide resistance is shown in Fig. 6. The G143A mutation in cytochrome *b* is highly advantageous under QoI fungicide selection, but this is not the case in all genetic backgrounds. Some fungi have an intron in cytochrome *b* for which codon 143 forms part of the splicing recognition site: in these cases, the nucleotide substitution encoding G143A removes that splice site, and therefore is lethal. In these fungi, the moderately resistant F129L substitution (or in some cases, G137R) is selected instead (Sierotzki et al., 2007).

More complex epistatic interactions have been identified between the multiple co-occurring *CYP51* mutations in *Zymoseptoria tritici*. At least 33 different mutations have been reported, with up to nine mutations in a single isolate (Huf et al., 2018). Whilst the accumulation of multiple mutations generally leads to stepwise increases in resistance levels, this is not always strictly additive, resulting in more complex, rugged fitness landscapes. For example, when heterologously expressed in yeast, the S524T *ZtCYP51* substitution conferred a resistance factor of 9 for epoxiconazole relative to the wild-type *Z. tritici* gene, and the L50S/Y461S double mutant had a resistance factor of 6, but the triple

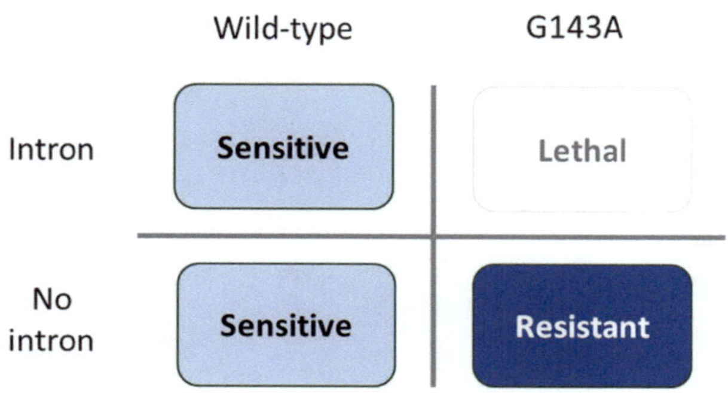

Figure 6 Epistatic interaction between codon 143 genotype and intron presence/absence in fungal cytochrome b, under selection by QoI fungicides (darker shade = higher fitness).

mutant had a resistance factor of 95 (Fig. 7a): this is 1.8 times the resistance factor expected from a purely additive combination, indicating a synergistic effect of positive magnitude epistasis. In contrast, for prothioconazole, S524T alone had a resistance factor of 33, the L50S/Y461S had a resistance factor of 2, but the triple mutant had a resistance factor of 28, indicating antagonism or diminishing returns (Fig. 7b) (Cools et al., 2011). In the field, codon 459–461 changes evolved before S524T emerged and before prothioconazole was introduced: it is likely that the accessibility of different evolutionary pathways depends on the order in which different azoles were introduced.

Epistatic interactions also affect fitness penalties associated with resistance mutations. When heterologously expressed in yeast, the I381V substitution in *Zt CYP51* was lethal alone or in combination with L50S, but the addition of Y461H restored enzyme function (Cools et al., 2010). This demonstrates how the accessibility of evolutionary pathways is also affected by functional constraints (Fig. 8).

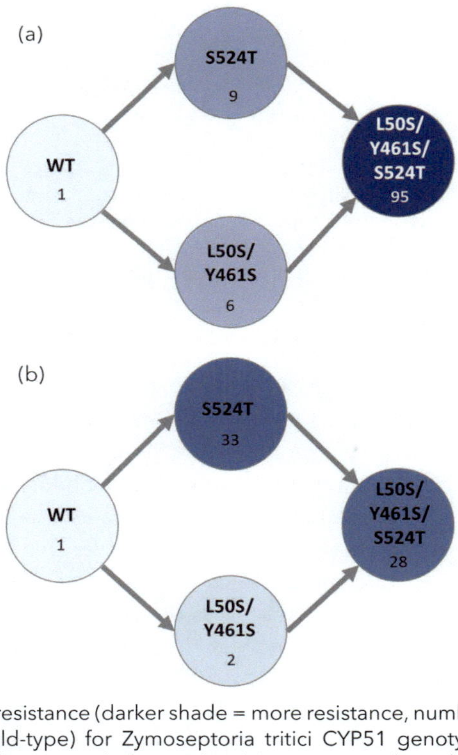

Figure 7 Fungicide resistance (darker shade = more resistance, numbers show resistance factor relative to wild-type) for Zymoseptoria tritici CYP51 genotypes heterologously expressed in yeast, for two azole fungicides: (a) epoxiconazole and (b) prothioconazole. Data from Cools et al. (2011).

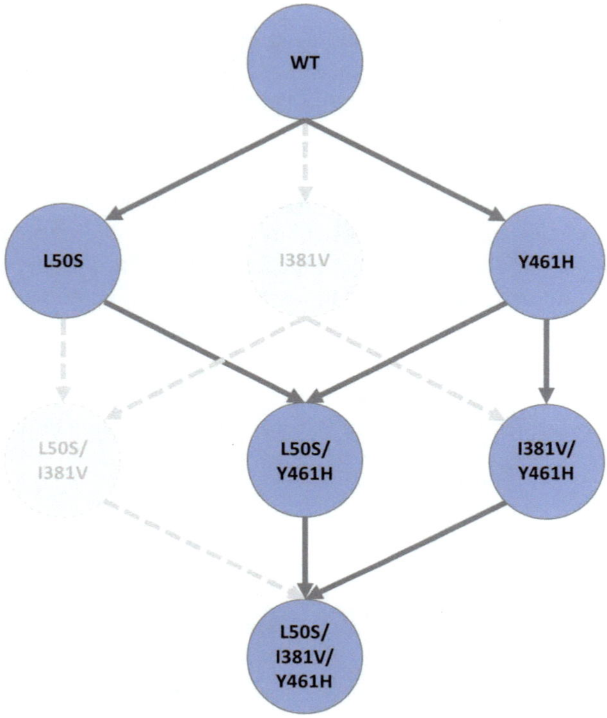

Figure 8 Evolutionary accessibility of different pathways to a multi-mutation fungicide-resistant haplotype. *Zymoseptoria tritici CYP51* variants were heterologously expressed in yeast and tested for viability. Greyed-out circles = non-viable genotypes; dashed lines = inaccessible pathways. Data from Cools et al. (2010).

6.3 Impact on risk assessments, monitoring and management

Resistance risk assessments involve predicting resistance evolution; therefore, the predictability of evolution will affect how accurate those assessments can be. The most predictable cases have a simple adaptive landscape, with one or few mutations likely to recur from lab to field and across multiple pathogen species (Hawkins and Fraaije, 2021). Where the adaptive landscape is more complex, evolution will be less predictable: this may indicate lower overall resistance risk, due the absence of a single highly resistant mutant, but mutation-specific predictions, such as the exact resistance level or cross-resistance pattern, may not be reliable if alternative mutations evolve in other populations. It may not be possible to predict which of a range of lab-selected mutants are likely to emerge in the field, or further data may be needed regarding the relative fitness of different mutants in different conditions (Hawkins and Fraaije, 2016).

Therefore, where resistance risk assessments are being used to inform management guidelines, it is important not only to make predictions but also to understand the level of confidence or uncertainty in those predictions. For example, if dose rates or mixing partners are selected based on the resistance levels and cross-resistance patterns of predicted mutations, it is important to consider the risk of different mutations with different properties being selected in some field populations. This risk will be mitigated if the management strategy does not rely entirely on fungicide mixtures, but also includes host resistance and cultural control. If monitoring programmes are set up to detect the predicted mutations, there is a risk that others could be missed, either because discriminatory doses for phenotypic testing have been set too high, or because allele-specific molecular diagnostics are used but a different resistant allele evolves instead. This risk can be reduced by also conducting EC_{50} testing for some isolates, so any shift in sensitivity will be detected, whatever the molecular basis.

7 Conclusion and future trends

Resistance is an ongoing challenge in crop protection, and is likely to remain so, whatever form future crop protection measures may take. Practical decisions in fungicide development and deployment need to consider the durability of protection against evolving pathogen populations, and not just the immediate protection of the current crop. Resistance development is ultimately an evolutionary process, so improved evolutionary understanding can lead to better-informed decision making at every stage.

Predicting the evolution of resistance against new fungicides is important to assess the risk of resistance, to determine the necessary levels of mitigation (EPPO, 2015). Mutagenesis and *in vitro* selection studies are now widely used in risk assessments (Scalliet et al., 2012; Fouché et al., 2022), but more detailed knowledge of evolutionary factors such as fitness penalties could improve predictions of which lab mutations will be important in the field (Hawkins and Fraaije, 2021). In future, predicting the resistance risk earlier in product development would make it possible to identify lower-risk leads earlier in the pipeline, or even to identify lower-risk targets for the rational design of inhibitors. Ideal targets would have low resistance risk, with a target site that is highly constrained and conserved in the pathogen (though preferably not in non-target organisms) so mutations are likely to be deleterious, and fungicide binding that is robust to possible changes in the binding site. The rational design of fungicides has been discussed for some time, for example by researchers identifying potential targets involved in pathogenicity (Cools and Hammond-Kosack, 2013), but computational advances in structural modelling, and new approaches such as RNAi which can inhibit targets at the nucleotide sequence level, mean this may now be closer to reality.

Understanding the level of risk and the likely nature of resistance, whether intermediate or highly resistant, compound-specific or generalised across different fungicides, enables more precise, tailored resistance management strategies to be designed. Resistance management should be in place from the first use of the fungicide, which means predicting potential evolutionary pathways of resistance before resistant genotypes emerge in the field, but also being aware of the different confidence levels in those predictions should evolution follow a more or less predictable path. Ongoing testing of optimal management strategies for current and new fungicides continues to use a combination of modelling and field testing of different management strategies, but future trends are likely to include greater use of *in vitro* experimental evolution which allows empirical testing of basic principles over far shorter time frames, and in far less space, than field trials (Ballu et al., 2021; Gutiérrez-Alonso et al., 2017).

Similarly, predicting the likely mechanisms of resistance but at the same time understanding the limitations of those predictions is important in planning resistance monitoring programmes. For example, molecular diagnostics can be designed for specific mutations, but there should also be some monitoring that would be able to detect any resistance due to different and unexpected mechanisms. Future trends include advances in sequencing technologies, enabling whole genes to be sequenced from a pathogen population sample so predictions of individual mutations within that gene will no longer be necessary for designing molecular diagnostics. Further developments in this area could include single-cell whole-genome sequencing, rather than limiting genotyping-by-sequencing to a single gene, although this will still require some prediction and subsequent functional validation of which genetic differences are likely to be resistance-associated.

Future developments in understanding the evolutionary biology of resistance will enable the development of lower-risk crop protection products and more evidence-based management guidelines, but it will still be prudent to include mitigations against unexpected evolutionary turns.

8 Where to look for further information

The following articles give an overview of resistance evolution and risk assessments:

- Lucas, J. A., Hawkins, N. J. and Fraaije, B. A. 2015. The Evolution of Fungicide Resistance. *Advances in Applied Microbiology*, 90, 29-92.
- van den Bosch, F., Oliver, R., Van Den Berg, F. and Paveley, N. 2014. Governing Principles Can Guide Fungicide-Resistance Management Tactics. *Annual Review of Phytopathology,* 52**,** 175-195.

- EPPO 2015. PP 1/213 (4) Resistance risk analysis. *EPPO Bulletin,* 45, 371-387.

Information on fungicide modes of action, resistance risk assessments, reported cases of resistance and updates on resistance in the field can be found at:

- Fungicide Resistance Action Committee, www.frac.info.
- Local and regional Resistance Action Groups, such as FRAG-UK, https://ahdb.org.uk/knowledge-library/the-fungicide-resistance-action-group-frag-uk.

9 References

AHDB. 2020. Spotlight on the loss of active substances [Online]. Agriculture and Horticulture Development Board, UK. Available at: https://ahdb.org.uk/knowledge-library/extension-of-authorisation-for-minor-use-eamu/spotlight-loss-active-substances [Accessed February 2020].

Allen, R. C., Engelstadter, J., Bonhoeffer, S., McDonald, B. A. and Hall, A. R. 2017. Reversing resistance: Different routes and common themes across pathogens. *Proceedings. Biological Sciences* 284(1863), 10.

Ballu, A., Despréaux, P., Duplaix, C., Dérédec, A., Carpentier, F. and Walker, A.-S. 2021. Delayed but generalist resistance in alternation: A trade-off shaped by the mode of action of antifungals, as revealed by experimental evolution in a plant pathogen. *bioRxiv*.

Barak, E. and Edgington, L. V. 1984. Cross-resistance of *Botrytis cinerea* to captan, thiram, chlorothalonil, and related fungicides. *Canadian Journal of Plant Pathology* 6(4), 318-320.

Baucom, R. S. and Mauricio, R. 2010. Defense to the herbicide RoundUp predates its widespread use. *Evolutionary Ecology Research* 12, 131–141.

Becher, R. and Wirsel, S. G. 2012. Fungal cytochrome P450 sterol 14α-demethylase (CYP51) and azole resistance in plant and human pathogens. *Applied Microbiology and Biotechnology* 95(4), 825–840.

Bell, G. and Maclean, C. 2018. The search for 'evolution-proof' antibiotics. *Trends in Microbiology* 26(6), 471–483.

Brent, K. J. and Hollomon, D. W. 2007a. *Fungicide Resistance in Crop Pathogens: How Can It Be Managed?* (2nd edn.). *FRAC Monograph*. CropLife International, Brussels, 56pp.

Brent, K. J. and Hollomon, D. W. 2007b. Fungicide resistance: The assessment of risk. *FRAC Monograph*. CropLife International, Brussels, 53pp.

Brunner, P. C., Stefanato, F. L. and McDonald, B. A. 2008. Evolution of the *CYP51* gene in *Mycosphaerella graminicola*: Evidence for intragenic recombination and selective replacement. *Molecular Plant Pathology* 9(3), 305–316.

Burke, M. K., Liti, G. and Long, A. D. 2014. Standing genetic variation drives repeatable experimental evolution in outcrossing populations of *Saccharomyces cerevisiae*. *Molecular Biology and Evolution* 31(12), 3228–3239.

Carter, H. E., Fraaije, B. A., West, J. S., Kelly, S. L., Mehl, A., Shaw, M. W. and Cools, H. J. 2014. Alterations in the predicted regulatory and coding regions of the sterol 14 alpha-demethylase gene (*CYP51*) confer decreased azole sensitivity in the oilseed rape pathogen *Pyrenopeziza brassicae*. *Molecular Plant Pathology* 15(5), 513–522.

Chen, S. N., Luo, C. X., Hu, M. J. and Schnabel, G. 2016. Fitness and competitive ability of *Botrytis cinerea* isolates with resistance to multiple chemical classes of fungicides. *Phytopathology* 106(9), 997–1005.

Chen, W. J., Delmotte, F., Richard-Cervera, S., Douence, L., Greif, C. and Corio-Costet, M. F. 2007. At least two origins of fungicide resistance in grapevine downy mildew populations. *Applied and Environmental Microbiology* 73(16), 5162–5172.

Chong, P., Essoh, J. N., Arango Isaza, R. E., Keizer, P., Stergiopoulos, I., Seidl, M. F., Guzman, M., Sandoval, J., Verweij, P. E., Scalliet, G., Sierotzski, H., De Lapeyre De Bellaire, L., Crous, P. W., Carlier, J., Cros, S., Meijer, H. J. G., Peralta, E. L. and Kema, G. H. J. 2021. A world-wide analysis of reduced sensitivity to DMI fungicides in the banana pathogen *Pseudocercospora fijiensis*. *Pest Management Science* 77(7), 3273–3288.

Cook, N. M., Chng, S., Woodman, T. L., Warren, R., Oliver, R. P. and Saunders, D. G. O. 2021. High frequency of fungicide resistance-associated mutations in the wheat yellow rust pathogen *Puccinia striiformis* f. sp. *tritici*. *Pest Management Science* 77(7), 3358–3371.

Cools, H. J. and Fraaije, B. A. 2008. Are azole fungicides losing ground against Septoria wheat disease? Resistance mechanisms in *Mycosphaerella graminicola*. *Pest Management Science* 64(7), 681–684.

Cools, H. J. and Hammond-Kosack, K. E. 2013. Exploitation of genomics in fungicide research: Current status and future perspectives. *Molecular Plant Pathology* 14(2), 197–210.

Cools, H. J., Hawkins, N. J. and Fraaije, B. A. 2013. Constraints on the evolution of azole resistance in plant pathogenic fungi. *Plant Pathology* 62, 36–42.

Cools, H. J., Mullins, J. G. L., Fraaije, B. A., Parker, J. E., Kelly, D. E., Lucas, J. A. and Kelly, S. L. 2011. Impact of recently emerged sterol 14α-demethylase (CYP51) variants of Mycosphaerella graminicola on azole fungicide sensitivity. *Applied and Environmental Microbiology* 77(11), 3830–3837.

Cools, H. J., Parker, J. E., Kelly, D. E., Lucas, J. A., Fraaije, B. A. and Kelly, S. L. 2010. Heterologous expression of mutated eburicol 14 alpha-demethylase (CYP51) proteins of Mycosphaerella graminicola to assess effects on azole fungicide sensitivity and intrinsic protein function. *Applied and Environmental Microbiology* 76(9), 2866–2872.

Corkley, I., Fraaije, B. and Hawkins, N. 2022. Fungicide resistance management: Maximising the effective life of plant protection products. *Plant Pathology* 71(1), 150–169.

Croll, D. and McDonald, B. A. 2017. The genetic basis of local adaptation for pathogenic fungi in agricultural ecosystems. *Molecular Ecology* 26(7), 2027–2040.

Dalla Lana, F., Paul, P. A., Godoy, C. V., Utiamada, C. M., Da Silva, L. H. C. P., Siqueri, F. V., Forcelini, C. A., Jaccoud-Filho, D. S., Miguel-Wruck, D. S., Borges, E. P., Juliatti, F. C., Campos, H. D., Nunes, J., Carneiro, L. C., Canteri, M. G., Ito, M. F., Meyer, M. C., Martins, M. C., Balardin, R. S., Furlan, S. H., Carlin, V. J. and Del Ponte, E. M. 2018. Meta-analytic modeling of the decline in performance of fungicides for managing soybean rust after a decade of use in Brazil. *Plant Disease* 102(4), 807–817.

Davidse, L. C. and Flach, W. 1977. Differential binding of methyl benzimidazol-2-yl carbamate to fungal tubulin as a mechanism of resistance to this antimitotic agent in mutant strains of *Aspergillus nidulans*. *Journal of Cell Biology* 72(1), 174–193.

Del Sorbo, G., Schoonbeek, H. J. and De Waard, M. A. 2000. Fungal transporters involved in efflux of natural toxic compounds and fungicides. *Fungal Genetics and Biology: FG and B* 30(1), 1–15.

Delye, C., Deulvot, C. and Chauvel, B. 2013. DNA analysis of herbarium specimens of the grass weed Alopecurus myosuroides reveals herbicide resistance pre-dated herbicides. *PLoS ONE* 8(10), e75117.

Di Rago, J. P., Coppee, J. Y. and Colson, A. M. 1989. Molecular basis for resistance to myxothiazol, mucidin (strobilurin A), and stigmatellin. Cytochrome b inhibitors acting at the center o of the mitochondrial ubiquinol-cytochrome c reductase in *Saccharomyces cerevisiae*. *The Journal of Biological Chemistry* 264(24), 14543–14548.

Diaz-Trujillo, C., Chong, P., Cordovez, V., Guzmán, M., De Wit, P. J., Stergiopoulos, I., Meijer, H. J., Isaza, R. E. A., Scalliet, G. and Sierotzki, H. 2016. A new resistance mechanism to DMI fungicides in the fungal banana black Sigatoka pathogen *Pseudocercospora fijiensis* is driven by increased expression of *PfCYP51* through multiple promoter repeats. PhD Thesis. *The Origin, Versatility and Distribution of Azole Fungicide Resistance in the Banana Black Sigatoka Pathogen Pseudocercospora fijiensis*, Wageningen University.

Dooley, H., Shaw, M. W., Spink, J. and Kildea, S. 2016. Effect of azole fungicide mixtures, alternations and dose on azole sensitivity in the wheat pathogen *Zymoseptoria tritici*. *Plant Pathology* 65(1), 124–136.

Dowling, M. E., Bryson, P. K., Boatwright, H. G., Wilson, J. R., Fan, Z., Everhart, S. E., Brannen, P. M. and Schnabel, G. 2016. Effect of fungicide applications on *Monilinia fructicola* population diversity and transposon movement. *Phytopathology* 106(12), 1504–1512.

Elderfield, J. A. D., Lopez-Ruiz, F. J., van den Bosch, F. and Cunniffe, N. J. 2018. Using epidemiological principles to explain fungicide resistance management tactics: Why do mixtures outperform alternations? *Phytopathology* 108(7), 803–817.

EPPO. 2015. PP 1/213 (4) Resistance risk analysis. *EPPO Bulletin* 45(3), 371–387.

Espeby, L. Å., Fogelfors, H. and Milberg, P. 2011. Susceptibility variation to new and established herbicides: Examples of inter-population sensitivity of grass weeds. *Crop Protection* 30(4), 429–435.

Estep, L. K., Torriani, S. F. F., Zala, M., Anderson, N. P., Flowers, M. D., McDonald, B. A., Mundt, C. C. and Brunner, P. C. 2015. Emergence and early evolution of fungicide resistance in North American populations of *Zymoseptoria tritici*. *Plant Pathology* 64(4), 961–971.

Fernandez-Ortuno, D., Tores, J. A., De Vicente, A. and Perez-Garcia, A. 2008. Mechanisms of resistance to QoI fungicides in phytopathogenic fungi. *International Microbiology: The Official Journal of the Spanish Society for Microbiology* 11(1), 1–9.

Förster, H., Luo, Y., Hou, L. and Adaskaveg, J. E. 2022. Mutations in *Sdh* gene subunits confer different cross-resistance patterns to SDHI fungicides in *Alternaria alternata* causing Alternaria Leaf spot of almond in California. *Plant Disease* 106(7), 1911–1918. DOI: 10.1094/PDIS-09-21-1913-RE.

Fouché, G., Michel, T., Lalève, A., Wang, N. X., Young, D. H., Meunier, B., Debieu, D., Fillinger, S. and Walker, A. S. 2022. Directed evolution predicts cytochrome b G37V target site modification as probable adaptive mechanism towards the QiI fungicide fenpicoxamid in *Zymoseptoria tritici*. *Environmental Microbiology* 24(3), 1117–1132.

Fraaije, B. A., Cools, H. J., Fountaine, J., Lovell, D. J., Motteram, J., West, J. S. and Lucas, J. A. 2005. Role of ascospores in further spread of QoI-resistant cytochrome b alleles (G143A) in field populations of *Mycosphaerella graminicola*. *Phytopathology* 95(8), 933–941.

FRAC. 2020. List of plant pathogenic organisms resistant to disease control agents [Online]. Fungicide Resistance Action Committee. Available at: https://www.frac.info/docs/default-source/publications/list-of-resistant-plant-pathogens/list-of-first-confirmed-cases-of-plant-pathogenic-organisms-resistant-to-disease-control-agents_05_2020.pdf?sfvrsn=7073499a_2 [Accessed 01/06/2020].

FRAC SDHI Working Group. 2022. *Minutes of the 2022 Meeting with Recommendations for 2022*. Fungicide Resistance Action Committee.

Frenkel, O., Cadle-Davidson, L., Wilcox, W. F. and Milgroom, M. G. 2015. Mechanisms of resistance to an azole fungicide in the grapevine powdery mildew fungus, *Erysiphe necator*. *Phytopathology* 105(3), 370–377.

Fungicide Resistance Action Committee. 2021. *FRAC Code List 2021: Fungal Control Agents Sorted by Cross Resistance Pattern and Mode of Action*. CropLife International.

Gambhir, N., Kamvar, Z. N., Higgins, R., Amaradasa, B. S. and Everhart, S. E. 2021. Spontaneous and fungicide-induced genomic variation in *Sclerotinia sclerotiorum*. *Phytopathology* 111(1), 160–169.

Gisi, U., Waldner, M., Kraus, N., Dubuis, P. H. and Sierotzki, H. 2007. Inheritance of resistance to carboxylic acid amide (CAA) fungicides in *Plasmopara viticola*. *Plant Pathology* 56(2), 199–208.

Grimmer, M. K., van den Bosch, F., Powers, S. J. and Paveley, N. D. 2014. Evaluation of a matrix to calculate fungicide resistance risk. *Pest Management Science* 70(6), 1008–1016.

Grimmer, M. K., van den Bosch, F., Powers, S. J. and Paveley, N. D. 2015. Fungicide resistance risk assessment based on traits associated with the rate of pathogen evolution. *Pest Management Science* 71(2), 207–215.

Gutiérrez-Alonso, O., Hawkins, N. J., Cools, H. J., Shaw, M. W. and Fraaije, B. A. 2017. Dose-dependent selection drives lineage replacement during the experimental evolution of SDHI fungicide resistance in *Zymoseptoria tritici*. *Evolutionary Applications* 10(10), 1055–1066.

Hahn, M. 2014. The rising threat of fungicide resistance in plant pathogenic fungi: *Botrytis* as a case study. *Journal of Chemical Biology* 7(4), 133–141.

Hamamoto, H., Hasegawa, K., Nakaune, R., Lee, Y. J., Makizumi, Y., Akutsu, K. and Hibi, T. 2000. Tandem repeat of a transcriptional enhancer upstream of the sterol 14 alpha-demethylase gene (*CYP51*) in *Penicillium digitatum*. *Applied and Environmental Microbiology* 66(8), 3421–3426.

Hawkins, N. J., Bass, C., Dixon, A. and Neve, P. 2019. The evolutionary origins of pesticide resistance. *Biological Reviews of the Cambridge Philosophical Society* 94(1), 135–155.

Hawkins, N. J., Cools, H. J., Sierotzki, H., Shaw, M. W., Knogge, W., Kelly, S. L., Kelly, D. E. and Fraaije, B. A. 2014. Paralog re-emergence: A novel, historically contingent mechanism in the evolution of antimicrobial resistance. *Molecular Biology and Evolution* 31(7), 1793–1802.

Hawkins, N. J. and Fraaije, B. A. 2016. Predicting resistance by mutagenesis: Lessons from 45 years of MBC resistance. *Frontiers in Microbiology* 7, 1814.

Hawkins, N. J. and Fraaije, B. A. 2018. Fitness penalties in the evolution of fungicide resistance. *Annual Review of Phytopathology* 56, 339–360.

Hawkins, N. J. and Fraaije, B. A. 2021. Contrasting levels of genetic predictability in the evolution of resistance to major classes of fungicides. *Molecular Ecology* 30(21), 5318–5327.

Heaney, S. P., Hall, A., Davies, S. A. and Olaya, G. 2000. Resistance to fungicides in the QoI-STAR cross-resistance group: Current perspectives. In: *The BCPC Conference - Pests - And Diseases 2000*, pp. 755–762.

Hermisson, J. and Pennings, P. S. 2005. Soft sweeps: Molecular population genetics of adaptation from standing genetic variation. *Genetics* 169(4), 2335–2352.

Hobbelen, P. H. F., Paveley, N. D. and van den Bosch, F. 2014. The emergence of resistance to fungicides. *PLoS ONE* 9(3), e91910.

Huf, A., Rehfus, A., Lorenz, K. H., Bryson, R., Voegele, R. T. and Stammler, G. 2018. Proposal for a new nomenclature for *CYP51* haplotypes in *Zymoseptoria tritici* and analysis of their distribution in Europe. *Plant Pathology* 67(8), 1706–1712.

Jorgensen, L. N., Matzen, N., Heick, T. M., Havis, N., Holdgate, S., Clark, B., Blake, J., Glazek, M., Korbas, M., Danielewicz, J., Maumene, C., Rodemann, B., Weigand, S., Kildea, S., Bataille, C., Brauna-Morzevska, E., Gulbis, K., Ban, R., Berg, G., Semaskiene, R. and Stammler, G. 2020. Decreasing azole sensitivity of *Z. tritici* in Europe contributes to reduced and varying field efficacy. *Journal of Plant Diseases and Protection* 15, 287–301.

Katan, T., Elad, Y. and Yunis, H. 1989. Resistance to diethofencarb (NPC) in benomyl-resistant field isolates of *Botrytis cinerea*. *Plant Pathology* 38(1), 86–92.

Kelly, S. L., Arnoldi, A. and Kelly, D. E. 1993. Molecular genetic analysis of azole antifungal mode of action. *Biochemical Society Transactions* 21(4), 1034–1038.

Khan, A. I., Dinh, D. M., Schneider, D., Lenski, R. E. and Cooper, T. F. 2011. Negative epistasis between beneficial mutations in an evolving bacterial population. *Science* 332(6034), 1193–1196.

Kildea, S., Marten-Heick, T., Grant, J., Mehenni-Ciz, J. and Dooley, H. 2019. A combination of target-site alterations, overexpression and enhanced efflux activity contribute to reduced azole sensitivity present in the Irish *Zymoseptoria tritici* population. *European Journal of Plant Pathology* 154(3), 529–540.

Klosowski, A. C., De Mio, L. L. M., Miessner, S., Rodrigues, R. and Stammler, G. 2016. Detection of the F129L mutation in the cytochrome b gene in *Phakopsora pachyrhizi*. *Pest Management Science* 72(6), 1211–1215.

Koenraadt, H. and Jones, A. L. 1993. Resistance to benomyl conferred by mutations in codon 198 or codon 200 of the beta-tubulin gene of *Neurospora crassa* and sensitivity to diethofencarb conferred by codon 198. *Phytopathology* 83, 850–854.

Koenraadt, H., Somerville, S. C. and Jones, A. L. 1992. Characterisation of mutations in the beta-tubulin gene of benomyl-resistant field strains of *Venturia inaequalis* and other plant pathogenic fungi. *Phytopathology* 82(11), 1348–1354.

Kretschmer, M., Leroch, M., Mosbach, A., Walker, A. S., Fillinger, S., Mernke, D., Schoonbeek, H. J., Pradier, J. M., Leroux, P., De Waard, M. A. and Hahn, M. 2009. Fungicide-driven evolution and molecular basis of multidrug resistance in field populations of the grey mould fungus *Botrytis cinerea*. *PLoS Pathogens* 5(12), e1000696.

Kristoffersen, R., Heick, T. M., Müller, G. M., Eriksen, L. B., Nielsen, G. C. and Jørgensen, L. N. 2020. The potential of cultivar mixtures to reduce fungicide input and mitigate fungicide resistance development. *Agronomy for Sustainable Development* 40(5), 36.

Lalève, A., Fillinger, S. and Walker, A. S. 2014. Fitness measurement reveals contrasting costs in homologous recombinant mutants of *Botrytis cinerea* resistant to succinate dehydrogenase inhibitors. *Fungal Genetics and Biology: FG and B* 67, 24–36.

Leroux, P. and Walker, A. S. 2011. Multiple mechanisms account for resistance to sterol 14 alpha-demethylation inhibitors in field isolates of *Mycosphaerella graminicola*. *Pest Management Science* 67(1), 44–59.

Ma, Z., Proffer, T. J., Jacobs, J. L. and Sundin, G. W. 2006. Overexpression of the 14α-demethylase target gene (CYP51) mediates fungicide resistance in *Blumeriella jaapii*. *Applied and Environmental Microbiology* 72(4), 2581–2585.

Ma, Z. H., Yoshimura, M. A. and Michailides, T. J. 2003. Identification and characterization of benzimidazole resistance in *Monilinia fructicola* from stone fruit orchards in California. *Applied and Environmental Microbiology* 69(12), 7145–7152.

Mair, W. J., Thomas, G. J., Dodhia, K., Hills, A. L., Jayasena, K. W., Ellwood, S. R., Oliver, R. P. and Lopez-Ruiz, F. J. 2020. Parallel evolution of multiple mechanisms for demethylase inhibitor fungicide resistance in the barley pathogen *Pyrenophora teres* f. sp. *maculata*. *Fungal Genetics and Biology: FG and B* 145, 103475.

Malandrakis, A. A., Markoglou, A. N., Nikou, D. C., Vontas, J. G. and Ziogas, B. N. 2011. Molecular diagnostic for detecting the cytochrome b G143S-QoI resistance mutation in *Cercospora beticola*. *Pesticide Biochemistry and Physiology* 100(1), 87–92.

Martin, C. H. and Wainwright, P. C. 2013. Multiple fitness peaks on the adaptive landscape drive adaptive radiation in the wild. *Science* 339(6116), 208–211.

McDonald, B. A. and Linde, C. C. 2002. Pathogen population genetics, evolutionary potential, and durable resistance. *Annual Review of Phytopathology* 40, 349–379.

McDonald, B. A., Suffert, F., Bernasconi, A. and Mikaberidze, A. 2022. How large and diverse are field populations of fungal plant pathogens? The case of *Zymoseptoria tritici*. *Evolutionary Applications* 15(9), 1360–1373.

McDonald, M. C., Renkin, M., Spackman, M., Orchard, B., Croll, D., Solomon, P. S. and Milgate, A. 2019. Rapid parallel evolution of azole fungicide resistance in Australian populations of the wheat pathogen *Zymoseptoria tritici*. *Applied and Environmental Microbiology* 85(4), e01908-18.

Miao, J., Cai, M., Dong, X., Liu, L., Lin, D., Zhang, C., Pang, Z. and Liu, X. 2016. Resistance assessment for oxathiapiprolin in *Phytophthora capsici* and the detection of a point mutation (G769W) in *PcORP1* that confers resistance. *Frontiers in Microbiology* 7, 615.

Mikaberidze, A. and McDonald, B. A. 2015. Fitness cost of resistance: Impact on management. In: Ishii, H. and Hollomon, D. W. (Eds.), *Fungicide Resistance in Plant Pathogens: Principles and a Guide to Practical Management*. Springer, Tokyo, Japan.

Mikaberidze, A., McDonald, B. A. and Bonhoeffer, S. 2014. Can high-risk fungicides be used in mixtures without selecting for fungicide resistance? *Phytopathology* 104(4), 324–331.

Nikou, D., Malandrakis, A., Konstantakaki, M., Vontas, J., Markoglou, A. and Ziogas, B. 2009. Molecular characterization and detection of overexpressed C-14 alpha-demethylase-based DMI resistance in *Cercospora beticola* field isolates. *Pesticide Biochemistry and Physiology* 95(1), 18–27.

Oliver, R. P. 2014. A reassessment of the risk of rust fungi developing resistance to fungicides. *Pest Management Science* 70(11), 1641–1645.

Omrane, S., Sghyer, H., Audéon, C., Lanen, C., Duplaix, C., Walker, A. S. and Fillinger, S. 2015. Fungicide efflux and the MgMFS1 transporter contribute to the MDR phenotype in *Zymoseptoria tritici* field isolates. *Environmental Microbiology* 17(8), 2805–2823.

Perry, J. A. and Wright, G. D. 2013. The antibiotic resistance 'Mobilome': Searching for the link between environment and clinic. *Frontiers in Microbiology* 4, 138.

Rehfus, A., Matusinsky, P., Strobel, D., Bryson, R. and Stammler, G. 2019. Mutations in target genes of succinate dehydrogenase inhibitors and demethylation inhibitors in *Ramularia collo-cygni* in Europe. *Journal of Plant Diseases and Protection* 126(5), 447–459.

Rehfus, A., Miessner, S., Achenbach, J., Strobel, D., Bryson, R. and Stammler, G. 2016. Emergence of succinate dehydrogenase inhibitor resistance of *Pyrenophora teres* in Europe. *Pest Management Science* 72(10), 1977–1988.

Rehfus, A., Strobel, D., Bryson, R. and Stammler, G. 2018. Mutations in *sdh* genes in field isolates of *Zymoseptoria tritici* and impact on the sensitivity to various succinate dehydrogenase inhibitors. *Plant Pathology* 67(1), 175–180.

Russell, P. E. 2004. Sensitivity baselines in fungicide resistance research and management. *FRAC Monograph*. CropLife International, Brussels, 56pp.

Samaras, A., Ntasiou, P., Myresiotis, C. and Karaoglanidis, G. 2020. Multidrug resistance of *Penicillium expansum* to fungicides: Whole transcriptome analysis of MDR strains reveals overexpression of efflux transporter genes. *International Journal of Food Microbiology* 335, 108896.

Scalliet, G., Bowler, J., Luksch, T., Kirchhofer-Allan, L., Steinhauer, D., Ward, K., Niklaus, M., Verras, A., Csukai, M., Daina, A. and Fonné-Pfister, R. 2012. Mutagenesis and functional studies with succinate dehydrogenase inhibitors in the wheat pathogen *Mycosphaerella graminicola*. *PLoS ONE* 7(4), e35429.

Schnabel, G. and Chen, F. 2013. Fungicide-induced mutagenesis in *Monilinia fructicola* and implications for resistance management. *Phytopathology* 103, 128–128.

Schnabel, G. and Jones, A. L. 2001. The 14 alpha-demethylase (*CYP51A1*) gene is overexpressed in *Venturia inaequalis* strains resistant to myclobutanil. *Phytopathology* 91(1), 102–110.

Schoustra, S. E., Debets, A. J. M., Slakhorst, M. and Hoekstra, R. F. 2006. Reducing the cost of resistance; experimental evolution in the filamentous fungus *Aspergillus nidulans*. *Journal of Evolutionary Biology* 19(4), 1115–1127.

Schroeder, W. T. and Provvidenti, R. 1969. Resistance to benomyl in powdery mildew of cucurbits. *Plant Disease Reporter* 53, 271–275.

Sierotzki, H., Frey, R., Wullschleger, J., Palermo, S., Karlin, S., Godwin, J. and Gisi, U. 2007. Cytochrome b gene sequence and structure of *Pyrenophora teres* and *P. tritici-repentis* and implications for QoI resistance. *Pest Management Science* 63(3), 225–233.

Stalder, L., Oggenfuss, U., Mohd-Assaad, N. and Croll, D. 2023. The population genetics of adaptation through copy number variation in a fungal plant pathogen. *Molecular Ecology* 32(10), 2443–2460. DOI: 10.1111/mec.16435.

Standish, J. R., Brenneman, T. B. and Stevenson, K. L. 2019. Quantifying the effects of a G137S substitution in the cytochrome bc(1) of *Venturia effusa* on azoxystrobin sensitivity using a detached leaf assay. *Plant Disease* 103(5), 841–845.

Steinhauer, D., Salat, M., Frey, R., Mosbach, A., Luksch, T., Balmer, D., Hansen, R., Widdison, S., Logan, G., Dietrich, R. A., Kema, G. H. J., Bieri, S., Sierotzki, H., Torriani, S. F. F.

and Scalliet, G. 2019. A dispensable paralog of succinate dehydrogenase subunit C mediates standing resistance towards a subclass of SDHI fungicides in *Zymoseptoria tritici*. *PLoS Pathogens* 15(12), e1007780.

Suemoto, H., Matsuzaki, Y. and Iwahashi, F. 2019. Metyltetraprole, a novel putative complex III inhibitor, targets known QoI-resistant strains of *Zymoseptoria tritici* and *Pyrenophora teres*. *Pest Management Science* 75(4), 1181–1189.

Sun, B., Zhu, G., Xie, X., Chai, A., Li, L., Shi, Y. and Li, B. 2022. Double mutations in succinate dehydrogenase are involved in SDHI resistance in *Corynespora cassiicola*. *Microorganisms* 10(1), 132.

Thomas, A., Langston Jr., D. B., Walcott, R. R., Gitaitis, R. D. and Stevenson, K. L. 2014. Evidence for fungicide-resistant seed-borne inoculum for gummy stem blight of watermelon. *Seed Science and Technology* 42(1), 92–96.

Torriani, S. F. F., Brunner, P. C., McDonald, B. A. and Sierotzki, H. 2009. QoI resistance emerged independently at least 4 times in European populations of *Mycosphaerella graminicola*. *Pest Management Science* 65(2), 155–162.

Trubenová, B., Krejca, M. S., Lehre, P. K. and Kötzing, T. 2019. Surfing on the seascape: Adaptation in a changing environment. *Evolution; International Journal of Organic Evolution* 73(7), 1356–1374.

Ulber, L., Nordmeyer, H. and Zwerger, P. 2013. Resistance risk assessment within herbicide authorisation - A call for sensitivity data. *Pest Management Science* 69(2), 160–164.

Van Den Berg, F., Paveley, N. D. and van den Bosch, F. 2016. Dose and number of applications that maximize fungicide effective life exemplified by *Zymoseptoria tritici* on wheat - A model analysis. *Plant Pathology* 65(8), 1380–1389.

van den Bosch, F., Oliver, R., Van Den Berg, F. and Paveley, N. 2014. Governing principles can guide fungicide-resistance management tactics. *Annual Review of Phytopathology* 52, 175–195.

van den Bosch, F., Paveley, N., Shaw, M., Hobbelen, P. and Oliver, R. 2011. The dose rate debate: Does the risk of fungicide resistance increase or decrease with dose? *Plant Pathology* 60(4), 597–606.

Weinreich, D. M., Delaney, N. F., Depristo, M. A. and Hartl, D. L. 2006. Darwinian evolution can follow only very few mutational paths to fitter proteins. *Science* 312(5770), 111–114.

Wright, S. 1932. The roles of mutation, inbreeding, crossbreeding, and selection in evolution. In: *Proceedings of the Sixth International Congress on Genetics*, pp. 355–366.

Yamashita, M. and Fraaije, B. 2018. Non-target site SDHI resistance is present as standing genetic variation in field populations of *Zymoseptoria tritici*. *Pest Management Science* 74(3), 672–681.

Young, D. H., Spiewak, S. L. and Slawecki, R. A. 2001. Laboratory studies to assess the risk of development of resistance to zoxamide. *Pest Management Science* 57(11), 1081–1087.

Zhan, J. and McDonald, B. A. 2013. Field-based experimental evolution of three cereal pathogens using a mark-release-recapture strategy. *Plant Pathology* 62, 106–114.

Chapter 8

The role of Extension in fungicide resistance management

Guido Schnabel, Clemson University, USA; and Phillip M. Brannen, University of Georgia, USA

1 Introduction

The basic components and principles of Extension and its interplay with resistance management are found in many systems throughout the world (i.e. Australia, New Zealand, USA, many countries in Europe, etc.). In this chapter, we review these basic principles as observed in our interactions within the US Extension system, but they are in fact prevalent in other government-supported outreach and fruit producer education programs elsewhere.

In the United States, the land-grant university system is charged with the task of serving the agricultural community with established and newly developed information. Land-grant universities and colleges are mostly public universities, but some are private schools (e.g. Cornell University). These institutions of higher education are designated by a state to receive the benefits of the Morrill Acts of 1862 and 1890. These acts granted land-grant institutions with federally controlled land to teach practical agriculture, science, and more. All land-grant institutions have a three-fold mission: teaching, research, and Extension/outreach; this makes them unique as compared to other institutions of higher learning.

When the Smith Lever Act was established in 1914, the act dictated that 'cooperative agricultural Extension work shall consist of the giving of instruction and practical demonstrations in agricultural and home economics to persons not

http://dx.doi.org/10.19103/AS.2022.0116.08

attending or resident in said colleges in the several communities and imparting to such persons information on said subjects through field demonstrations, publications, and otherwise.' To paraphrase, one of the senators at the time stated that 'the purpose of this bill is to help the tillers of the land to discover the hidden riches of the soil, to devise methods of cultivation which will lessen the burden of farm life by shortening the hours of drudgery and render more productive the land. Its splendid purpose is to improve the man, enlarge his mental horizon, and give intelligent direction to his efforts.' Though the means of developing and disseminating information may have changed, the mission has not. The U.S. Cooperative Extension Service exists today to solve the problems of producers - serving the agricultural community through directed research and technology transfer.

Land-grant universities typically fund multiple Extension agents responsible for serving the residents of their state. These agricultural agents typically hold a BS or MS degree in a field related to agricultural sciences, travel to farms, interact with producers, and support local agricultural businesses with know-how and various activities. Extension specialists (including the authors of this chapter) are typically PhD-level faculty members of the university, often with commodity-specific responsibilities in research and Extension. Many advise MS and PhD students and may also teach classes for undergraduate and graduate students. The role of Extension personnel in agriculture is twofold:

1. Agents and specialists typically relay current research-based knowledge to producers (established technology transfer).
2. In addition, they must react to emerging issues, develop research-based solutions, and work to re-educate farmers relative to new knowledge/technology required to address such issues. To reiterate, Extension agents and specialists take the following actions:

    - They identify unusual, novel problems - starting at specific or multiple farms.
    - They study these problems and suggest modified actions to solve the problems.
    - They monitor outcomes of modified actions and develop/recommend adjustments as needed.

Extension personnel follow the age-old axiom of rational thought and the scientific process when addressing new and emerging issues. There is no place where this is more true than when addressing resistance development in plant pathogens. Something that works today may not work tomorrow. Pathogen populations are constantly changing and adapting to their environment. That includes farmer practices. For example, every fungicide spray changes the dynamics of the pathogen populations. Multi-site fungicides knock down the pathogen population without regard to phenotype or genotype, whereas the application of single-site fungicides select for genotypes that can survive this environmental stress. Depending on the farm practices, this selection may be accelerated if few or no anti-resistance management strategies are applied; these include use restrictions, strategic placements, and tank mixtures or rotations of products with different modes of action, as designated by Fungicide Resistance Action Committee (FRAC) codes. Consequently, there is a constant need for a review of the efficacy of Extension recommendations (e.g. fungicides or management schemes). This requires field observations by farmers and Extension workers, diligent research to confirm resistance in pathogen populations, farmer training on principles of fungicide resistance management, and distribution of new information to farmers so that they can adjust their management methods – a near constant feedback loop.

The greatest contributions of the Extension Service to fungicide resistance management are producer education, design of suitable resistance management strategies that fit location-specific circumstances, and aiding in implementation of these strategies. The primary goal of these efforts is the prevention of resistance outbreaks in the first place, but when prevention efforts fail, a secondary goal is the rapid identification of field resistance, to be followed by recommendations for fungicide program adjustments. The following three case studies are given as good examples of recent responses of Extension in the Southeast (USA) to resistance development. They cover resistance development in two fungi, one of which is obligate, and an example of the development of new technology (an app for iPhone and Android devices) designed to help producers more easily navigate and group active ingredients and trade names of multi-site and single-site fungicides, fine tune spray programs, and avoid resistance development.

2 Case study 1: Establishment of a fungicide resistance management program for gray mold control of strawberry

Gray mold of strawberry drives the spray program for strawberry producers in the southern United States. The pathogen, *Botrytis cinerea*, comes in every fall through latent infections of leaves on transplants that are used in plasticulture production (Oliveira et al., 2017; Peres, 2015; Schnabel et al., 2015); these introduced populations mix with local populations spreading from decaying plant tissue. Resistant fungal strains can quickly dominate the population due to the selection of fungicide-resistant phenotypes imported from nurseries and local populations (Schnabel et al., 2015; Fernández-Ortuño et al., 2014).

To generate healthy planting stock, fungicide applications are initiated in strawberry nurseries; once plants are transplanted to production fields, fungicide applications continue in the fall and spring to protect leaves, flowers, and fruit from infection. Cold weather and use of row covers for frost/freeze protection precludes winter use of fungicides, but depending on location, fungicides can be applied when conditions are warm and generally conducive to infection and disease development. Because of the selection pressure applied in nurseries and production farms in any given season, there is no assurance that control will be satisfactory when fungicides at risk for resistance development are applied to protect ripening fruit. Resistance to multiple FRAC codes is often already present in fields (Fernández-Ortuño et al., 2014), and lack of control by these FRAC codes can occur even if they have not yet been applied that season.

Over a period of two years (2010–2011), producers started reporting increased preharvest and postharvest gray mold on strawberry crops. After discussing possible causes with Extension agents and specialists, including the historic use of all IPM tools available, fungicide resistance emerged as a strong culprit for the loss of fruit to gray mold disease. Research indicated the presence of *B. cinerea* populations with resistance to commonly used fungicides (Fernández-Ortuño and Schnabel, 2012), and the need for modified disease management practices was evident. The situation entailed not only a change of spray materials and spray strategy but also greater producer knowledge of fungicide resistance evolution and pesticide chemistry.

A resistance monitoring program was developed to provide producers with a location-specific fungicide resistance profile. We assessed sensitivity to seven different classes of fungicides in early spring using isolates from frost-damaged flowers. The testing required a minimum of six but most often ten samples (isolates) per field. The small sample size was not sufficient for a scientific population analysis, but it was a number that the research lab could handle, and it made sample collection feasible for producers and Extension agents. The resistance profile of this relatively small sample size indicated the presence of resistance, and it educated producers about the importance of FRAC code principles. If resistance was present in this small sample size, it was likely widespread in the field and on the farm. The producer was then advised which FRAC codes to best rotate in the spray program for that year. Often the resistance profile was repeated later in the season for another potential adjustment.

Resistance profiling was conducted as described previously (Fig. 1) (Li et al., 2014). Briefly, flowers with a black torus were submitted to the lab, petals were removed, and the tori were surface sterilized and incubated for 2–3 days in a moist chamber. Spores emerging from conidiophores were carefully transferred with a toothpick to fungicide-amended growth medium, and mycelial growth

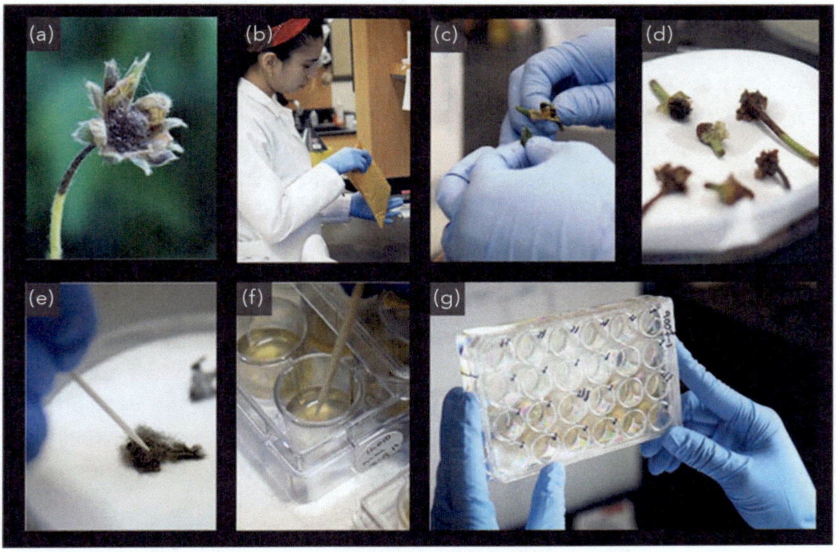

Figure 1 Processing profile samples: (a) flowers from the field were (b) received (c) sepals were removed and flowers were sterilized then (d) placed in moist chamber (e) spores were collected from the surface of blossoms and (f) transferred to a discriminatory dose of fungicide-amended media (g) growth was rated and results sent to the producer.

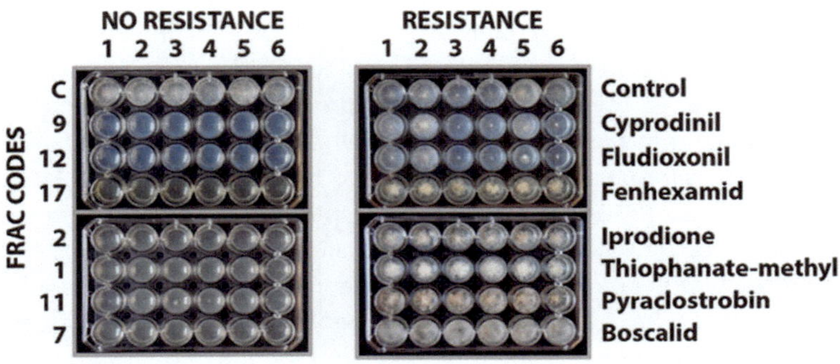

Figure 2 An example of resistance profile plates containing six sensitive isolates (left) and six resistant isolates (right). FRAC codes and fungicides used for testing are also provided.

was assessed in relation to an untreated control (Fig. 2). Resistance profile results and disease and fungicide resistance management recommendation were shared with the grower and Extension agent. Since the sample size was too low for detailed location-specific recommendations, the recommendations were generalized. Also, the recommendations explained that the resistance profiles were likely to change with each subsequent application of a site-specific product.

During the life of the monitoring program (2012–2019), profile resistance testing was conducted on 5359 individual *B. cinerea* isolates from 50 farms over 16 states (Alabama, Arkansas, California, Connecticut, Delaware, Florida, Georgia, Louisiana, Maryland, Michigan, Missouri, North Carolina, Ohio, Pennsylvania, South Carolina, and Virginia). Producers and specialists provided overwhelmingly positive feedback with statements such as 'This service is not good.... It is excellent,' 'thank you for the screen results of the fungicides on *Botrytis*. It is very informative and helps in deciding on spray materials...we have a better chance at efficacy!' and 'it (the resistance profile service) has saved me a lot of money personally from not buying a chemical that I would have bought otherwise. It is a great service you provide.' Producers also benefited from this program by learning to use the FRAC code system for resistance management.

While *B. cinerea* was by far the most common species isolated from flowers, some cultures did not match the description of this species. Nucleotide sequence analysis of G3PDH, NEP2, NEP1, HSP60, and HBP gene regions revealed some isolates belonged to *B. fragariae*, *B. caroliniana*, and *B. mali* species (Dowling et al., 2017a; Dowling and Schnabel, 2017b; Fernández-Ortuño et al., 2012). For simplicity, and because most isolates collected were *B. cinerea*, this chapter only reports results for *B. cinerea*.

One of the most striking initial results was the high frequency of resistance to thiophanate-methyl (Fig. 3). Almost 70% of isolates were resistant to thiophanate-methyl in 2012 and this level was maintained across experimental years despite producers discontinuing use of FRAC 1 fungicides in 2013. This illustrates the selective power of this fungicide and stability of this phenotype – even in the absence of selection pressure. Later, we confirmed that thiophanate-methyl was present in virtually every multi-fungicide-resistant phenotype. Therefore, resistance continued to be selected over the course of the following seasons, not directly by any application of thiophanate-methyl, but rather indirectly by 'selection by association' (Hu et al., 2016). Another striking result was the rapid increase of resistance to quinone outside inhibitor (QoI) fungicides (FRAC 11) for the first three years and the maintenance of high resistance frequencies for the remainder of the monitoring program (Fig. 3). Although FRAC 11 fungicides are not technically 'botryticides', they do have suppressive action on sensitive isolates and are therefore used in tank mixtures with FRAC 7 (succinate dehydrogenase inhibitor [SDHI]) products to help slow down selection for this fungicide class. Our results clarified that this resistance management strategy was no longer working and that the SDHIs were completely 'unprotected' regarding resistance selection. Furthermore, spraying the FRAC 7/11 mixture for gray mold control had become ineffective and potentially detrimental in the absence of other pathogens controlled by QoI fungicides, such as powdery mildew, leaf spots, or anthracnose.

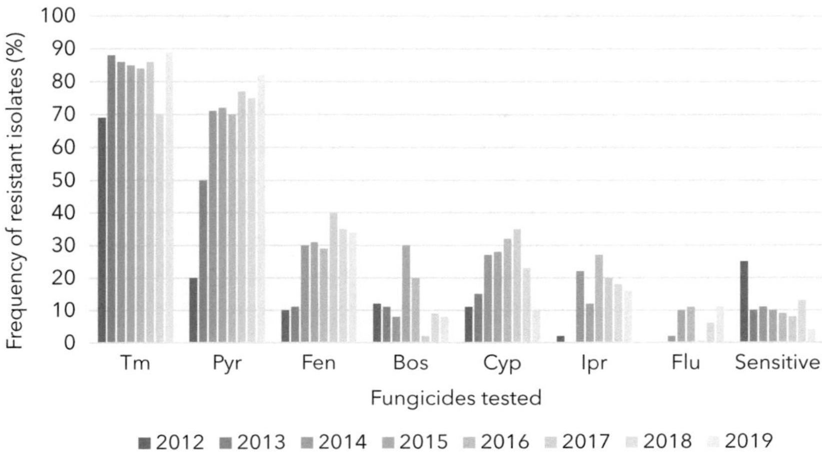

Figure 3 Frequency of resistance to the fungicides Thiophanate-methyl (Tm), Pyraclostrobin (Pyr), Fenhexamid (Fen), Boscalid (Bos), Cyprodinil (Cyp), Iprodione (Ipr), and Fludioxonil (Flu) in the 8 years that strawberry flower samples were profiled.

Resistance frequencies to fenhexamid, cyprodinil, iprodione, and boscalid were lower than those of thiophanate-methyl or QoI fungicides and remained relatively stable over the 8 years of monitoring (Fig. 3). Frequencies ranged from 10% (or less) to peaks of up to 40%. For some products, such as fenhexamid, the resistance frequencies quickly rose but then plateaued at levels of 30–40%. For others (i.e. cyprodinil, boscalid, and iprodione), the frequencies rose but then dropped back down to acceptable levels (10–20%) – where a single use of the product might still do more good than harm in managing the disease. For fenhexamid and iprodione, the management of resistance frequencies may have been due to a tighter restriction of the number of applications per season (one) compared to label recommendations; the stable and low frequency of resistance to boscalid can perhaps be explained by the unique and inconsistent action of FRAC 7 fungicides. FRAC 7 products continue to be applied multiple times per season. However, new FRAC 7 products have been released that do not necessarily select for the same resistance genotypes. For example, boscalid and fluxapyroxad select for isolates with H277Y/R/L mutations in the *sdh*B gene, but fluopyram and isofetamid do not. It is therefore possible that the introduction of next-generation SDHIs such as fluopyram and isofetamid helped de-select isolates with resistance to boscalid due to incomplete cross resistance (Stammler et al., 2015).

Fludioxonil continues to be a producer favorite with its high efficacy against gray mold and anthracnose. The latter can be so explosive and devastating that it can wipe out an entire crop if conditions are suitable. Fludioxonil use has increased significantly in southern strawberry fields, with up to five applications per season. However, resistance frequencies remained low (at about 10%) over the course of this testing. It is still unclear whether isolates resistant to fludioxonil experience a fitness penalty. If they do, it could explain the observed results. The fungicide is sold in combination with cyprodinil (FRAC 9), which may also contribute to resistance management. Resistance to cyprodinil has remained around 10–20%. We do believe, however, that it is of utmost importance to tightly regulate the frequency of use of this combination product. The current, regional recommendation by Extension specialists is to use it no more than twice per season, rather than the maximum allowed by the label of 4–5 applications (depending on dose) and to rotate this product with multi-sites or with other FRAC codes.

In contrast to our concerns at the beginning of this study, the frequency of multi-fungicide-resistant isolates has stabilized over time (Fig. 4). At first, we saw a dramatic increase in frequency of isolates with resistance to 3 and 4 FRAC codes (3CCR, 4CCR) and emergence of 5CCR, 6CCR, and 7CCR isolates. However, the number of 5CCR isolates has returned to low levels in recent years, and frequencies of 6 and 7CCR did not increase for our sampled isolates. One possible explanation is that the monitoring program recommendations were successful and resulted in stricter control of the number of applications per season for each FRAC code, a larger focus on using multi-site fungicides such as thiram and captan, and avoidance of FRAC codes that were projected to be of little use due to very high resistance frequencies. Another is that the more chemical classes an isolate becomes resistant to, the more likely it is for fitness

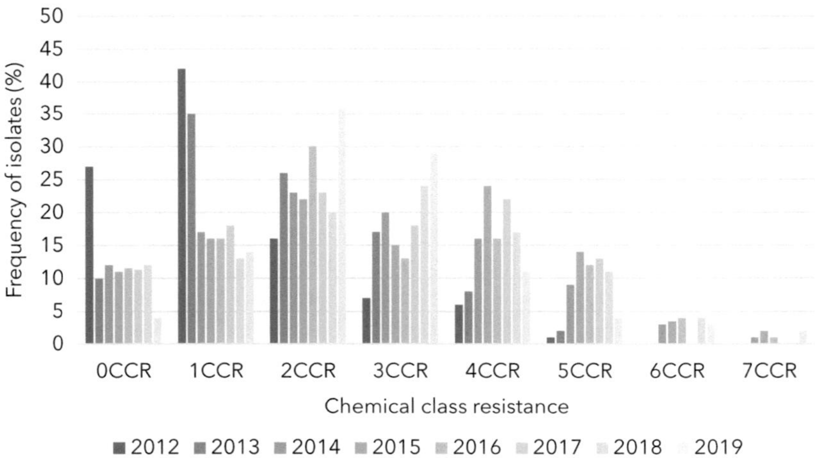

Figure 4 Frequency of isolates with differing chemical class resistance (CCR), where 0CCR refers to isolates sensitive to all 7 chemical classes tested and 7CCR refers to isolates with resistance to all 7 chemical classes tested in the 8 years that strawberry flower samples were profiled.

penalties to occur. A previous study conducted under controlled conditions indicated substantial competitive disadvantages of 5CCR and 7CCR isolates when compared to 0CCR isolates (Chen et al., 2016).

Though this program was highly successful, an Extension specialist can only be a 'service provider' when funding and time is sufficiently available to run the program. Reliable, accurate, and timely response during the sampling timeframe, even for just one commodity, can severely hamper the Extension specialist's ability to redirect efforts to other priorities. Therefore, specialists may not have time to run an established program indefinitely, and this was the case for this profile system in strawberries. However, one solution to this problem is to commercialize the activity within the land-grant institution. Addressing this need, the University of Georgia established a fee-for-service resistance-testing service at the Plant Molecular Diagnostic Lab in Tifton, Georgia (USA). The success of the resistance-testing service will largely depend on the quality and timeliness of the service provided and the willingness of producers to pay for the service. Private industry could also take on this same task, but profitability may be in question.

In conclusion, this case study covered a response to a producer-reported lack of control efficacy in the field. An Extension program was established that monitored resistance in *B. cinerea* to fungicides, identified potential weaknesses in existing spray programs, increased producer awareness of the need for resistance management practices, modified grower behavior, and led to a more sustainable IPM approach for gray mold management. The program has been transitioned and continues to be maintained within the land-grant system.

3 Case study 2: The role of Extension in identification and management of resistance in the obligate pathogen *Plasmopara viticola*

Obligate fungal pathogens, those that cannot be cultured, pose some difficulty when trying to determine field resistance. As an example, grapevine downy mildew (GDM), caused by the oomycete pathogen *Plasmopara viticola*, is prevalent in numerous viticultural regions on *Vitis vinifera* and other grape species and hybrids (Gessler et al., 2011; Kassemeyer et al., 2015). GDM damages early-season fruit, though ontogenic resistance occurs as fruit matures (Kennelly et al., 2005). Damage to leaves can be significant as well, resulting in substantial defoliation that can result in yield loss, poor fruit quality on harvested fruit, and winter injury or death to stressed vines. Frequent precipitation and warm temperatures encourage downy mildew development, and when grapes are grown in climates that provide these conditions, total crop loss can incur if vines are not managed through the use of prophylactic treatments with chemical fungicides.

Due to the season-long infection potential of *P. viticola* and other grape pathogens observed throughout Georgia (USA), vineyard managers employ as many as 17 sprays in any given crop year. Because the number of active ingredients with different modes of action (MOA) are limited for GDM, multiple applications of fungicides with the same MOA are often required, making resistance development likely (Gisi and Sierotzki, 2008). Based on numerous reports from other states and other countries (Blum et al., 2010; Collina et al., 2005; Corio-Costet et al., 2011; Gisi et al., 2000; Gisi et al., 2002; Gisi et al., 2007; Heaney et al., 2000), we knew that *P. viticola* could readily develop resistance to multiple fungicide classes. With the help of Extension agents, we reviewed spray programs of several producers who were having issues with GDM management; even with strong spray programs, wine grape producers were observing significant disease levels, and resistance development, although unproven, was suspected. Fungicide classes of particular concern were: (1) phenylamides (PA; e.g. mefenoxam), (2) quinone outside inhibitors (e.g. azoxystrobin), and (3) carboxylic acid amides (CAA; e.g. mandipropamid).

To address the concerns relative potential resistance issues, plans were initiated that included bringing on a graduate student and coordinated efforts with a mycologist (fungal populations expert), an epidemiologist, and other experts on resistance management. This is a great example of how Extension can utilize and collaborate with the other missions of a land-grant institution, teaching and research, as well as the entire national system of Extension and research, to quickly solve problems for agricultural producers. In short, we were able to assess resistance development of *P. viticola* in Georgia wine grape vineyards through a combination of resistance tests of *P. viticola* populations (bioassays and PCR lab tests) and field trials (research station and on-farm field trials with Extension agents). From these efforts, we confirmed that resistance to QoI fungicides was widely present in *P. viticola* populations and the degree to which the lab results matched real-world field resistance.

The ultimate goal of these Extension efforts was to provide information to grape producers that would allow them to tailor spray programs that would take advantage of fungicides that are still active, as opposed to those that no longer provide efficacy. Starting in 2015, we collected GDM isolates for various research projects, to include resistance assessment. For sample collection, Extension agents were instrumental to the success of our efforts, as they provided liaison with wine grape producers, and they provided samples from producer sites. With the collected isolates, bioassays, and PCR tests were conducted in 2017 and 2018 to address whether fungicide resistance was present, and to the degree possible, what the genetic cause of the resistance might be (Campbell et al., 2020). Based on assessments over years, we confirmed shifts to QoI resistance in *P. viticola* populations. In field trials with multiple MOA fungicides at both research and producer locations, we were further able to confirm fungicide field failures because of resistant populations of *P. viticola* (Campbell et al., 2021). Again, local Extension agents provided liaison to find sites for research, as well as helping in the conduct of the field trials. As resistance information was developed, producers were informed through numerous outlets (e.g. newsletters, blogs, meetings, etc.) as new information became available; Extension efforts were instrumental in providing continuous feedback to producers, and as a result, producers modified their spray programs - resulting in less GDM and much healthier fruit and vines.

Both bioassays and PCR detection of genetic determinants of resistance can be used to identify resistance in an obligate pathogen (Brent and Hollomon, 2007). Bioassays are not dependent on any single molecular basis for resistance and are thus more encompassing and arguably more accurate - potentially identifying either qualitative or quantitative resistances for which the cause is or is not known. However, when the resistance determinants (e.g. specific point mutations that confer resistance) are known, PCR assays are highly reliable in rapidly identifying field resistance. For example, in the case of the G143A mutation that confers qualitative resistance and provides cross resistance to most QoI fungicides, PCR may be more than adequate (Baudoin et al., 2008). Based on our experience in Georgia surveys (Fig. 5), having conducted both bioassays and PCR, the G143A mutation will likely be the first and most common type of resistance observed in *P. viticola* (Campbell et al., 2020). Additional resistance determinants may be of interest if G143A is not detected. However, if the G143A mutation has been established in a population, for all practical purposes, no other QoI resistance determinant really matters any longer, as complete resistance to the QoIs will be observed at this site without regard, due to the qualitative nature of QoI resistance. Therefore, the use of PCR alone is likely sufficient when the most common resistance determinants are known. This is particularly important when rapid turnaround times are important, such as in the role of Extension providing rapid results where fungicide field failures are starting to occur within a season.

One question that often arises when testing for field resistance is, 'How many samples should an Extension agent provide to a lab to determine resistance?' While it is true that the use of a few isolates can't tell you much about resistance distribution, the collective analysis of isolates from a region will provide a good indication of whether resistance is of concern, and the degree to which it is of concern. From an Extension point of view, and academic concerns aside, producers need fast answers to existential problems. The Extension Service needs to provide those answers within the limitations of the system and the real-world costs of conducting trials. Given, more samples are always better up to a point, but when virtually 100% of isolates are showing QoI

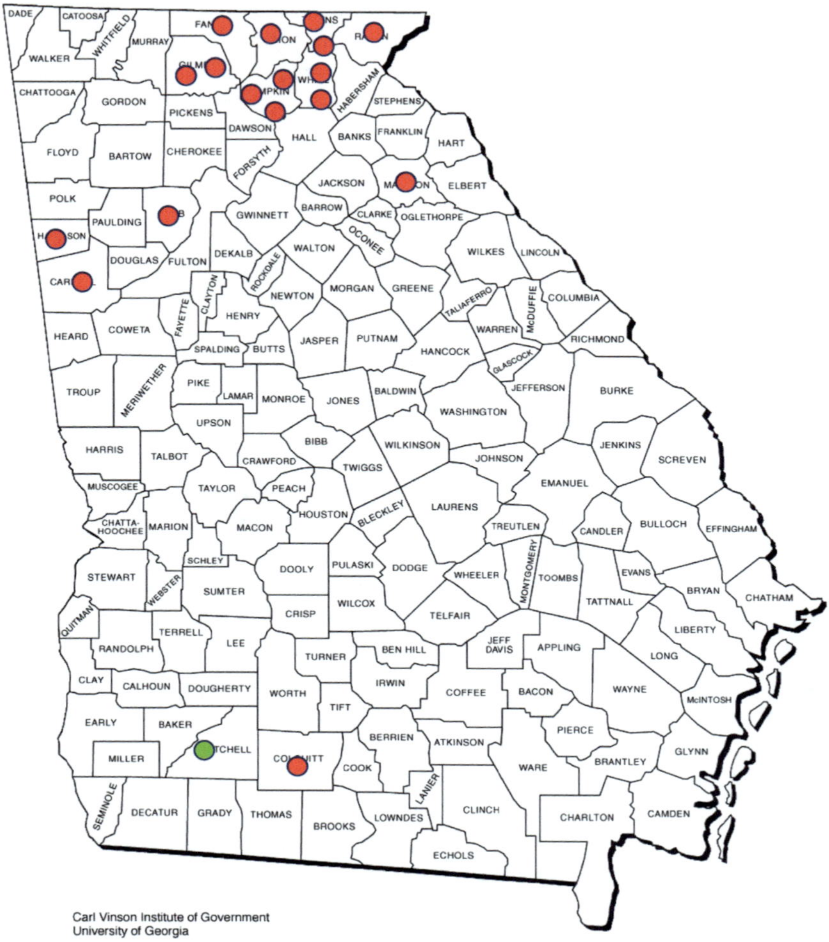

Figure 5 Results of Extension survey conducted to review QoI resistance of Plasmopara viticola isolates from Georgia wine grape vineyards (2015–2018; USA). Red dots = resistance to QoI fungicides; green dot = sensitive to QoI fungicide.

resistance across a region, without regard to sample numbers achieved, there is most likely a severe resistance issue that needs to be relayed to producers (Baudoin et al., 2008; Campbell et al., 2020). One must balance the scientific need for large sample numbers versus logic and the limitations of the number of samples Extension agents can provide. While too few samples can lead to false conclusions, we would maintain that one may still determine resistance issues with relatively few samples, say 10–20, from individual locations (Baudoin et al., 2008; Colcol and Baudoin, 2016; Corio-Costet, 2015).

As a result of resistance monitoring and field trials conducted by Extension personnel in Georgia, we now know that resistance of *P. viticola* to the QoI fungicides is widespread in Georgia, and that these fungicides no longer control downy mildew in most locations. In addition, we established which other classes are still active. Therefore, Georgia producers have received critical information they need to prevent application of fungicides that no longer have activity. The work with the Extension Service also educated producers across the state about the need to take resistance management practices seriously. They learned these measures do affect the bottom line of a farm operation, provide better disease control, and produce grapes with higher wine quality. This is a great example of how Extension can respond to the needs of producers, helping them solve their problems. However, resistance surveys are simply a snapshot in time; without continued testing on an annual basis, one can't be sure that resistance to another fungicide class has not occurred. With an obligate pathogen such as downy mildew, PCR testing on an annual basis for specified mutations is possible, and these lab tests can be utilized to determine whether field resistance is likely present. However, due to cost and difficulty of execution, bioassays and field trials to confirm PCR results are unlikely to be practical for annual assessments with obligate pathogens. Most land-grant universities and colleges also provide diagnostic services, but few provide diagnostic services for determining resistance development in obligate or other pathogens. This is a critical need, and in particular, molecular diagnostics for resistance development should be incorporated into diagnostic services provided by Extension.

In Georgia, due to extensive field surveys of pathogens for resistance development, we now have known resistance in multiple grape pathogens, of which GDM is a critical one based on our environment. With the advent of resistance to multiple pathogens, we realized that producers were confused as to how to develop efficacious spray programs, especially as related to resistance management (Oliver et al., 2021). Considering the intense and complicated nature of resistance management and resistance development in Georgia, we initiated spray program workshops specifically for wine grape producers in 2020. These workshops were modifications of those developed for the Washington State University Viticulture Certificate Extension Program (personal communication M. M. Moyer) in the Pacific Northwest. These day-long workshops allow producers, Extension agents, and Extension specialists to develop spray programs that will address all grape pathogens, as well as insects. The producers provide each entry for each phenological growth stage and anticipated fungicide application, all while being advised by Extension personnel. At the end of each day, producers have a complete example spray program that will address all pathogens and insects, while considering resistance management and the likely presence of resistance in particular pathogens or insects. Many producers simply adopt this spray program as their own. Anecdotally, we have observed great success with this effort – with disease levels being reduced dramatically from year to year under similar environmental constraints. Feedback from producers has been incredibly positive relative to these meetings and their results.

4 Case study 3: Development of a novel Extension tool to navigate pesticides and promote fungicide resistance principles

Staying up to date on fungicides and their potential to develop resistance is not easy. Traditional Extension tools such as bulletins, books, websites, etc. are often outdated from the time they are published or posted; in addition, they are not always readily accessible to producers while in the field. Around the time when *B. cinerea* fungicide resistance was suspected as the culprit of increased gray mold in strawberries, the first author of this chapter had a discussion with a young, third-generation farmer in South Carolina. He proudly declared that he had developed an excel spreadsheet that listed active ingredients, trade names, and FRAC codes of fungicides available to him for his crops. This farmer was trying to make sense of the increasing pesticide choices and how to best fit them into a program that is effective and sustainable, while also providing resistance management. Extension agriculture specialists strive to provide up-to-date production information for producers to increase productivity in a sustainable and environmentally responsible way. However, based on this conversation, it was clear that the available options had failed to convey the necessary information to this young producer, as well as many others.

A more creative way was needed to fill that void. Mobile smartphones had increased in number, and they were also widely utilized for communication, entertainment, and increased productivity. The first agricultural smartphone apps were being developed for pest and disease identification, crop disorders, agricultural news, disease risk assessment, weather forecasts, and much more. The advantages of smartphones for farmers were obvious. Smartphones are communication devices, and even technology-averse producers are now seeing the benefits of using a hand-held mini-computer/phone. Helpful information can be conveyed through software applications (apps), and it can be retrieved at any time and virtually at any place, even in remote orchards.

In response to producer needs, we developed MyIPM, a free for download app designed, created, and maintained by numerous Extension specialists (Schnabel et al., 2015). It features a variety of different aspects of disease and pest management for commercial fruit producers in the form of text, high resolution pictures, interactive tables, and audio. MyIPM integrates different pest and disease management approaches including pest and disease diagnostics, chemical and nonchemical management options, and tools for insecticide and fungicide resistance management. It features key elements of regional spray guides, current knowledge of control, and a picture library for diagnostic purposes. The app is designed for producers, field advisors, and specialists and is available in the Google Play Store for Android devices and in the Apple App Store for iOS devices.

From a software perspective, MyIPM can be divided into two main components: the application and an external server. The application consists of all the relevant code that allows the app to run, as well as an internal database that stores information local to the device. The server includes an external database that holds the most up-to-date version of the app's information, as well as PHP and HTML scripts which allow a user (specialist) to edit the external database, and for an individual app to update its data.

The relationship between the app and the internal database can be seen as a model-view-controller. This design pattern divides a user interface into three components: a model, a view, and a controller. The view is everything the user sees in the application, like tables or user interface elements such as buttons or menus. The controller defines how the user interacts with the application. For example, it determines when a user presses a button and what logic should be executed when the event (the button was pressed) occurs. This information is then sent to the model. The model contains all the data for a particular view and alerts its associated views and controllers when they need to be updated. In MyIPM, each update to the model is accompanied by a query to the application's internal database. For example, every time a user requests a piece of information, such as selecting 'Strawberry' from the application's main menu, the application will load the relevant information from the database, in this case, a list of diseases that affect strawberries and an image of each disease. The app then updates the view and associated controller to reflect this new data.

The server can best be explained though the client-server model. In this architectural style, there is a central server that stores information and provides a variety of services which can be requested by the client. For context, this model is the basis for how websites are accessed on the internet. Our server provides two main services. The first provides an authoring tool for the information stored in the external database. This tool provides a way for an administrative user to add, edit, and delete the information that defines the application. For example, a user with access to the authoring tool can dynamically add a new fruit or disease to the app without modifying the source code of the application. The second service connects the external database to each individual instance of the application. Using PHP scripts, the server provides a way for the mobile application to ask if any changes have been made to the external database. If so, the app will download the new information and integrate it into its own internal database, where it can be seen by the user. Ultimately, this allows for real-time changes to IPM and resistance management information as it becomes available - readily available in a user-friendly form for producer use immediately after each specialist update.

5 Conclusion and future trends in research

Extension specialists and agents have played an integral part in proactively relaying resistance management information to fruit producers - generally through face-to-face meetings, farm visits, phone calls, newsletter articles, and a myriad of traditional venues. The emphasis is on 'proactively' relaying resistance management information, because preventing resistance-related disease outbreaks by applying preventative resistance management practices is the key to success. Though traditional outlets for information and technology transfer will continue to be utilized, the expansion of the internet and other electronic outlets will allow for more interactive apps, podcasts, computer meetings, etc. as we move forward. Even with the best preventative efforts, resistance will continue to be a challenge for producers, and Extension will continue to play a critical role in helping producers understand and manage resistance.

Of equal importance, we realize that the information provided to producers in one year will not necessarily reflect the resistance situation in the following one. Resistance is a continual 'moving target,' and a near constant feedback loop is needed for both producer education and resistance testing from the orchard, vineyard, or field. Lab and field methods for rapid profiling of resistant pathogen populations are now available or are being developed for numerous pathosystems. When combined with up-to-date resistance management recommendations, such as provided by the MyIPM app, these new tools can provide critical aids to producer decision matrices concerning which, how, and how often products can be used when developing spray programs - on a yearly or within-year basis.

Unfortunately, there are no federally funded resistance monitoring programs in place. Rather, resistance management programs are rather piecemeal, and they generally arise following field failures due to local resistance development in a particular commodity within a state or region - putting out fires. This is costly to producers and not particularly efficient; funding and establishment of national resistance-testing and management programs could further the prevention of major disease outbreaks and help manage resistance-related problems throughout US fields and orchards.

As indicated, Extension agents and specialists are problem solvers, of which resistance is just one problem that needs their near constant attention. They have two goals associated with resistance management: (1) provision of non-biased, accurate, and current information to producers relative to resistance management best practices and (2) rapid identification of resistance when it occurs, to be quickly followed by new recommendations – the feedback loop. The long-time, historical relationship and trust observed between Extension and producers is critical to technology transfer, and if producers trust the source of the information, they are generally more willing to address rapid changes required for resistance management. To reiterate, the strong working relationship between Extension and producers educates both parties, continues to build trust, and generates the foundation for change. The original intent of and mission of Extension is accurately portrayed in resistance management; established and new information must be generated and transferred to action, and Extension agents and specialists continue to be instrumental for both.

6 References

Baudoin, A., Olaya, G., Delmotte, F., Colcol, J. F. and Sierotzki, H. (2008). QoI resistance of *Plasmopara viticola* and *Erysiphe necator* in the mid-Atlantic United States, *Plant Health Prog.* 9(1), 25. doi: 10.1094/PHP-2008-0211-02-RS.

Blum, M., Waldner, M. and Gisi, U. (2010). A single point mutation in the novel *PvCesA3* gene confers resistance to the carboxylic acid amide fungicide mandipropamid in *Plasmopara viticola*, *Fung. Gen. Biol.* 47, 499–510.

Brent, K. J. and Hollomon, D. W. (2007). Fungicide resistance in crop pathogens: How can it be managed?, FRAC Monograph No. 1 (second, revised edition). Available at: https://www.frac.info/docs/default-source/publications/monographs/monograph-1.pdf, Online publication.

Campbell, S. E., Brannen, P. M., Scherm, H. and Brewer, M. T. (2020). Fungicide sensitivity survey of *Plasmopara viticola* populations in Georgia vineyards, *Plant Health Prog.* 21(4), 256–261.

Campbell, S. E., Brannen, P. M., Scherm, H., Eason, N. and MacAllister, C. (2021). Efficacy of fungicide treatments for *Plasmopara viticola* control and occurrence of strobilurin field resistance in vineyards in Georgia, USA, *Crop Prot.* 139, 1–6.

Chen, S. N., Luo, C. X., Hu, M. J. and Schnabel, G. (2016). Fitness and competitive ability of *Botrytis cinerea* isolates with resistance to multiple chemical classes of fungicides, *Phytopathology* 106(9), 997–1005.

Colcol, J. F. and Baudoin, A. B. (2016). Sensitivity of *Erysiphe necator* and *Plasmopara viticola* in Virginia to QoI fungicides, Boscalid, quinoxyfen, thiophanate methyl, and mefenoxam, *Plant Dis.* 100(2), 337–344.

Collina, M., Landi, L., Guerrini, P., Branzanti, M. B. and Brunelli, A. (2005). QoI resistance of *P. viticola* in Italy: Biological and quantitative real-time PCR approaches. In: Dehne, H. W., Gisi, U., Kuck, K. H., Russell, P. E. and Lyr, H. (Eds.), *Modern Fungicides and Antifungal Compounds* (Vol. IV). BCPC & International Reinhardsbrunn Symposium, Alton, UK, pp. 81–88.

Corio-Costet, M. F. (2015). Monitoring resistance to obligate pathogens by bioassays relating to field use: Grapevine powdery and downy mildews. In: Ishii, H. and

Hollomon, D. W. (Eds.), *Fungicide Resistance in Plant Pathogens*. Springer, Tokyo, pp. 251–279.

Corio-Costet, M. F., Dufour, M. C., Cigna, J., Abadie, P. and Chen, W. J. (2011). Diversity and fitness of *Plasmopara viticola* isolates resistant to QoI fungicides, *Eur. J. Plant Pathol.* 129(2), 315–329.

Dowling, M. E., Hu, M. J. and Schnabel, G. (2017a). Identification and characterization of *Botrytis fragariae* isolates on strawberry in the United States, *Plant Dis.* 101(10), 1769–1773.

Dowling, M. E. and Schnabel, G. (2017b). First report of *Botrytis mali* causing gray mold on strawberry in the United States, *Plant Dis.* 101(6), 1034.

Fernández-Ortuño, D., Chen, F. and Schnabel, G. (2012). Resistance to pyraclostrobin and Boscalid in *Botrytis cinerea* isolates from strawberry fields in the Carolinas, *Plant Dis.* 96(8), 1198–1203.

Fernández-Ortuño, D., Grabke, A., Bryson, P. K., Amiri, A., Peres, N. A. and Schnabel, G. (2014). Fungicide resistance profiles in *Botrytis cinerea* from strawberry fields of seven southern U.S. states, *Plant Dis.* 98(6), 825–833.

Fernández-Ortuño, D., Li, X. P., Wang, F. and Schnabel, G. (2012). First report of gray mold of strawberry caused by *Botrytis caroliniana* in North Carolina, *Plant Dis.* 96(6), 914.

Gessler, C., Pertot, I. and Perazzolli, M. (2011). *Plasmopara viticola*: A review of knowledge on downy mildew of grapevine and effective disease management, *Phytopathol. Mediterr.* 50, 3–44.

Gisi, U., Chin, K. M., Knapova, G., Küng Färber, R., Mohr, U., Parisi, S., Sierotzki, H. and Steinfeld, U. (2000). Recent developments in elucidating modes of resistance to phenylamide, DMI and strobilurin fungicides, *Crop Prot.* 19(8–10), 863–872.

Gisi, U. and Sierotzki, H. (2008). Fungicide modes of action and resistance in downy mildews, *Eur. J. Plant Pathol.* 122(1), 157–167.

Gisi, U., Sierotzki, H., Cook, A. and McCaffery, A. (2002). Mechanisms influencing the evolution of resistance to Qo inhibitor fungicides, *Pest Manag. Sci.* 58(9), 859–867.

Gisi, U., Waldner, M., Kraus, N., Dubuis, P. H. and Sierotzki, H. (2007). Inheritance of resistance to carboxylic acid amide (CAA) fungicides in *Plasmopara viticola*, *Plant Pathol.* 56(2), 199–208.

Heaney, S. P., Hall, A. A., Davies, S. A. and Olaya, G. (2000). Resistance to fungicides in the QoI-STAR cross resistance group: Current perspectives. *Proc. BCPC Conf. Pests Dis.* 2, 755–762.

Hu, M. J., Cox, K. D. and Schnabel, G. (2016). Resistance to increasing chemical classes of fungicides by virtue of "selection by association" in *Botrytis cinerea*, *Phytopathology* 106(12), 1513–1520.

Kassemeyer, H. H., Gadoury, D. M., Hill, G. and Wilcox, W. F. (2015). Downy mildew. In: Wilcox, W. F., Gubler, W. D. and Uyemoto, J. K. (Eds.), *Compendium of Grape Diseases, Disorders, and Pests* (2nd edn.). APS Press, St Paul, MN, pp. 46–52.

Kennelly, M. M., Gadoury, D. M., Wilcox, W. F., Magarey, P. A. and Seem, R. C. (2005). Seasonal development of ontogenic resistance to downy mildew in grape berries and rachises, *Phytopathology* 95(12), 1445–1452.

Li, X., Fernández-Ortuño, D., Chen, S., Grabke, A., Luo, C. X., Bridges, W. C. and Schnabel, G. (2014). Location-specific fungicide resistance profiles and evidence for stepwise accumulation of resistance in *Botrytis cinerea*, *Plant Dis.* 98(8), 1066–1074.

Oliveira, M. S., Amiri, A., Zuniga, A. I. and Peres, N. A. (2017). Sources of primary inoculum of *Botrytis cinerea* and their impact on fungicide resistance development in commercial strawberry fields, *Plant Dis.* 101(10), 1761–1768.

Oliver, C. L., Cooper, M. L., Lewis Ivey, M. L., Brannen, P. M., Miles, T. D., Mahaffee, W. F. and Moyer, M. M. (2021). Assessing the United States grape industry's understanding of fungicide resistance mitigation practices, *Am. J. Enol. Vitic.* 72(2), 181–193.

Peres, N. A. (2015). Florida plant disease management guide: Strawberry. Available at: http://edis.ifas.ufl.edu/pg056 (accessed 8 August 2016).

Schnabel, G., Hu, M. J. and Fernández-Ortuño, D. (2015). Monitoring resistance by bioassay: Relating results to field use using culturing methods. In: Ishii, H. and Hollomon, D. W. (Eds.), *Fungicide Resistance in Plant Pathogens*. Springer Press, Tokyo, pp. 251–280.

Schnabel, G., Pargas, R., Hu, M. J. and Edison, G. (2015). A new smartphone app for disease and resistance management. APS Annual Meeting, Pasadena, CA. Poster.

Stammler, G., Wolf, A., Glaettli, A. and Klappach, K. (2015). Respiration inhibitors: Complex II. In: Ishii, H. and Hollomon, D. W. (Eds.), *Fungicide Resistance in Plant Pathogens: Principles and a Guide to Practical Management*. Springer, Japan, Tokyo.

Chapter 9

Key challenges in developing new fungicides

Gregory M. Kemmitt, Corteva Agriscience™, UK

1 Introduction

The Earth's population is predicted to rise around 30% in the next two decades from its current 7.8 billion individuals to between 9 and 10 billion by 2050, peaking at 11 billion by 2100 (United Nations, 2019). This statistic must be set in the context that only 5 billion hectares or 38% of the Earth's land surface is used for agricultural purposes of which one third is crop land, and the remaining two thirds are pasture for livestock grazing (FAOSTAT, 2020). Within the cropland segment (approximately 1.6 billion hectares), 10% is set aside to permanent crops such as vineyards, orchards, oil palm and cocoa plantations. Today, the global food system feeds nearly 8 billion people, which is a testimony to its productivity, although more than 800 million people remain malnourished (National Academy of Sciences, 2021). The dynamic of an increasing population, coupled with a finite area of productive arable land, means that the amount of arable land per capita has fallen continuously over the period 1990–2019 from approximately 0.9 hectares per capita to 0.6 hectares per capita (FAOSTAT, 2019). The world needs to grow more food from the currently available land and limited water supplies whilst making modern agriculture more sustainable with respect to impact on environment. By 2050 global agricultural production may need to increase 60–110% (FAO, 2009; Tilman et al., 2011). However, there remains a concern that yield increases are not currently on track to meet

http://dx.doi.org/10.19103/AS.2022.0116.09

these needs, (Ray et al., 2013). A number of studies estimating crop losses due to pests including insects, weeds and plant pathogens have been published and Oercke (2006) reported estimated global crop losses due to pathogens in the key staple crops of wheat, rice, maize, potatoes and soybeans of 15.6, 13.5, 9.4, 21.2 and 11%, respectively. A more recent analysis by Savary et al. (2019) reported an impact of around 20–30% across the same 5 staples when looking at losses due to pests. Post-harvest losses where plant pathogens play a significant role have been variously estimated at up to 10% worldwide in durable commodities such as cereals, oilseeds and pulses rising to around 25% in perishable crops in the tropics (Narayanasamy, 2006). It has been postulated that crop yields would be around half their current levels if no crop protection measures were implemented (Leadbeater and Gisi, 2010). Crops with increased yields may be associated with a higher vulnerability to diseases and pests. It has been estimated that potential loss of wheat yield due to disease, increases from less than 10% at an attainable yield of 2 tonnes/ha to over 20% in a production system where inputs are optimized to achieve a yield of 12 tonnes/ha (Oercke, 2006).

The use of fungicides to provide effective economic control of plant diseases has been an integral component of crop production systems since their widespread introduction from the 1940s onwards. The Fungicide Resistance Action Committee (FRAC) (FRAC Code List, 2022) reports a total of around 250 fungicidal materials sorted by cross resistance pattern and mode of action into 77 individually coded groups. Despite this apparent cornucopia of fungicide products available on the market today there is still a clear need for new and innovative fungicides due to the impact of factors such as a need for proactive resistance management, high reliance on a limited number of modes of action, a dynamic regulatory environment causing a rapidly accelerating loss of products, grower/customer expectations for more sustainable solutions, and new or spreading crop diseases.

Discovering, developing and registering new fungicidal solutions be they synthetic chemistry or biological based is a costly and time-consuming business. A significant proportion of development costs for a new fungicide are driven by the extensive suite of regulatory studies required to demonstrate safety to humans and the environment. These high costs might be seen as a potential barrier to entry of new solutions, but the author would argue that as the regulatory bar is raised higher, manufacturers are obliged to look for new and innovative solutions with more benign toxicological profiles, lower use rates thus reducing pesticide load in the environment and molecules with the flexibility to be integrated into different delivery systems such as targeted application, seed treatment, trunk injection etc.; all of which may help to reduce operator and non-target organism exposure. This can be seen as a positive development in an era where government led regulation, NGO scrutiny and public perception means agriculture, like many other industries, is actively working towards developing solutions to improve sustainability and reduce negative impact on the environment.

Manufacturers invested in significant discovery based efforts to develop new fungicidal solutions are typically seeking to uncover innovative new fungicides with novel modes of action (MOAs) and if possible, inherently low resistance risk. This is particularly important as a new MOA in the grower's disease control toolbox will, in theory, enhance the opportunity to manage resistance risk. Rotation or mixing of molecules with different MOAs within a spray programme is a central pillar of managing risk. Developing a molecule with a new MOA may bring additional challenges in achieving the correct balance between efficacy versus the target organism and safety to non-target species, parameters which are more clearly understood by both manufacturers and regulators for existing chemical classes of fungicides e.g. Quinone Outside Inhibitors (QoIs), Sterol Biosynthesis Inhibitors (SBIs) and Succinate Dehydrogenase Inhibitors (SDHIs) etc. Indeed, the multiple examples of commercialized molecules from the three aforementioned classes is eloquent testament to the industry's ability to take established (in some cases decades old) MOAs and to continue to innovate within the group by looking to improve such characteristics as spectrum, systemicity, intrinsic efficacy, use rate etc.

With the continuing critical need for new fungicidal solutions, it is imperative that agrochemical companies maintain strong lead generation programmes to discover and develop the next generation of sustainable products. A lead molecule can broadly be defined as having an activity profile and degree of novelty that warrant a more significant investment in exploring the potential of the opportunity (Loso et al., 2017). Lead generation is the growth 'engine' for any agrochemical company aiming to maintain a robust pipeline of new high-value products. The challenge is how to discover these innovative solutions in a timely, cost-effective manner.

No doubt fungicides will remain a vital component of modern agricultural production systems for the foreseeable future. However their development and use within agronomic production systems will likely need to be adapted and grounded within a more holistic and sustainable system of food production driven by consumer demand and regulatory agency mediated government policy in many regions of the globe. Not forgetting that such solutions also need to address the growers fundamental need for tools which allow them to profitably produce healthy safe food. These over-arching dynamics will likely require a more multidisciplinary approach to the challenge of effective plant disease control where synthetic fungicides will be but one component (Jeger et al., 2021).

2 Past and present fungicide development: a story of success

Humankind has been battling against pests and diseases of plants grown for nutritional purposes since the practice of agriculture first began in the Fertile Crescent around 5000BC. One of the first recorded uses of a pesticidal material is around 4500 years ago by the Sumerians who used sulphur compounds to control insects and mites (Umetsu and Shirai, 2020). Around 1000 BC, Homer also reported use of sulphur for fumigating crops (Williams and Cooper, 2004). The Romans prayed to the god of mildew, Robigus, in an attempt to avert fungal diseases of their crops (Money, 2007). Prevost (1807) demonstrated that bunt of wheat (causal agent *Tilletia caries*) could be controlled by application of copper sulphate to wheat seed. The nineteenth century saw the concept of promotion of chemical disease control begin to gain traction with individuals specifically publishing recommendations such as use of sulphur for the control of peach powdery mildew (Robertson, 1824) and lime sulphur for grape powdery mildew control (Kenrick, 1833). Bordeaux mixture for the control of grape downy mildew was first announced by Millardet in 1885 (Agrios, 1988) and versions of this material are still in use today.

From the 1940s onwards, large chemical companies started to manufacture and sell protectant fungicidal materials such as the dithiocarbamates e.g. (Mancozeb, Nabam, Thiram etc.) and the phthalimides e.g. (Captan, Folpet). The 1970s to the end of the twentieth century saw a significant increase in the diversity of fungicidal chemistry being developed with the introduction of more systemic lower use rate materials and the commercial birth of key chemical classes such as anilinopyrimidines, carboxylic acid amides, benzimidazoles, phenylamides, triazoles, imidazoles, dicarboxamides and morpholines. The first two decades of the twenty-first century have seen continued innovation and commercialization of fungicidal solutions including the introduction of major groups such as the QoIs and SDHIs.

In spite of the challenges involved in fungicide development it is evident that industry has been very successful in discovering and bringing new fungicides to the market over the past 80 years. For a comprehensive chronological list of fungicide introductions from 1940 up to 2015 Russell (2005) and Leadbeater (2015) are suggested as useful references. Table 1 lists a more recent chronology of fungicide development based on published ISO common names for the years including 2015 up to the present (Q4 2021 at the time of writing). Twenty three compounds have been announced in this 7 year time frame which compares very favourably with the earlier 7 year window (2008–2014) where 27 new ISO common names were assigned for new fungicidal compounds, (BCPC Compendium of Pesticide common names https://pesticidecompendium.bcpc.org/index_new_frame.html ;accessed

Table 1 ISO common names of fungicides announced from 2015 onwards and their mode of action where reported[a]

Year	ISO common name	Mode of action (in parenthesis – FRAC MOA code where assigned)
2015	Pydiflumetofen	SDHI (7)
2016	Ipfentrifluconazole	DMI (3)
	Mefentrifluconazole	DMI (3)
	Quinofumelin	Unknown[b] (NC)
	Dichlobentiazox	Host plant defence induction (salicylate related) (P08)
	Fenpicoxamid	QiI (21)
2017	Fluindapyr	SDHI (7)
	Isoflucypram	SDHI (7)
	Aminopyrifen	GWT-1[d] (NC)
	Inpyrfluxam	SDHI (7)
	Pyrapropoyne	SDHI[c] (NC)
	Florylpicoxamid	QiI (21)
	Fluopimomide	Delocalization of spectrin-like proteins (43)
	Ipflufenoquin	Unknown[b] (NC)
	Metyltetraprole	QoI (subgroupA) (11A)
	Pyridachlometyl	Novel anti-tubulin MOA[e] (NC)
2018	Fluoxapiprolin	OSBPI (49)
2020	Cyclobutrifluram	SDHI (7)
	Flubeneteram	SDHI[c] (NC)
2021	Flufenoxadiazam	Unknown[b] (NC)
	Flumetylsulforime	Unknown[b] (NC)
	Fluoxytioconazole	DMI (3)
	Metarylpicoxamid	QiI (21)

[a]Based on information publicly reported by BCPC in form of an Index of New ISO Common names (https://pesticidecompendium.bcpc.org/index_new_frame.html).
[b]NC = no MOA code currently assigned by FRAC. Probable histone deacetylase inhibitor (Winter et al., 2020).
[c]MOA code not currently classified by FRAC but assumed to be SDHI based on structure.
[d]MOA reported by Hatomoto et al. (2019).
[e]MOA reported by Matsuzaki et al. (2020).

October 20th 2021). This represents a relatively healthy rate of continuing introduction of new actives onto the market.

At first glance, the more than 250 compounds representing 77 separate groupings as classified by FRAC would suggest that growers have abundant tools at their disposal. However, a high level analysis of sales data shows that just three key classes of fungicides, namely the SBIs, QoIs (strobilurins) and SDHIs account for around 58% of the total market (Table 2). The evident

Table 2 Estimated sales of fungicides in 2020 ($US in million)[a] as a function of chemical class or group

Chemical class/group	Global sales US$ in millions
DMIs (triazoles)	4880
QoIs (strobilurins plus others)	3355
SDHIs	1548
Dithiocarbamates (i.e. mancozeb and related)	1313
Copper	718
Benzimidazoles	549
Chlorothalonil	484
Other multi-sites	402
Carboxylic acid amides	329
Phenylamides	288
Amines	223
Melanin biosynthesis inhibitors (MBIs)	207
Anilinopyrimidines	158
Biologicals	149
Phosphonates	145
Dicarboxamides	129
QiIs	115
Phenyl pyrroles	109
Other fungicides	1581
Total	16721

[a]Values calculated based on Corteva agriscience analysis using internal sales panel data and (Kynetec, 2020) market analysis.

commercial success of these three classes of fungicides is a function of industry's ability to continue to innovate within a known area of chemistry due to a detailed understanding of the inherent associated regulatory risks and the fact that looking for holes within competitors' patents which can be exploited is on balance more likely to yield success. The SDHIs are a good example of where a known MOA (the first example was Carboxin, introduced in 1966) has subsequently undergone significant innovation with an associated broadening of spectrum and use patterns as well as differences in cross resistance patterns (Sierotzki and Scallet, 2013; Yamashita and Fraaije, 2018). Today there are 24 SDHIs listed in FRAC code group 7.

Some sectors of the fungicide market are well served with a wide range of effective and diverse MOAs. A good example are the solutions currently on the market for control of diseases caused by the *Phytophthora* genus of Oomycetes where at least ten different MOAs including multi-site chemistry are registered for use. Many of these MOAs are still fully effective against key

Phytophthora species since resistance has been slow to develop with the exception of phenylamides (Hermann et al., 2019) and more recently Fluazinam (Schepers et al., 2018). A diverse toolbox of effective molecules allows growers to fully rotate and mix different MOA's and thus practise effective resistance risk management within their spray programmes.

Other sectors of the fungicide market are in more need of innovation and additional chemical diversity. A good example is the control of the key foliar disease of wheat in Western Maritime regions of Europe i.e. *Zymoseptoria tritici*. SBIs, QoIs and SDHIs are all registered for use on this pathogen however resistance is widespread to QoIs in Western maritime wheat growing regions (Garnault et al., 2019; Kildea et al., 2019), SBI sensitivity has shifted significantly in the European Union (EU) population with an associated loss in performance in some regions (Jørgensen et al., 2021) and mutants with reduced sensitivity to SDHIs are also present in the field (Rehfus et al., 2020). The situation is further compounded by the recent loss of chlorothalonil due to a non-renewal decision by the EU commission meaning growers have lost a valuable multi- site tool for managing resistance risk. Pythium control is another area where additional solutions are desperately needed. Historically Phenylamides have been the only truly effective solutions on this important pathogen and resistance has become a significant issue (Hermann et al., 2019). Picarbutrazox introduced in 2013 has been a welcome additional tool for Pythium control but further solutions are still required. (Watanabe et al., 2019)

A wide variety of approaches exist for the generation of new leads, with varying levels of demonstrated success. Some of the more commonly used approaches involve competitor inspired lead generation, diversity screening, natural product inspired sources of new leads and bioactive hypothesis testing (Loso et al., 2017). One traditional approach to discovering new fungicide leads has involved high throughput screening of many tens of thousands of randomly sourced molecules annually in the hope of turning up interesting potential leads which could then be optimized by groups of chemists via a systematic exploration of the structure activity relationship of the molecule of interest. Estimates of numbers of molecules screened in order to eventually put one product onto the market vary, but a figure of 160,000 has been reported based on a survey of five major manufacturers (McDougall, 2016). Success is a function of the number of molecules which can be pushed through the screens in a finite period of time and much thought has been given to how this process can be streamlined and optimized via the use of targeted assays, robotics and other bespoke approaches which most companies would consider proprietary.

Structure or mechanism based design is an iterative process using a multidisciplinary approach which is established in the pharmaceutical industry and has played an important role in the development of a number of drugs. In theory designing molecules to specifically inhibit fungal target proteins at a specific novel target site would help to avoid undesirable effects in target plants and mammals. However the approach has had no success in the agrochemical development arena to the author's knowledge.

Each different lead generation approach has its own merits as well as inherent drawbacks and a related level of risk. While risks can be mitigated in a variety of different ways depending on the approach, a common theme for enhancing success is an effective interface between biology and chemistry, which is essential for hypothesis-driven research. Managing risk, and ultimately the long-term viability of a lead generation programme merits the utilization of a portfolio of approaches for the generation of new leads.

Innovation and diversity in the development of the fungicidal toolbox available to growers has brought many benefits to modern agricultural production systems, including improved safety to humans, non-target organisms and the environment. When considering market segments and targets for development all major R&D companies have their own research focused targets which dictate the strategies and the types of products they are looking to develop. Since it is potentially easier and by implication lower risk to innovate around existing MOAs and classes of chemistry (competitor inspired lead generation) than it is to discover and register a truly novel MOA there is a potential head wind slowing innovation with respect to the delivery of a diversity of MOAs in the chemical toolbox of solutions available to the grower.

This is not to say that companies are not actively trying to develop new MOAs which are registerable, active against key pathogen targets and with an associated low resistance risk. Being first to market with a novel MOA or chemistry is likely to have a significant advantage (Sparks et al., 2018). First to market with a new class of fungicide also has the opportunity to capture market space and set the stage for grower perception and use. The first company into a new area of chemistry will spend a large amount of time and money to address any issues which may negatively impact the ability to register the molecule. This will inevitably leave a path for others to follow and it is typical practice for the same company to generate an initial patent estate for the chemistry in order to protect freedom to operate. Patents from companies have increased in complexity in an effort to raise the barrier for competitors seeking to break into the same chemistry space. This is clearly only partially successful as a strategy as evidenced by the large number of molecules developed within the same chemistry space by numerous competitor companies, see the SDHI example mentioned earlier in this section.

Table 3 lists some of the characteristics of an ideal fungicide. These ideal requirements are easy to articulate on paper, but to address them all in a single molecule is an extremely challenging task in reality. Discovering and developing new molecules with high intrinsic potency against target fungi whilst having negligible or preferably no adverse effects on non-target

Table 3 Combination of desired properties of the 'Ideal' fungicide

Efficacy	Environmental profile	Flexibility of use	Economics
Highly selective to target.	Low use rate.	Low acute and chronic toxicity.	Patentable area with freedom to operate.
Safe to crop.	Low soil mobility.	>3 year shelf life.	Viable cost of manufacture.
Low resistance risk.	Rapid degradation in environment.	Easy to formulate in wide range of matricies.	Affordability to grower, positive impact on return on investment.
Broad spectrum.	No metabolites of toxicological concern.	Compatible in tank mix with other materials including biologicals.	Sustainable manufacturing process with low environmental impact.
Long residual plus curative action.	Low toxicity to beneficials.	Can be applied via range of equipment including drones.	
Good rainfastness.	Low/no residues on harvested components.	Low drift.	
MOA not shared by antifungals used in human medicine.	Compatibility with integrated pest management programmes.	Effective over wide range of temperatures and environmental conditions.	
Novel MOA.	Low toxicity to aquatic organisms.	Effective over wide range of water hardness and pH.	
Good redistribution within target plant (protection of new or unsprayed growth).	Low toxicity to birds and mammals.		

plants, beneficial organisms and humans and other mammals is an exercise in threading the needle. Ideally, the target site will be unique to fungi and thus avoid some of the aforementioned selectivity issues. Low use rate is one of the properties listed in Table 3 and it is reasonable to point out that the industry has made some progress in this area in certain market segments with newer molecule introductions having effective use rates which are significantly lower than the older molecules in the segment. Lowering the quantity of synthetic pesticide going into the environment is clearly a key desire of both consumers and regulators and is directly aligned to the goal of sustainable pesticide use. Table 4 illustrates this positive trend by comparing the chronology of the introduction years and the associated use rates of a number of important oomycete actives introduced over the past 60 years.

Table 4 Use rates of some important oomycete fungicides

Active ingredient	Approximate year of ISO common name assignment or introduction	Global average label rate (gain/ha)
Mancozeb	1961	1500–2000
Chlorothalonil	1964	1200–1700
Fosetyl Al.	1977	1500–2000
Dimethomorph	1988	225
Fluazinam	1992	200
Mefenoxam	1996	250–1000
Zoxamide	1999	160
Cyazofamid	2000	80–90
Mandipropamid	2004	150
Fluopicolide	2004	100
Amisulbrom	2005	100
Ametoctradin	2009	240–300
Oxathiapiprolin	2012	12–15

3 Challenges facing fungicide development and registration

Pesticide development is a complex, time-consuming and expensive exercise. From 1995 to 2014 the estimated research and development expenditure of bringing a new active ingredient to market rose from US$152 million to US$286 million (McDougall, 2016). These figures were based on data reported from five major manufacturers and represent an increase of 88% over the time period surveyed. The same report quotes a figure of US$2625 million of total R&D expenditure across 11 surveyed companies which equated to 5.4% of their agrochemical sales. Unfortunately published figures for development costs in the period 2014 to the present day are not available, but it is certain that they have continued to rise over time. High development costs mean that target markets need to be of sufficient size and value in order for companies to justify the expenditure of bringing new fungicides to the market as well as the significant sums which are then required to maintain registrations over the natural life cycle of the product.

Due to the cost and complexity of molecule development the cost benefit ratio naturally prioritizes development for key market segments such as Asian soybean rust, Western European cereals, Oomycetes, rice and selected segments of the specialty crop market (vegetables, grapevines, pome etc.). One consequence of this approach is that smaller value markets or minor crops may miss out on new introductions leaving growers with fewer effective options. A number of organizations such as IR-4 in the US specifically target this issue by working with major manufacturers to share costs on extending registrations into minor crops or by supporting specialty registrations on major crops.

Time to development is another significant headwind for manufacturers to contend with. In 1995 it took an average of 8.3 years to bring a new molecule to market, this had risen to 11.3 years by 2014 (McDougall, 2016). Even a new formulated product containing known registered actives can take up to 4–5 years to gain registration in some regions of the world. It is certain that a large part of the increase in development time is being driven by more stringent and complex regulatory requirements. Long development times result in erosion of a companies' period of patent exclusivity based on the initial filing for composition of matter. This is a significant risk since the period of patent protection is a critical period over which companies maximize their opportunity to recoup development costs and generate a return on their investment.

Increasing development times is important since the Net Present Value (NPV) of a business case supporting molecule development is highly time sensitive, even a one year delay in obtaining registration for a new active or product can have a significant negative impact on NPV. Another important element of having long lead times before reaching market is that manufacturers need to become proficient in predicting the future. A great deal can happen in the market over an 11 year period. Will resistance drastically alter over time to the MOA they are working on rendering it obsolete? Uncertainty about whether certain actives will be re-registrable due to changes in regulation over the development period can have an impact on selection of molecules to partner in co-formulated combinations or more than one active. If one of the partners in the mixture does not get re-registered, then the product is lost from the market. This would be particularly painful if it were to occur during the development phase of the product resulting in significant losses due to sunk costs. Indeed, the latter situation particularly in Europe has seen manufacturers developing a greater preponderance of so called 'co-packs' where two separate actives are delivered in separate containers which need to be used together in the spray tank. However this solution has its downsides including increased numbers of containers needing recycling and the potential for misuse of the product due to growers inadvertently splitting the packs and using the actives separately thus rendering any built in resistance management worthless.

The number of companies with active pesticide discovery programmes has decreased significantly over the past 50 years as a result of numerous corporate mergers. This drastic consolidation has been driven by numerous factors including but not limited to expansion into more profitable opportunities in the seed market via biotechnology, soaring development and regulatory costs driven by more stringent hurdles to bring products to market and intellectual property rights protection making vertical integration into downstream industries a necessity. Some NGO observers have seen this trend as a negative which is designed to accelerate corporate control and consolidate and monopolize global, regional and local food systems (Hass Institute, 2018).

With fewer companies actively engaged in looking for new MOAs one would expect the pace of introduction to have been diminishing over time. An analysis of ISO common name approvals for new fungicides since 1990 to the present day reveals the following trends (Fig. 1); 109 new fungicide ISO common names were issued (some of the latter molecules have yet to come to market as commercial products and not all of the molecules have become commercially successful) of which at the time of introduction 82 belonged to existing MOAs whilst 27 represented either a new or unknown MOA. It is assumed for the purposes of this exercise that the unknown MOA molecules do likely represent a new MOA. Despite the best efforts of manufacturers MOA cannot always be elucidated during development of a new molecule. Looking at 10 year increments the rate of introduction of new fungicide actives has remained relatively stable with 37 actives in the period 1990–2000, 33 actives 2001–2010 and 37 actives from 2011 to present. From 1990 to 2000 10 molecules with a new or unknown MOA were developed with this figure dropping to 9 in the period 2001 to 2010 and further decreasing to 7 from 2011 to the present day. This does suggest that novelty or innovation as measured purely by introduction of new MOAs into the market has been slowly decreasing over the past 20 years.

An increasingly stringent regulatory environment has been partially responsible for driving up the costs of new active ingredient development as well as the removal of a number of key fungicide products from the market place. The switch in the EU from an active ingredient approvals process based on risk based to hazard based criteria (Directive 1107/2009) has resulted in the loss of a number of key products including (but not limited to) Myclobutanil, Propiconazole, Epoxiconazole, Cyproconazole, Fenpropimorph, Picoxystrobin,

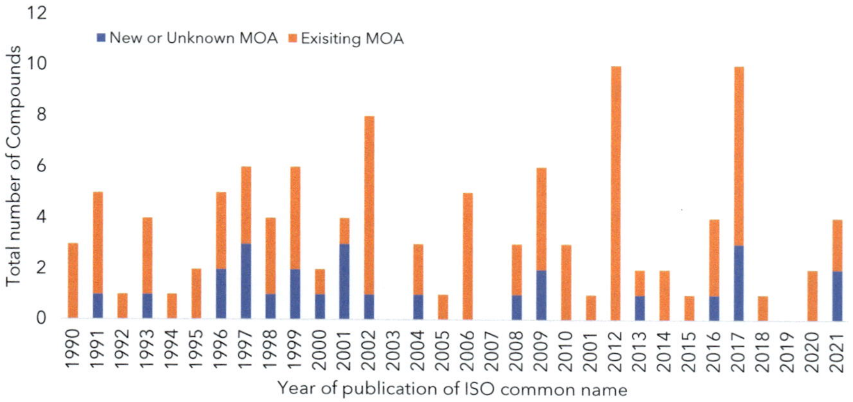

Figure 1 Chronology (1990–2021) of new Fungicide ISO common name introductions.

Mancozeb and Chlorothalonil. As can be seen from this list of actives the EU's criteria for assessing endocrine disrupting properties (Andersson et al., 2018) has accounted for the loss of registration of a number of azole fungicides in the region due to the potential of this class of compounds to impact the endocrine systems of different organisms (Taxvig et al., 2007). It has been estimated by Bryson and Brix (2018) that around 70% of the fungicide products registered in the EU contain an active ingredient which may be prohibited due to cut-off criteria or listed as a candidate for substitution. Such a loss/restriction of solutions from the grower's toolbox would have a profound impact on disease control in the EU.

Regulatory pressure on the agrochemical industry is certain to increase in the future. The EU has one of the most conservative regulatory systems for pesticide approvals worldwide which has its genesis in a move towards an approval process based on the precautionary principle as opposed to science based risk assessment. The recently published European Green Deal contains within its framework Farm to Fork and Biodiversity strategies within which plans are outlined for an updated implementation of the current Sustainable use Directive (Directive 2009/128/EC) committing the EU Commission to take actions to reduce by 50% the use and risk of chemical pesticides by 2030 and reduce by 50% the use of the more hazardous pesticides by 2030. (EU Commission report, 2020). Pesticides including fungicides are seen in some quarters as incompatible with the philosophy of moving towards a more sustainable system of growing food (Lykogianni et al., 2021) although the reality is that it is likely the latter goal will need to be built on a multidisciplinary approach where synthetic fungicides will still play a role.

Policy maker, consumer and NGO perception of pesticide use continues to frame the environment in which the crop protection industry needs to operate when developing new solutions for disease control. Fungicide and insecticide use in fruit and vegetable production inevitably results in residues on the harvested produce, and this is yet another complex challenge faced by manufacturers. Many large food retailers impose their own secondary approval standards on the number and levels of residues they will accept on produce. These are typically more restrictive than the Maximum Residue Limits (MRLs) originally set by the regulatory agencies when approving the registration of a product. These standards may drive the growers' choice of product and spray timing if they want to be able to supply key retailers. Loss of registration of a fungicide in the EU due to the molecule triggering cut-off criteria related to human health concerns will result in the associated MRL for the molecule being reduced to the Limit of Quantification. This will have a significant impact on fresh produce distributors importing into the EU since growers outside the EU producing for export may still be able to use the same fungicide in their spray programmes. Manufacturers when faced with this situation for their products need to work with and provide advice to growers on how they adjust use patterns for the molecule and post harvest interval in order to be in a non-detectable residue scenario. Typically, this requires a thorough understanding of the molecule's residuality on and in the target crop via conducting extensive in-field reverse decline studies under a range of environmental conditions and in different geographic regions.

Modern fungicides and antimycotic pharmaceuticals interact with target receptors or enzymes via the same molecular recognition process. In some cases a homologous enzyme or receptor is addressed and this has given rise to different actives from the same chemical class which have found commercial utility in both sectors, the azoles being the classic example here. One unintended consequence of this situation has been the rise of concerns about the use of triazoles in an agricultural setting selecting for resistant strains of *Aspergillus fumigatus* (Verweij et al., 2016). The fungus is ubiquitous in soil and decaying matter and is an opportunistic pathogen in humans with the ability to cause life threatening invasive aspergillosis in immunodeficient and immunocompromised patients. Azoles are an important clinical treatment for this condition, and azole resistant isolates of *A. fumigatus* have been found in patients around the globe.

The spread of these isolates globally has come to be regarded as a potential global human health concern (Verweij et al., 2020). A recent Chinese paper has linked prevalence of resistant *A. fumigatus* isolates in agricultural fields to triazole application (Cao et al., 2021). Composted waste from the flower bulb production industry is a reported hot spot for selection of azole resistant *A. fumigatus* (Gisi, 2014, 2021). Other research in this area suggests the role of agricultural triazole use in resistance selection to *A. fumigatus* requires additional investigation (Barber et al., 2020, Fraaije et al., 2020). Similar concerns have been raised for other fungi including *Candida* species which are exposed to triazoles in agricultural settings (Gisi, 2021). Unsurprisingly, regulatory agencies have had their attention drawn to this issue, and the area of concern around pesticides and pharmaceuticals sharing the same target site in fungi and any potential associated crossover risk of resistance will need to be carefully considered by the crop protection industry in future development plans.

Considering the regulatory complexities faced by manufacturers when developing new fungicides, much critical thinking and experimental design goes into effectively screening at an early stage in the discovery process for undesirable traits in molecules which would likely prevent registration. The earlier this is detected, the greater the potential savings in time and money. Late detection of a significant issue resulting in a molecule development programme being terminated may result in millions of dollars lost due to sunk R&D costs. The challenge is to operate a predictive safety screening process which is calibrated at a level of sensitivity which filters out problem molecules whilst not rejecting promising new chemistry. The latter point is important here since early stage lead molecules with undesirable properties can potentially have such issues rectified via further synthetic modification of the structure.

4 Fungicide resistance: a driver for innovation

Fungicide resistance is the process by which repeated exposure of a fungal population to the selection pressure of a fungicide results in the emergence of individuals within the population with reduced sensitivity to the fungicide being used to control them. Once the frequency of these resistant individuals within the population reaches a critical level, field failure of the fungicide will be expressed as a lack of effective control. Fungicide resistance can have negative implications for both growers and manufacturers with potential loss of yield and yield quality due to poor disease control and loss of sales due to a shortened life cycle of the product on the market. Fungicide resistance is a recurring issue in the management of a large number of agronomically important plant diseases (Stevenson et al., 2019). The phenomenon of resistance developing to multiple MOAs in the same pathogen is also an additional challenge, with *Botrytis cinerea* being a notorious example (Weber, 2011).

Fungicide resistance management strategies using a diverse range of MOAs are common practice in an attempt to delay or avoid the development of resistance and the subsequent commercial impact should field resistance arise. These strategies are integral components of the broader concept of promoting effective product stewardship and sustainable use. A thorough assessment of the perceived resistance risk for a new fungicidal molecule along with an appropriate risk management strategy designed to mitigate the likelihood of resistance developing during commercial use are key components of the efficacy section of the dossier which is submitted to the authorities when seeking a new approval. The primary aim of this analysis and any resultant management strategies is to understand the theoretical level of risk and prolong the effective life of the fungicide. This is done via a detailed look at the potential mechanisms of resistance and an analysis of the fitness penalties associated with any identifiable mutations conferring resistance (Hawkins and Fraaije, 2018), as well as the impact of agronomic factors associated with the crop pathogen complex. Modelling of occurrence of resistance to better facilitate its management can also play an important role here (van den Bosch et al., 2014).

A joint publication by the Resistance Action Committees (Nauen et al., 2008) has previously recommended that facilitating effective implementation of a sustainable resistance management strategy for a particular crop and pest complex requires no less than three effective pesticidal MOAs be considered for a particular crop and target pest in individual geographies, of which one should ideally be from an accepted low risk category. Four or more MOAs are desirable. They further state that in crops where season long application programmes are typical the use of less than four MOAs to control a crop and pathogen and pest situation is considered extremely risky and therefore unsustainable from an effective disease and pest management perspective as well as from a resistance management perspective. Five or more MOAs are desirable. The unfortunate reality is that effective resistance management strategies traditionally built around mixing different MOAs, or blocking or alternating MOAs within multiple application spray programmes, are not always possible depending on molecule availability in a particular crop and region. Not all products are registered in all regions of the world or effective on all pathosystems. See previous comments in Section 2 of this chapter regarding the available diversity of fungicidal solutions for growers looking to control *Zymoseptoria tritici* and *Pythium* as two examples. Another significant impact on the grower's ability to effectively rotate different MOAs within a spray programme are the secondary standards imposed by the food chain as mentioned in Section 3 of this chapter. A number of large fruit and vegetable importers/distributors mandate no more than 3 detectable pesticide residues on produce they are supplying to large retail customers. This can directly limit the number of different molecules a grower can use in their programme particularly if they are using both insecticides and fungicides.

As previously mentioned in the introduction to this chapter, the 2022 FRAC Code list catalogues around 250 fungicidal materials grouped into 77 different code groupings based on cross resistance behaviour. At first glance this seems like a more than adequate armoury of fungicidal solutions for the grower seeking to control disease in their crops. However if we correlate total global fungicide sales with resistance risk (based on FRAC classification of individual actives) (Table 5), we see a more nuanced picture of the actual situation. High and medium risks compounds accounted for close to 70% of annual global sales in 2020 whilst medium to low and low risk compounds represented a mere 28% of the market. This number appears to have remained unchanged since a similar analysis was conducted and reported 6 years ago (Leadbeater, 2015).

This lack of true diversity of MOAs in the market as measured by product use is a reflection of the huge commercial success of just three key classes of fungicides; the C-14 demethylase inhibitors (majority are triazoles), QoIs and SDHIs, due to their effectiveness, breadth of registrations and popularity with growers. These three target sites have been deeply mined by multiple manufacturers and numerous examples exist in the market place. The FRAC code list reports 24 SDHIs (group 7), 20 QoIs (group 11) and 37 C-14 demethylase inhibitors (group 3). Unsurprisingly given their frequency of use

Table 5 2020 global fungicide; sales ($US in million) by FRAC code group resistance risk classification

Resistance risk[a]	Number of FRAC code groups[a]	Number of molecules	Global sales 2020[b]
High	5	27	4289
High to medium	5	24	1896
Medium	12	49	5410
Medium to low	11	30	1167
Low	29	13	3433
Unknown	18	74	514
Total		217	16709

[a]Taken from (FRAC Code List, 2022).
[b]Values calculated based on Corteva agriscience analysis using internal sales panel data and (Kynetec, 2020) market analysis.

and associated resistance risk (medium to high) there are numerous reports in the literature of resistance having been detected to these groups of fungicides in a wide range of economically important plant pathogens (FRAC, 2020).

Site-specific fungicides targeting a narrowly defined binding site on a target protein are typically at much higher risk of resistance development than multi-site molecules (Brent and Hollomon, 2007). Traditionally multi-sites have had high use rates and protectant only activity, in some instances uptake into plant tissue causes phytotoxicity hence the need to keep the material on the plant surface. Activity on multiple biochemical steps (common to many living organisms) and high use rates in the environment have resulted in multi-site fungicides coming under increasing regulatory pressure with the recent loss in the EU of two important molecules used extensively in spray programmes for managing resistance risk namely - Mancozeb and Chlorothalonil. A new generation of low resistance risk multi-site materials with low use rates and a favourable regulatory profile would be a significant breakthrough.

Certainly, the impact of the phenomenon of resistance development to fungicides with a higher risk single site MOA, and the fact that three key MOAs (C14-demethylase inhibitors, QoIs and SDHIs) account for very nearly 60% of current annual fungicide sales is a strong motivational force for manufacturers to continue to innovate in bringing new MOAs to the market and to increase the diversity of chemistry in the growers' disease control toolbox. Preferably, these new molecules will represent a MOA and associated target site with a low resistance risk.

5 Fungicide development: a look into the future

It is highly probable that increasing regulatory complexity and its associated raised barrier to entry into the market, along with a clear desire by society to accelerate continuing reduction of pesticide use, will continue to frame the future landscape within which companies seeking to introduce new fungicides will need to operate. In the future, as more biocontrol agents, plant defence elicitors and possible microbiome manipulations become available, and reduce the reliance on synthetic fungicides, a new understanding of the role of plant pathogens and the diseases they cause in whole cropping systems and in the provision of ecosystem services may be needed, (Jeger et al., 2021).

Agriculture is under pressure to become more sustainable by lowering its environmental impact, its levels of synthetic inputs and its carbon footprint. Growers are looking for innovative solutions from manufacturers to help them address these challenges head on whilst allowing them to continue to operate profitably. Synthetic fungicides are more likely to be seen in the future as just one component of a truly integrated approach to disease control. This will likely involve a coordinated use of multiple tools such as decision support systems for optimal spray timing and placement, disease surveillance and in field detection and diagnostics, varietal resistance (via traditional breeding, gene editing and transgenic approaches), biologicals and biostimulants.

Global sales of biofungicides were reported to have reached US$1.3 billion for 2019 (Markets and Markets, 2020). This accounts for approximately 8% of the total crop protection market spend on disease control in 2019. The predicted Compound Annual Growth Rate (CAGR) of the biofungicides segment for the period 2020–25 is 15.9% which is 5.8× the predicted CAGR for conventional synthetic fungicide products (2.7%) over the same time period (AgbioInvestor Report, 2021). Traditionally, the majority of biofungicide sales have been in the fruit and vegetables segments, although their use in row crops is beginning to grow. It is clear that biologicals are gaining in importance due to their perceived potential for a direct role in sustainable agricultural production systems. The majority of major R&D companies involved in pesticide development now have active divisions dedicated to the discovery and development of biological solutions including biofungicides. Their portfolios of biofungicide solutions are certain to grow in the future as a function of internal discovery and continued active acquisition of third party solutions.

Key here will be addressing and evaluating the integration of new fungicidal products with other components of the disease control strategy during the development phase and not as an afterthought. In the future, manufacturers may be in a position to offer truly integrated solutions to growers involving a combination of the correct germplasm, trait and seed treatment plus foliar spray programme tailored to the specific needs and agronomic situation of the customer. Some of this thinking is already well developed with a good example being the recent proposal to leverage the heterogeneity of cultivar mixtures to mitigate against selection for fungicide resistance, against Septoria leaf blotch in wheat (Kristofersen et al., 2020). Unfortunately this approach may be challenging where growers are looking to satisfy particular grain quality parameters required by millers for high quality bread making flour.

Traditionally biologicals have been positioned at the end of a spray programme on fruits and vegetables replacing synthetic fungicides for the last one or two sprays in order to manage detectable residue levels of the latter. In addition, the concept of mixing or alternating biologicals with synthetic fungicides in order to lower use rates of the latter is an area being explored for various crop pathosystems (Ons et al., 2020). Tailoring fungicide dose to the level of disease risk i.e. lower rates applied on crop varieties with higher disease tolerance or in regions where the climate is not favourable for strong epidemics is a simple practice which ties in well with sustainable fungicide use (Ritchie et al., 2018).

Precision agriculture is a knowledge-based technical management system, where sensing, information technologies, and mechanical systems enable sub-field crop management that can help optimize farm profits and minimize agriculture's impact on the environment. This technology has been developed over several decades through private sector and university research efforts, with new technologies and applications seeming to appear year on year. The rapid rise of precision farming technologies aligned to the use of remote sensing to detect disease in the field has the potential to open up new areas of research where the use of fungicides within a wider assessment of cultural practices for managing disease can be examined (Kitchen, 2008). Airborne and satellite imagery have been used for detecting and mapping cotton root rot, a destructive soilborne fungal disease, in cotton fields. Prescription maps derived from the imagery have been coupled with variable rate and site-specific fungicide application resulting in effective control of the disease without the need to prophylactically treat the entire field (Yang, 2020).

Fungicide application via drones or Unmanned Aerial Vehicles (UAVs) has seen a significant increase in market acceptance over the past decade. UAVs can operate at low altitudes and varying speeds using low water volumes thus making them suitable for use in scenarios where ground application is challenging or impossible and where precision application is required. Adoption in South East Asia, Japan and China has been particularly rapid and in 2020 over 450 million hectares of crops were treated by UAV in China alone, (Gallup and Barbosa, 2020). It seems certain that development of fungicides and formulations which can be effectively utilized via UAV application systems will continue to be a key area of focus moving forward.

The future for fungicide discovery and development may require manufacturers to make improvements in how they implement the intersection of chemistry, biochemistry, molecular biology, formulation science and genomics in order to bring forward so called 'designer' fungicides that address the hurdle of regulation and work effectively in an integrated approach with other tools. Is it possible to design fungicidal molecules with low resistance risk and a benign regulatory profile? Using natural products as a source of inspiration for developing novel or innovative new fungicide leads is one area which probably requires more attention from manufacturers in the future. Screening a wide range of biodiverse sources of organisms has the potential to provide a rich source of inspiration for new fungicide leads. A good example is the soil microbiome where we find microorganisms equipped with an arsenal of unique biosynthetic enzymes producing bioactive secondary metabolites allowing them to engage in chemical warfare as they compete with other microorganisms in their environmental niches (Spraker et al., 2018). It should be noted here that natural products may have toxicological profiles which make them unsuitable for development as fungicides, for example they may be uncouplers, carcinogens or teratogenic. Any natural product screening effort requires an effective set of controls to screen out molecules with undesirable properties at an early stage.

The recent launch of the fungicide Fenpicoxamid, a first new target site MOA for control of *Zymoseptoria tritici* in wheat in nearly 20 years is a good example of a natural product inspired solution. The molecule is based on an antibiotic UK-2A produced by a soil borne *Streptomyces* sp. 517-02 (Owen et al., 2017). UK-2A is produced by industrial scale fermentation and then a single step post fermentation modification is conducted in order to generate Fenpicoxamid. This modification is done in order to impart UV stability to the molecule thus increasing residuality as well as lowering use rate. More interesting is the fact that Fenpicoxamid converts back to the fungicidally active natural product UK-2A when in the presence of fungal or plant tissue. When one considers a recent analysis based on ecological scaling laws that suggests only 0.001% of all microbial life on earth is known to science (Locey and Lennon, 2016) it is difficult not to contemplate what treasure trove of novel fungicidal natural products remain undiscovered.

The plant microbiome has been intensively studied over the last decade including in food crops (Ding et al., 2019). It has been understood for a period of time now that the plant-associated phytobiome (phyllosphere, endophyte, and rhizosphere microbiome), can affect directly or indirectly disease development (Rabiey et al., 2019; Tsolakidou et al., 2019). Could molecules be developed in the future which interact at this level and prevent disease development? Plant defence stimulation is another area which probably warrants additional attention in the future. At present the number of commercially available disease control materials categorized by FRAC as stimulating host defence induction is somewhat limited.

The impact of climate change and its related effects on the emergence or re-emergence of new plant diseases and their spread to new locations will likely pose new challenges going forward (Gottwald et al., 2019; Carvajal-Yepes et al., 2019). This will continue to drive the need for adaptability within the crop protection industry with repurposing of existing solutions or development of new regionally integrated disease control strategies where synthetic fungicides will continue to play an important role.

6 Conclusion

The One Health concept proposes that in the long term, people's health and that of the environment are indivisible (National Academy of Sciences, 2021). The application of a One Health approach is seen as critical for achieving the UN 2030 Agenda for Sustainable Development (United Nations, 2015). This has direct implications for agriculture. People need nutritious food to remain healthy and agriculture does have the potential to cause negative environmental impact. The challenge of how the world will address the need to feed an extra 2 billion mouths by 2050 whilst re-tooling some modern agricultural practices to align with a vision for a more sustainable system with lower environmental impact is a huge task which will involve many individuals and organizations in the private, public and non-profit sectors. Only by greater cooperation and coordination between key players will this be possible.

Synthetic pesticides, including fungicides, will likely be a component of disease control strategies for the foreseeable future, but what seems certain is that their future continued use and development will evolve within an ever more demanding environment of regulatory pressure and consumer acceptance. Certainly biological solutions will continue to grow in importance and relevance. This dynamic needs to be seen by manufacturers involved in fungicide discovery and development as an opportunity to innovate and adapt to the new paradigm rather than as an intrinsic threat to their business model. There will be significant opportunities for companies who correctly identify and develop any future disruptive plant disease control technologies, and many growers are actively looking for more sustainable solutions which will continue to allow them to farm profitably. Section 2 of this chapter alluded to the remarkable success of the agrochemical industry over the past 80 years in developing increasingly innovative and effective plant disease control materials. It can only be hoped that the story of the next 80 years will be equally as successful.

7 Where to look for further information

- Leadbeater, A. and Gisi, U. (2010). "The challenges of chemical control of plant diseases". In U. Gisi, I. Chet and M. L. Gullino (Eds), *Recent Developments in Management of Plant Diseases. Plant Pathology in the 21st Century*, DOI 10.10007/978-1-4020-8804-9_1, Springer Science + Business media B.V. 2010.

- Leadbeater, A. (2015). Recent developments and challenges in chemical disease control. *Plant. Protect. Sci.,* 51(4), 163–169.
- Russell, P. E. (2005). A century of fungicide evolution. *J. Agric. Sci,* 143, 11–25.
- Jeger, M., Beresford, R., Bock, C., Brown, N., Fox, A., Newton, A., Vicent, A., Xu, X. and Yuen, J. (2021). Global challenges facing plant pathology: multidisciplinary approaches to meet the food security and environmental challenges in the mid-twenty-first century. *CABI Agric. Biosci.* 2:20 https://doi.org/10.1186/s43170-021-00042-x.
- Umetsu, N. and Shirai, Y. (2020). "Development of novel pesticides in the 21st century." *J. Pestic. Sci.,* 45(2), 54–74.

8 References

AgbioInvestor. (2021). Market insight crop protection and traded seed market 2025 forecast. Available at: www.agbioinvestor.com.

Agrios, G. N. (1988). *Plant Pathology*. San Diego, CA: Academic Press.

Andersson, N., Arena, M., Auteri, D., Barmaz, S., Grignard, E., Kienzler A., Lepper, P., Lostia, A. M., Munn, S., Parra Morte, J. M., Pellizzato, F., Tarazona, J., Terron, A. and Van der Linden, S. (2018). Guidance for the identification of endocrine disruptors in the context of Regulations (EU) no 528/2012 and (EC) no 1107/2009. *EFSA J.* 16(6), 5311. doi: 10.2903/j.efsa.2018.5311.

Barber, A. E., Riedel, J., SaeOng, T., Kang, K., Brabetz, W., Panagiotou, G., Deising, H. B. and Kurzai, O. (2020). Effects of agricultural fungicide use on *Aspergillus fumigatus* abundance, antifungal susceptibility, and population structure. *mBio* 11(6), e02213-20. doi: 10.1128/mBio.02213-20.

BCPC Compendium of Pesticide Common Names. Index of new ISO common names. Available at: https://pesticidecompendium.bcpc.org/index_new_frame.html.

Brent, K. J. and Hollomon, D. W. (2007). Fungicide resistance: The assessment of risk. FRAC Monograph no 2 (2nd rev. edn.). Available at: https://www.frac.info/docs/default-source/publications/monographs/monograph-2.pdf?sfvrsn=749d419a_8.

Bryson, R. J. and Brix, H. D. (2018). Challenges and prospects for fungicidal control of wheat diseases. In: Oliver, R. (Ed.), *Integrated Disease Management of Wheat and Barley*. UK: Burleigh Dodds Science Publishing Limited, pp. 219–234.

Cao, D., Wang, F., Yu, S., Dong, S., Wu, R., Cui, N., Ren, J., Xu, T., Wang, S., Wang, M., Fang, H. and Yu, Y. (2021). Prevalence of azole-resistant *Aspergillus fumigatus* is highly associated with Azole fungicide residues in the fields. *Environ. Sci. Technol.* 55(5), 3041–3049. doi: 10.1021/acs.est.0c03958.

Carvajal-Yepes, M., Cardwell, K., Nelson, A., Garrett, K. A., Giovani, B., Saunders, D. G. O., Kamoun, S., Legg, J. P., Verdier, V., Lessel, J., Neher, R. A., Day, R., Pardey, P., Gullino, M. L., Records, A. R., Bextine, B., Leach, J. E., Staiger, S. and Tohme, J. (2019). A global surveillance system for crop diseases. *Science* 364(6447), 1237–1239.

Code List, F. (2022). Fungal control agents sorted by cross resistance pattern and mode of action (including coding for FRAC Groups on product labels). Available at: https://

www.frac.info/docs/default-source/publications/frac-code-list/frac-code-list-2022--final.pdf?sfvrsn=b6024e9a_2.

Ding, L. J., Cui, H. L., Nie, S. A., Long, X. E., Duan, G. L. and Zhu, Y. G. (2019). Microbiomes inhabiting rice roots and rhizosphere. *FEMS Microbiol. Ecol.* 95(5), fiz040.

EU Commission Report. (2020). "Report from the Commission to the European Parliament and the Council: On the experience gained by Member States on the implementation of national targets established in their National Action Plans and on progress in the implementation of Directive 2009/128/EC on the sustainable use of pesticides." SANTE/11319/2017-EN CIS. Available at: europa.eu.

FAO. (2009). How to feed the world 2050. *Glob. Agric. Towards. High-Level Expert Forum* (Vol. 2050). Rome, 12–13 October 2009. Available at: https://www.fao.org/fileadmin/templates/wsfs/docs/Issues_papers/HLEF2050_Global_Agriculture.pdf.

FAOSTAT. (2019). Land use statistics and indicators: Global, regional and country trends 1990–2019. Available at: https://www.fao.org/3/cb6033en/cb6033en.pdf.

FAOSTAT. (2020). Sustainable food and agriculture. Land use in agriculture by the numbers. Available at: https://www.fao.org/sustainability/news/detail/en/c/1274219/.

Fraaije, B., Atkins, S., Hanley, S., Macdonald, A. and Lucas, J. (2020). The Multi-Fungicide resistance status of *Aspergillus fumigatus* populations in Arable soils and the wider European environment. *Front. Microbiol.* 11, 599233. doi: 10.3389/fmicb.2020.599233.

FRAC. (2020). List of first confirmed cases of plant pathogenic organisms resistant to disease control agents. Available at: https://www.frac.info/docs/default-source/publications/list-of-resistant-plant-pathogens/list-of-first-confirmed-cases-of-plant-pathogenic-organisms-resistant-to-disease-control-agents_05_2020.pdf?sfvrsn=7073499a_2.

Gallup, C. and Barbosa, R. (2020). Rapid adoption of drone application technology in Asia-Pacific: A story of grower demands diagnostics of plant diseases 22 defining new management techniques. In: *APS Workshop of Unmanned Aerial Vehicles for making plants healthy - Do we have a winner?* 3 December 2020, Minneapolis, USA. Available at: https://www.apsnet.org/meetings/mtngwshops/phworkshops/Pages/default.aspx.

Garnault, M., Duplaix, C., Leroux, P., Couleaud, G., Carpentier, F., David, O. and Walker, A. S. (2019). Spatiotemporal dynamics of fungicide resistance in the wheat pathogen *Zymoseptoria tritici* in France. *Pest Manag. Sci.* 75(7), 1794–1807.

Gisi, U. (2014). Assessment of selection and resistance risk for demethylation inhibitor fungicides in *Aspergillus fumigatus* in agriculture and medicine: A critical review. *Pest Manag. Sci.* 70(3), 352–364. doi: 10.1002/ps.3664.

Gisi, U. (2021). Crossover between the control of fungal pathogens in medicine and the wider environment, and the threat of antifungal resistance. *Plant. Path* 71, 131–149. doi: 10.1111/ppa.13429.

Gottwald, T. R., Luo, W., Posny, D., Riley, T. and Louws, F. (2019). A probabilistic census-travel model to predict introduction sites of exotic plant, animal and human pathogens. *Phil. Trans. R. Soc. B Biol. Sci.* 374(1776), 20180260.

Haas Institute. (2018). The era of corporate consolidation and the end of competition. Available at: https://escholarship.org/content/qt100895ng/qt100895ng_noSplash_ba7c85156bcc6472adabfa93a6fd3e79.pdf?t=qc32is.

Hatomoto, M., Aizawa, R., Kobayashi, Y. and Fujimura, M. (2019). A novel fungicide aminopyrifen inhibits GWT-1 protein in glycosylphosphatidylinositol-anchor biosynthesis in *Neurospora crassa*. *Pestic. Biochem. Physiol.* 156, 1–8.

Hawkins, N. J. and Fraaije, B. A. (2018). Fitness penalties in the evolution of fungicide resistance. *Annu. Rev. Phytopathol.* 56, 339–360.

Hermann, D. C., McKenzie, D., Cohen, Y. and Gisi, U. (2019). Chapter 6: Phenylamides: Market trends for resistance evolution for important oomycete pathogens more than 35 years after product introduction (FRAC code 4). In: Stevenson, K. L., McGrath, M. T. and Wyenandt, C. A. (Eds.), *Fungicide Resistance in North America* (2nd edn.). St. Paul, MN: The American Phytopathological Society.

Jeger, M., Beresford, R., Bock, C., Brown, N., Fox, A., Newton, A., Vicent, A., Xu, X. and Yuen, J. (2021). Global challenges facing plant pathology: Multidisciplinary approaches to meet the food security and environmental challenges in the mid-twenty-first century. *CABI Agric. Biosci.* 2(1), 20. doi: 10.1186/s43170-021-00042-x.

Jørgensen, L. N., Matzen, N., Heick, T. M., Havis, N., Holdgate, S., Clark, B., Blake, J., Glazek, M., Korbas, M., Danielewicz, J., Maumene, C., Rodemann, B., Weigand, S., Kildea, S., Bataille, C., Brauna-Morževska, E., Gulbis, K., Ban, R., Berg, G., Semaskiene, R. and Stammler, G. (2021). Decreasing Azole sensitivity of *Zymoseptoria tritici* in Europe contributes to reduced and varying field efficacy. *J. Plant Dis. Prot.* 128(1), 287–301. doi: 10.1007/s41348-020-00372-4.

Kenrick, W. (1833). *The New American Orchardist*. Boston.

Kildea, S., Buar, D. E., Hutton, F., Rosa, Sdl, Welch, T. E. and Phelan, S. (2019). Prevalence of QoI resistance and mtDNA diversity in the Irish *Zymoseptoria tritici* population. *Ir. J. Agr. Food Res.* 58(1), 27–33.

Kitchen, N. R. (2008). Emerging technologies for real-time and integrated agriculture decisions. *Comput. Electron. Agric.* 61(1), 1–3.

Kristofersen, R., Heick, T. M., Müller, G. M., Eriksen, L. B., Nielsen, G. C. and Jørgensen, L. N. (2020). The potential of cultivar mixtures to reduce fungicide input and mitigate fungicide resistance development. *Agron. Sustain. Dev.* 40(5), 36. doi: 10.1007/s13593-020-00639-y.

Kynetec. (2020). Global market research in animal health and agriculture. Available at: https://www.kynetec.com/.

Leadbeater, A. (2015). Recent developments and challenges in chemical disease control – A review. *Plant Prot. Sci.* 51(4), 163–169.

Leadbeater, A. and Gisi, U. (2010). The challenges of chemical control of plant diseases. In: Gisi, U., Chet, I. and Gullino, M. L. (Eds.), *Recent Developments in Management of Plant Diseases: Plant Pathology in the 21st Century* . Springer Science+Business Media B.V. doi: 10.10007/978-1-4020-8804-9_1.

Locey, K. J. and Lennon, J. T. (2016). Scaling laws predict global microbial diversity. *Proc. Natl Acad. Sci. U. S. A.* 113(21), 5970–5975.

Loso, M. R., Garizi, N., Hegde, V. B., Hunter, J. E. and Sparks, T. C. (2017). Lead generation in crop protection research: A portfolio approach to agrochemical discovery. *Pest Manag. Sci.* 73(4), 678–685.

Lykogianni, M., Bempelou, E., Karamaouna, F. and Aliferis, K. A. (2021). Do pesticides promote or hinder sustainability in agriculture? The challenge of sustainable use of pesticides in modern agriculture. *Sci. Total Environ.* 795, 148625. doi: 10.1016/j.scitotenv.2021.1486250048-9697/.

Markets and Markets. (2020). Report. Agricultural microbial market – Global forecast to 2025. Available at: www.marketsandmarkets.com.

Matsuzaki, Y., Watanabe, S., Harada, T. and Iwahashi, F. (2020). Pyridachlometyl has a novel anti-tubulin mode of action which could be useful in anti-resistance management. *Pest Manag. Sci.* 76(4), 1393–1401.

McDougall, P. (2016). The Cost of New Agrochemical Product Discovery, development and Registration in 1995, 2000, 2005–8 and 2010–2014. R&D expenditure in 2014 and expectations for 2019. Available at: https://croplife.org/wp-content/uploads/2016/04/Cost-of-CP-report-FINAL.pdf.

Money, N. P. (2007). Chapter 6. In: *Triumph of the Fungi: A Rotten History*. Oxford: Oxford University Press.

Narayanasamy, P. (2006). Chapter 1: Postharvest pathogens and disease management. In: *Wiley Interscience*. Hoboken, NJ: John Wiley and Sons Inc., pp. 4–5.

National Academy of Sciences. (2021). *The Challenge of Feeding the World Sustainably: Summary of the US-UK Scientific Forum on Sustainable Agriculture*. Washington, DC: The National Academies Press. doi: 10.17226/26007.

Nauen, R., Leadbeater, A. and Thompson, A. (2008). Proposal on the revision of EU Directive 91/414. The impact on Resistance Management and Sustainable Crop Production. *Outlooks Pest Manag.* 19(4), 150–151. doi: 10.1564/19aug03.

Oercke, E.-C. (2006). Crop losses to pests. *J. Agric. Sci.* 144(1), 31–43.

Ons, L., Bylemans, D., Thevissen, K. and Cammue, B. P. A. (2020). Combining biocontrol agents with chemical fungicides for integrated plant fungal disease control. *Microorganisms* 8(12), 1930. doi: 10.3390/microorganisms8121930.

Owen, W. J., Yao, C., Myung, K., Kemmitt, G., Leader, A., Meyer, K. G., Bowling, A. J., Slanec, T. and Kramer, V. J. (2017). Biological characterization of fenpicoxamid, a new fungicide with utility in cereals and other crops. *Pest Manag. Sci.* 73(10), 2005–2016. doi: 10.1002/ps.4588.

Prevost, B. (1807). Memoire sur la cause immediate de la carie ou charbon des bles, et de plusieurs autres maladies des plantes, et sur les preservatifs de la carie. Paris: Phytopatholigcal Classics No 6. American Phyto-Pathological Society, p. 1939.

Rabiey, M., Hailey, L. E., Roy, S. R., Grenz, K., Al-Zadjali, M. A. S., Barrett, G. A. and Jackson, R. W. (2019). Endophytes vs tree pathogens and pests: Can they be used as biological control agents to improve tree health? *Eur. J. Plant Pathol.* 155(3), 711–729.

Ray, D. K., Mueller, N. D., West, P. C. and Foley, J. A. (2013). Yield trends are insufficient to double global crop production by 2050. *PLoS ONE* 8(6), e66428.

Rehfus, A., Katamadze, A., Strobel, D., Bryson, R. and Stammler, G. (2020). Evolution of SDHI adaptation in cereal pathogens in Europe. In: Deising Fraaije, H. B., Mehl, A., Sierotzki, O. E. C. and Stammler, G. (Eds.), *Modern Fungicides and Antifungal Compounds IX. Proceedings of the 19th International Reinhardsbrunn Symposium*.

Ritchie, F., Bain, R. A., Lees, A. K., Boor, T. R. W. and Paveley, N. D. (2018). Integrated control of potato late blight: Predicting the combined efficacy of host resistance and fungicides. *Plant Pathol.* 67(8), 1784–1791.

Robertson, J. (1824). On the mildew and some diseases incident to fruit trees. *Trans. Lond. Hortic. Soc.* 5, 175.

Russell, P. E. (2005). A century of fungicide evolution. *J. Agric. Sci.* 143(1), 11–25.

Savary, S., Willocquet, L., Pethybridge, S. J., Esker, P., McRoberts, N. and Nelson, A. (2019). The Global burden of pathogens and pests on major food crops. *Nat. Ecol. Evol.* 3(3), 430–439.

Schepers, H. T. A. M., Kessel, G. J. T., Lucca, F., Förch, M. G., van den Bosch, G. B. M., Topper, C. G. and Evenhuis, A. (2018). Reduced efficacy of fluazinam against *Phytophthora infestans* in the Netherlands. *Eur. J. Plant Pathol.* 151(4), 947–960.

Sierotzki, H. and Scallet, G. (2013). A review of current knowledge of resistance aspects for the next generation of succinate dehydrogenase inhibitor fungicides. *Phytopathology* 103(9), 880–887.

Spraker, J. E., Wiemann, P., Bacciie, J. A., Venkatesh, N., Schumacher, J., Schroeder, F. C., Sanchez, L. M. and Keller, N. P. (2018). Conserved responses in a war of small molecules between a plant-pathogenic bacterium and fungi. *mBio* 9(3), e00820-18. doi: 10.1128/mBio.00820-18.

Sparks, T. C., Hunter, J. E., Lorsbach, B. A., Hanger, G., Gast, R. E., Kemmitt, G. and Bryant, R. J. (2018). Crop protection discovery: Is being the first best? *J. Agric. Food Chem.* 66(40), 10337–10346.

Stevenson, K. L., McGrath, M. T. and Wyenandt, C. A. (2019). *Fungicide Resistance in North America* (2nd edn.). Saint Paul: APS Press.

Taxvig, C., Hass, U., Axelstad, M., Dalgaard, M., Boberg, J., Andeasen, H. R. and Vinggaard, A. M. (2007). Endocrine-Disrupting activities *in vivo* of the fungicides tebuconazole and epoxiconazole. *Toxicol. Sci.* 100(2), 464–473. doi: 10.1093/toxsci/kfm227.

Tilman, D., Balzer, C., Hill, J. and Befort, B. L. (2011). Global Food demand and the sustainable intensification of agriculture. *Proc. Natl Acad. Sci. U. S. A.* 108(50), 20260–20264. doi: 10.1073/pnas.1116437108.

Tsolakidou, M. D., Stringlis, I. A., Fanega-Sleziak, N., Papageorgiou, S., Tsalakou, A. and Pantelides, I. S. (2019). Rhizosphere-enriched microbes as a pool to design synthetic communities for reproducible beneficial outputs. *FEMS Microbiol. Ecol.* 95(10), fiz138.

Umetsu, N. and Shirai, Y. (2020). Development of novel pesticides in the 21st century. *J. Pestic. Sci.* 45(2), 54–74.

United Nations. (2015). Transforming our world: The 2030 agenda for sustainable development. Available at: https://www.un.org/ga/search/view_doc.asp?symbol=A/RES/70/1&Lang=E.

United Nations. (2019). Department of economic and social affairs: "World population prospects 2019." Available at: https://population.un.org/wpp/.

van den Bosch, F., Oliver, R., van den Berg, F. and Paveley, N. (2014). Governing principles can guide fungicide-resistance management tactics. *Annu. Rev. Phytopathol.* 52, 175–195.

Verweij, P. E., Chowdhary, A., Melchers, W. J. and Meis, J. F. (2016). Azole resistance in *Aspergillus fumigatus*: Can we retain the clinical use of mold-active antifungal azoles? *Clin. Infect. Dis.* 62(3), 362–368. doi: 10.1093/cid/civ885.

Verweij, P. E., Lucas, J. A., Arendrup, M. C., Bowyer, P., Brinkmann, A. J. F., Denning, D. W., Dyer, P. S., Fisher, M. C., Geenen, P. L., Gisi, U., Hermann, D., Hoogendijk, A., Kiers, E., Lagrou, K., Melchers, W. J. G., Rhodes, J., Rietveld, A. G., Schoustra, S. E., Stenzel, K., Zwaan, B. J. and Fraaije, B. A. (2020). The One Health Problem of Azole resistance in *Aspergillus fumigatus*: Current insights and future research agenda. *Fungal Biol. Rev.* 34(4), 202–214. doi: 10.1016/j.fbr.2020.10.003.

Watanabe, S., Urihara, I., Fujii, T., et al. (2019). *2019 Ghent Crop Protection Chemistry*, 3.26, Abstr. IUPAC.

Weber, R. W. S. (2011). Resistance of *Botrytis cinerea* to multiple fungicides in northern German small fruit production. *Plant Dis.* 95(10), 1263–1269.

Williams, J. S. and Cooper, R. M. (2004). The oldest fungicide and newest phytoalexin - A reappraisal of the fungitoxicity of elemental sulphur. *Plant Pathol.* 53(3), 263–279.

Winter, C., Fehr, M., Craig, I. R., Grammenos, W., Wiebe, C., Terteryan-Seiser, V., Rudolf, G., Mentzel, T. and Quintero Palomar, M. A. (2020). Trifluoromethyloxadiazoles: Inhibitors of histone deacetylases for control of Asian soybean rust. *Pest Manag. Sci.* 76(10), 3357–3368. doi: 10.1002/ps.5874.

Yamashita, M. and Fraaije, B. (2018). Non target site SDHI resistance is present as standing genetic variation in field populations of *Zymoseptoria tritici*. *Pest Manag. Sci.* 74(3), 672–681. doi: 10.1002/ps.4761.

Yang, C. (2020). Remote sensing and precision agriculture technologies for crop disease detection and management with a practical application example. *Engineering* 6(5), 528–532.

Part 2

Case studies: resistance in key groups of fungicides

Chapter 10

Understanding resistance to sterol biosynthesis inhibitor fungicides

Andreas Mehl, Bayer AG, Crop Science Division, Germany

1 Introduction

Sterols like ergosterol, ergosta-5,7-dien-3β-ol, and ergosta-5,24(24^1)-dien-3β-ol are essential components of fungal cell membranes regulating their permeability and stability. Consequently, inhibitors of key enzymes involved in fungal sterol biosynthesis have become very effective disease management tools (Loeffler et al., 1992; Shirane et al., 1996; Chapuis et al., 1996; Ziogas and Malandrakis 2015; Ishii et al., 2021).

Products applied for plant disease control containing fungicides that interfere with sterol biosynthesis, known as sterol biosynthesis inhibitors (SBIs), account today for around 40% of the global fungicide market. This makes SBI fungicides the most important fungicide class being used worldwide in a wide range of applications to control a variety of pathogens in many crops. According to the Fungicide Resistance Action Committee (FRAC), SBIs can be divided into four subclasses:

1. Demethylation inhibitors (DMIs),
2. Amines (formerly known as 'morpholines'),
3. Keto-reductase inhibitors (KRIs), and
4. Squalene-epoxidase inhibitors.

http://dx.doi.org/10.19103/AS.2022.0116.10

The latter are used only for pharmacological purposes (Table 1). This classification is based on the cross-resistance behaviour and common mode of action, rather than on the chemical structures of the active ingredients. According to the latest version of the FRAC Code List, revised in April 2022, 37 agricultural fungicides inhibit the C14 demethylase enzyme in the sterol biosynthesis pathway. These fungicides are therefore referred to as DMIs and are the most important group within the SBI fungicides (Fig. 1; Anonymous, 2022a).

SBI fungicides exhibit a broad spectrum of activity against a number of economically important pathogens belonging to the so-called true fungi (i.e. ascomycetes and basidiomycetes) in various crops. Examples of economically important ascomycetes affecting cereals include *Blumeria graminis* f. sp. *tritici*, *B. graminis* f. sp. *hordei*, *Zymoseptoria tritici*, *Pyrenophora* spp., *Ramularia collo-cygni* and *Rhynchosporium secalis*, while *Alternaria solani*, *Botrytis cinerea*, *Cercospora beticola*, *Erysiphe necator*, *Plenodomus lingam*, *Podosphaera leucotricha*, *Sclerotinia sclerotiorum*, *Venturia inaequalis* are examples of important pathogens on industrial and other crops. The most important basidiomycetes are rust fungi such as *Puccinia* spp. or *Phakopsora pachyrhizi*, and smut fungi such as *Ustilago* sp. Tebuconazole, prothioconazole and metconazole play important roles in controlling *Fusarium* species such as *Fusarium culmorum* and *F. graminearum*, which produce deoxynivalenol (DON) and other mycotoxins (Kang et al., 2001; Pirgozliev et al., 2002; Paul et al., 2008). Members of the oomycetes, e.g. species of the order Peronosporales (e.g. *Phytophthora* spp., *Plasmopara* spp.), lack full *de novo* sterol biosynthesis and thus are not sensitive to SBI fungicides (Warner et al., 1982; Mercer, 1984).

In terms of physicochemical properties, SBI fungicides are not phloem-mobile but exhibit loco-systemic activity (Büchel, 1986; Shephard, 1985). The

Table 1 Grouping of SBI fungicides in the FRAC classification (www.frac.info)

SBI class	I	II	III	IV
FRAC Codes	G1/Code 3	G2/Code 5	G3/Code 17	G4/Code 18
Group name	Demethylation inhibitors (DMIs)	Amines (formerly 'morpholines')	Keto-reductase inhibitors (KRIs)	Squalene-epoxidase inhibitors
Target in sterol biosynthesis	Sterol C14 demethylase (*erg11/cyp51*)	Δ^{14} Reductase and $\Delta^7 \rightarrow \Delta^8$ -isomerase (*erg24, erg2*)	3-Keto reductase, C4-demethylation (*erg27*)	Squalene-epoxidase (*erg1*)
Chemistry	Piperazines Pyridines Pyrimidines Imidazoles Triazoles	Morpholines Piperidines Spiroketalamines	Hydroxyanilides Amino-pyrazolinones	Thiocarbamates Allylamines

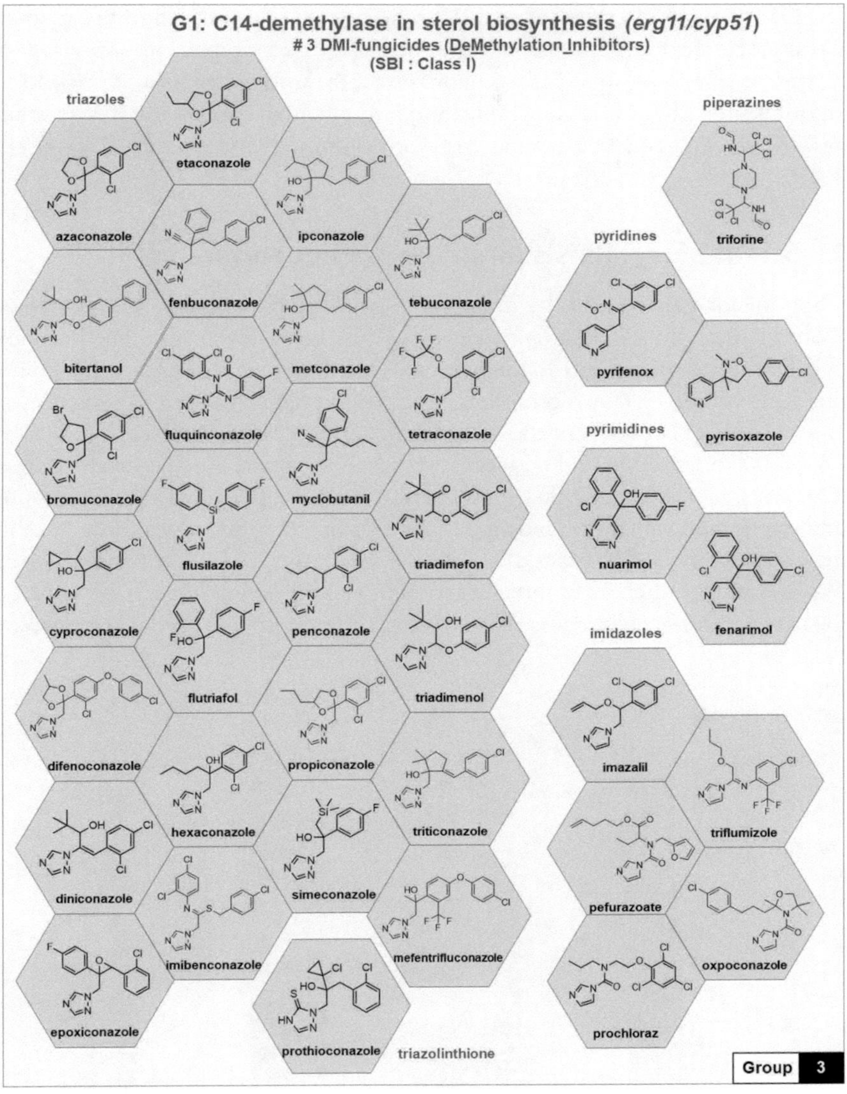

Figure 1 FRAC classification of DMI fungicides (FRAC Poster 2022, www.frac.info).

long-lasting protective effect of the most successful active ingredients is due to a high degree of systemicity. SBI fungicides can provide preventive, curative and eradicative activity, making them suitable for use as seed, foliar and/or ear treatment agents. In addition, positive side effects of SBI fungicides on plant health and growth have been noted (Kuck and Scheinpflug, 1986; Burden et al., 1989; Kuck et al., 2012).

Considering the number of SBI fungicides that have reached the market stage since their introduction in the 1960s, the economic importance and competitiveness of this class are proven. SBI fungicides have achieved a unique status due to the extremely long and effective life span of this specific mode of action, which is not the case for other fungicide classes (Kuck et al., 2012).

2 Sterol biosynthesis inhibitor market and trends

SBI fungicides are one of the most important fungicide classes with a market share in 2020 of around 40% considering their value on distributers' level for 'products used on ground' in the foliar and seed treatment segment, including the value of non-SBI mixing partners of products within ready-mixtures (Bayer internal market assessment based on panel data in 2022). With a total value of €7505 million in 2020, the economic importance of SBI fungicides is obvious (Fig. 2). Within DMI fungicides, triazoles dominate both in terms of market share and the number of compounds that have been continuously introduced to the market since the 1980s (Kuck et al., 2012; Mehl et al., 2019). Other members of the DMI group, such as piperazines, pyridines and pyrimidines, have not had the same commercial success as the triazoles. Therefore, the success of triazoles

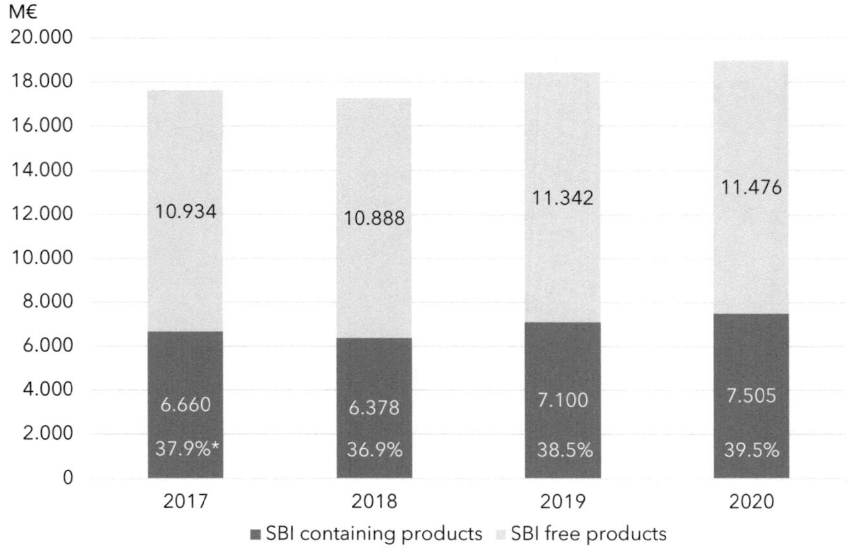

Figure 2 Development of global market share (in M€) of the SBI classes in agriculture according to their value on distributers' level for 'products used on ground' in the foliar and seed treatment segment, including the value of non-SBI partners of products within ready-mixtures (Bayer internal market assessment based on panel data in 2022). *SBI share of the total fungicide market.

is largely responsible for the success of SBI fungicides. One explanation for the steady increase in triazole sales over the years, despite the introduction of other broad-spectrum fungicides such as QoIs or SDHIs, could be the introduction of new mixtures and formulations containing triazoles. Triazoles are often the backbone of fungicide products.

In contrast to the year 2009, when Europe represented by far the most important regional market for SBIs and accounting for 53% of the total SBI market (Mehl et al., 2019), in 2020 region EMEA, covering Europe, the Middle East and Africa, had about the same share of SBI containing products as region Latin America, with 45.4% and 45.2%, respectively (Fig. 3). Considering that the most important crops for SBI fungicides are cereals and soybean, where DMI applications to control *P. pachyrhizi* have increased to a significant degree, particularly in Latin America, the reason for this regional distribution is obvious (Fig. 4).

Due to the expected continued massive use of DMI fungicides, selection pressure for sensitivity shifts will continue to increase, especially in crops and regions where the majority of key target pathogen populations have strong disruptive resistance to other important fungicide groups, such as the QoIs. Besides their broad-spectrum activity, the 'shifting' characteristic of DMI resistance development may have the strongest influence on the continued success of DMIs.

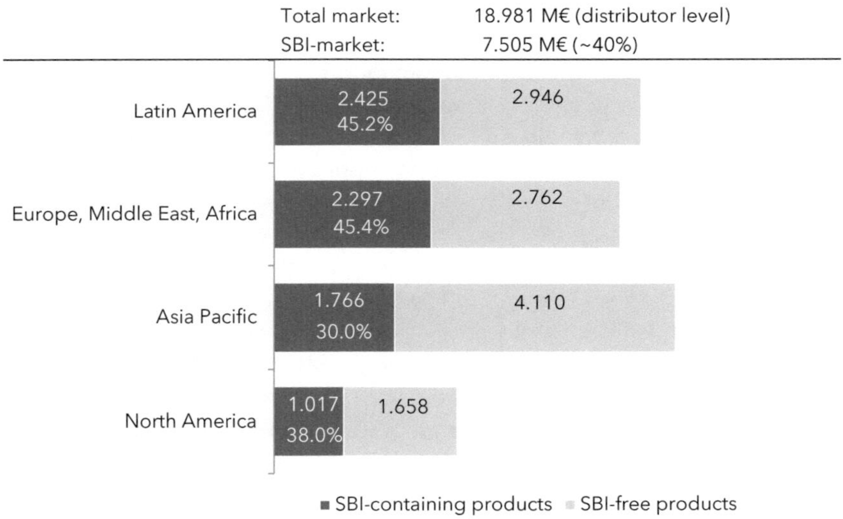

Figure 3 World fungicide/SBI market in agriculture by value and region in 2020 (in M€) according to their value on distributors' level for 'products used on ground' in the foliar and seed treatment segment, including the value of non-SBI partners of products within ready-mixtures (Bayer internal market assessment based on panel data in 2022).

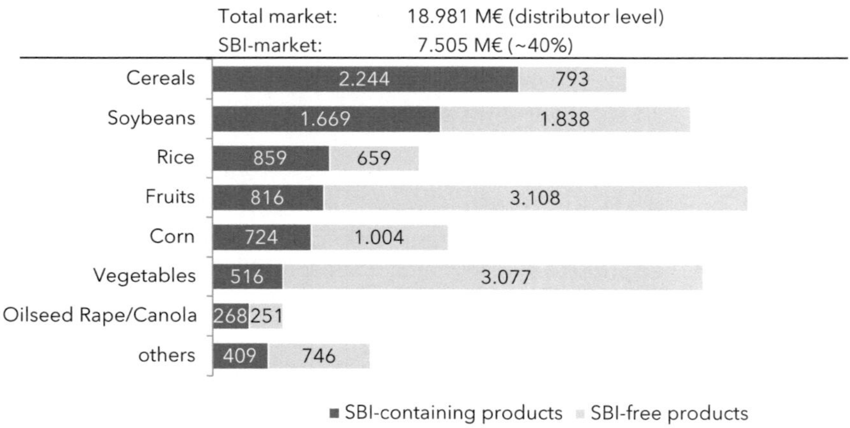

Figure 4 World fungicide/SBI market in agriculture by value and crop in 2020 (in M€) according to their value on distributors' level for 'products used on ground' in the foliar and seed treatment segment, including the value of non-SBI partners of products within ready-mixtures (Bayer internal market assessment based on panel data in 2022).

3 Short history of demethylation inhibitor fungicides

Until the 1960s, most fungicides offered only a limited spectrum of activity or protective action and showed no systemic effect. As a result, crop protection in important crop segments such as cereals was often inadequate. In the search for new solutions, imidazoles were discovered as pharmaceutical agents against fungal infections on human skin and were the first DMIs to be marketed (Büchel, 1986). This was the starting point for the development of antimycotics and fungicides. Since several comprehensive reviews of DMI fungicides are already available for the first generations of DMI fungicides (Kuck and Scheinpflug 1986; Kuck et al. 1995), only a brief overview of some important compounds is given here.

The breakthrough and the beginning of the DMI success story in agriculture was the introduction of triadimefon in 1976, the first systemic triazole (Kuck and Scheinpflug, 1986; Tomlin, 2011) which controlled at the time of market introduction very efficiently powdery mildew and rust, as well as various leaf spot diseases. Since then, nearly 50 SBI fungicides have been marketed, 37 of which belong to the DMI class. In the early 1980s, bitertanol and triadimenol were introduced to the market. Bitertanol was marketed as a foliar fungicide against *V. inaequalis* and *Pseudocercospora fijiensis* and simultaneously as a seed treatment fungicide against *Microdochium nivale* and *Tilletia* spp. Triadimenol was initially introduced exclusively as a systemic seed treatment fungicide in cereals but was later also developed for use as a foliar fungicide on monocotyledons and dicotyledons. The next new product was propiconazole, which had a broader spectrum of activity and provided exceptional leaf spot control in cereals and bananas. Not all DMIs were broad-spectrum products. Flusilazole became one of the leading fungicides in barley because of its particular strength in barley disease control. Tebuconazole, introduced in 1988, became a very successful fungicide worldwide because of its very broad spectrum of disease control. It was initially introduced in cereals but was later developed as a foliar fungicide for a wide range of crops including bananas, grapevines, oilseed rape, soybeans, pome and stone fruits. It was also marketed as a seed treatment fungicide in cereals against various smut and bunt diseases (Tomlin, 2011). A few years later, epoxiconazole was marketed as a broad-spectrum fungicide effective on a range of crops including cereals, coffee, rice, sugar beet and tea. Due to its long duration of action and pronounced control of leaf spot diseases and rusts in cereals, epoxiconazole gained an important role in the cereal fungicide market, both as a single treatment and as a combination product with mixing partners such as QoI fungicides. Another compound effective against a broad spectrum of pathogens is fluquinconazole, which was introduced a year after epoxiconazole in 1993. Fluquinconazole can be used as a foliar spray against diseases in cereals, soybeans, pome fruits or grapevines or as a seed treatment in cereals. Prothioconazole, being introduced in 2004, exhibits efficacy against a broad spectrum of diseases as well as pronounced strength against cereal, soybean and oilseed rape diseases and has therefore rapidly gained a prominent role in the global fungicide market. It occupies a special position among DMIs, not only because of its toxophore moiety, which differs from other DMI fungicides by an additional 'thione' at the third nitrogen atom but also, for example, because of the lack of cross-resistance to triazole- or prochloraz-resistant isolates of *Oculimacula acuformis* (Mauler-Machnik et al., 2002; Jautelat et al., 2004; Kuck and Mehl, 2004). In addition to foliar applications, cereal seed dressings containing prothioconazole are also marketed. In 2020, mefentrifluconazole was introduced as the latest innovation in SBI class I, representing a highly active ingredient from the new sub-class of isopropanol-triazole fungicides. It is registered for use on a variety of crops including maize, soybeans, potatoes, sugar beets, grapes, pome and stone fruit, oilseed rape, various cereal grains and legume vegetables. It is also approved for seed treatment of selected crops like wheat, soybean or maize and for use on non-residential turf and ornamentals.

4 Resistance risk and general resistance characteristics of sterol biosynthesis inhibitor fungicides

DMI fungicides are particularly effective against plant pathogens from genera of the Ascomycetes and Basidiomycetes. Although their strong intrinsic activity and frequent worldwide use lead to considerable selection pressure on fungal populations, the resistance risk of DMI fungicides (SBI class I) is generally classified as medium. For amine fungicides (SBI class II) and KRI fungicides (SBI class III), the resistance risk is even considered low to medium (Brent and Hollomon, 2007b, Anonymous, 2022a).

Since resistance to DMIs was first observed in the early 1980s, many reports of resistance have been published, but cases of complete loss of efficacy in the field are relatively rare due to an adapted resistance management strategy and the continued introduction of intrinsically more active compounds (Kuck et al., 2012). DMI fungicides typically exhibit a 'shifting-type' resistance characteristic resulting from a continuous stepwise polygenetic selection process in which multiple mutations must accumulate to reduce sensitivity. In contrast, the benzimidazoles or QoI fungicides exhibit a 'disruptive' resistance characteristic, as a single specific point mutation in the target gene is sufficient for complete resistance (Brent and Hollomon, 2007a; Sierotzki and Stammler, 2019; Olaya and Geddens, 2019). DMI resistance is thus characterised by a gradual loss of efficacy under field conditions. This comparably slow increase in resistance is caused by high selection pressure and can be provoked by low dose rates (Anonymous, 2022b), particularly if the frequency of applications is increased in order to compensate for any reduced performance from a lowered dose (Brent and Hollomon, 2007b). However, when selection pressure is reduced, a partial shift back to more sensitive pathogen populations has been observed in several cases (Kuck, 2002; Brent and Hollomon, 2007a). This partial reversion can be attributed to a fitness disadvantage associated with the mode of resistance.

In addition to this 'shifting' process, differences in sampling methods and bioassay variations cause sensitivity parameters to vary over time. Therefore, only stable trends that consistently leave the typical range can be considered sensitivities that have 'shifted'. The evolution of resistance is primarily driven by mutations in the DMI target-encoding *cyp*51 gene, but other mechanisms such as enhanced active efflux causing multidrug resistance and overexpression of the target-encoding *cyp*51 gene have also been considered (Cools et al., 2005b, 2010; Leroux et al., 2007; Leroux and Walker, 2011; Wei et al., 2020; Li et al., 2020; Zhang et al., 2020a; Zhang et al., 2021).

The observed changes in sensitivity are highly dependent on the frequency of fungicide applications and variation in selection pressure, as well as on the fungal species. In various horticultural and industrial crops resistance to DMI fungicides has been observed in numerous pathogens such as *C. beticola*, *E. necator*, *Monilinia fructicola*, *P. fijiensis*, or *V. inaequalis*. In cereals, for example, due to the long-term and intensive use of DMI fungicides, a large number of papers have been published in the last three decades showing resistance to individual active ingredients in pathogens such as *B. graminis* f.sp. *tritici*, *B. graminis* f.sp. *hordei*, *Oculimacula* spp., *Z. tritici*, *Pyrenophora teres*, *P. tritici-repentis*, *Puccinia striiformis* f.sp. *tritici*., *R. secalis* or *Fusarium* and *Microdochium* species (e.g. Blake et al., 2018; Cook et al., 2021; Cooke et al.,

2004; Cools et al., 2005a; Cristani and Gambogi, 1993; Ellwood et al., 2019; Garnault et al., 2019; Heick et al., 2020; Hellin et al., 2020; Jørgensen et al., 2021; Leroux and Marchegay, 1991; Leroux and Walker, 2011; Mavroedi and Shaw, 2005; Metcalfe et al., 2000; Sheridan et al., 1985). Because several of these publications described DMI resistance in laboratory mutants or focused on only a few strains within local trials, it is difficult to draw firm conclusions about the overall sensitivity status of individual pathogens on a continental, statewide or even regional scale based on these reports.

Due to their broad activity and strong efficacy towards many plant pathogens, DMI-containing products are used extensively for the control of agricultural fungal pathogens. When the extensive use of these compounds is combined with their single-site mode of action, the risk of sensitivity shifting is considerable and their use should be strictly regulated by resistance management guidelines.

'Shifting-type' resistance makes meaningful and effective resistance management somewhat easier. The development of new active ingredients with higher intrinsic activity can often compensate for the gradual loss of efficacy (Kuck, 2002). Classes I, II and III SBI fungicides generally exhibit cross-resistance. However, complete cross-resistance between individual members of each class has remained relatively rare, and efficacy has usually been maintained at a high level for decades despite the extensive use of these SBI classes. Although there have been a number of reports of sensitivity shifts, DMIs continue to be one of the most important fungicide classes due to strict resistance management guidelines and their exceptionally broad spectrum of disease control, representing a good example of the importance and success of sound resistance management strategies. Therefore, it is critical to continue to follow these resistance management guidelines to further maintain the efficacy of this important fungicide class in the future.

5 Mode of action and mechanisms of resistance

Studies on the mode of action of triarimol (Ragsdale and Sisler, 1973) showed for the first time that fungal sterol biosynthesis in *Ustilago maydis* is the target of this pyrimidine compound, which is chemically closely related to the commercial product fenarimol. Later, Ragsdale (1975) published that C14 demethylase is the major target site affected in sterol biosynthesis by triarimol. Currently, C14-demethylase (CYP51) is known to be the common target of over 35 agro-fungicides belonging to different chemical classes grouped under the name of demethylation inhibitors.

There are several detailed reviews of the ergosterol biosynthetic pathway of fungi and of fungicides that interfere with this pathway (Ragsdale, 1975; Mercer, 1984; Berg and Plempel, 1988; Burden et al., 1989). Therefore, only a brief overview is given here. Eleven enzymes are involved in the major steps of ergosterol biosynthesis, converting squalene to ergosterol (Fig. 5). Sterol C14 demethylase, encoded by *cyp*51 (also known as *erg*11 or lanosterol 14α-demethylase), is the target of all DMIs. It is localised in the microsomal

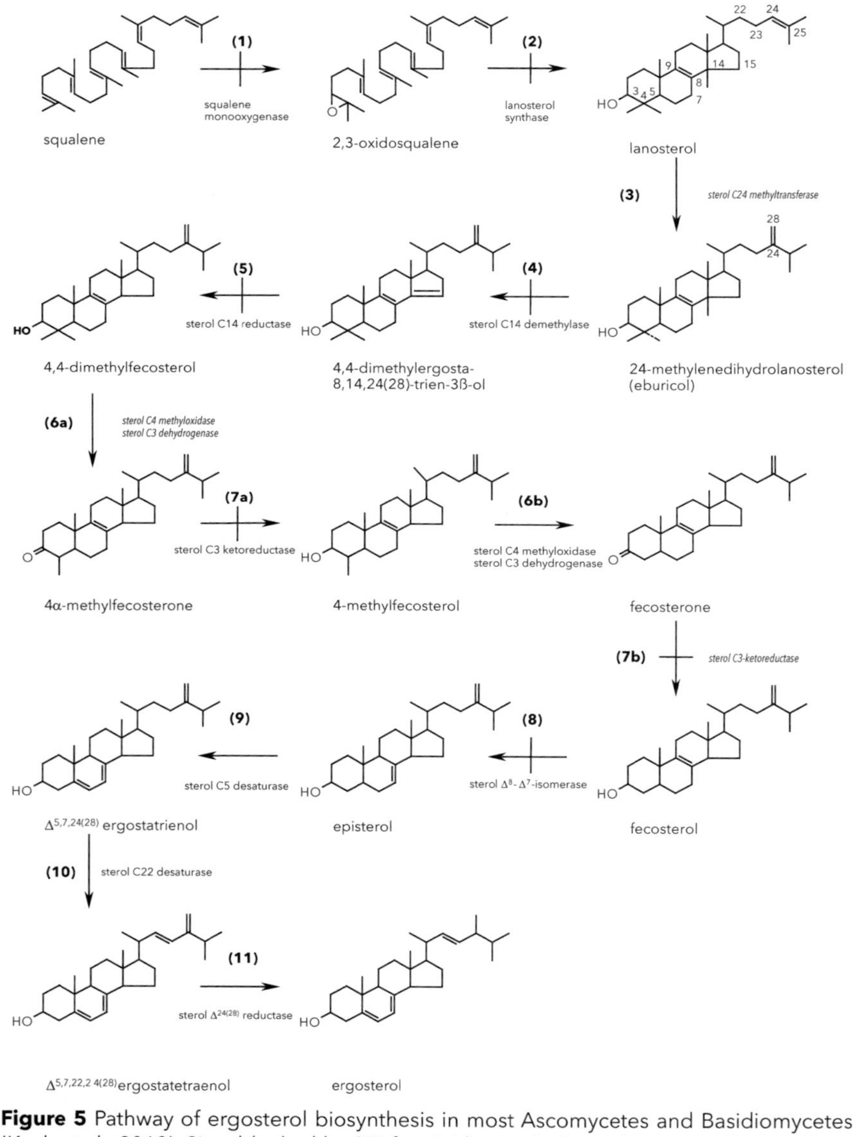

Figure 5 Pathway of ergosterol biosynthesis in most Ascomycetes and Basidiomycetes (Kuck et al., 2012). Sites blocked by SBI fungicides are indicated by crossed-out arrows. Further information is given in Table 2 and in the text.

Table 2 Enzymes involved in fungal sterol biosynthesis and targets of agricultural fungicides

Step	Enzyme	Gene	Other enzyme designations	Agricultural inhibitors
1	Squalene monooxygenase	*Erg1*	Squalene-epoxidase Oxidosqualene synthase	Target of G4 inhibitors like allylamines; side target of some amines (G2)
2	Lanosterol synthase	*Erg 7*	Oxidosqualene cyclase	Side target of some amines (G2)
3	Sterol C24 methyltransferase	*Erg6*	Sterol methyltransferase	
4	Sterol C14 demethylase	*Erg11*	*CYP51* Lanosterol 1α demethylase	Target of the DMI fungicides (G1)
5	Sterol C14 reductase	*Erg24*	Sterol Δ^{14} reductase	Main target of fenpropidin and spiroxamine (G2)
6	Sterol C4 methyloxidase	*Erg25*	-	
	Sterol C3 dehydrogenase	*Erg26*	Sterol C4 decarboxylase	
7	Sterol C3 ketoreductase	*Erg27*	-	Target of the KRI fungicides (G3)
8	Sterol Δ^8-Δ^7 isomerase	*Erg2*	Sterol C-8 isomerase	Main target of tridemorph (G2)/ secondary target of other amines (G2)
9	Sterol C5 desaturase	*Erg3*	C5 dehydrogenase	
10	Sterol C22 desaturase	*Erg5*	Ergosterol Δ22 desaturase	
11	Sterol $\Delta^{24(28)}$ reductase	*Erg4*	24-Methylene sterol (24(28))-reductase	

Step numbers are those shown in Figure 5 (Kuck et al., 2012).

membrane and catalyses the demethylation of eburicol, which is essential for the synthesis and orientation of sterols, near the beginning of the pathway. The primary biochemical consequences of impaired ergosterol biosynthesis have often been a matter of debate. Sisler and Ragsdale (1977) noted that a 'lack of ergosterol impedes the synthesis of new membranes and leads to deterioration of existing membranes'. In conjunction with these changes, fatty acid synthesis was observed to continue at a relatively high rate, resulting in a mismatch between fatty acid synthesis and utilisation for phospholipids. Further detailed studies have shown a variety of consequences of ergosterol depletion on the one hand and the accumulation of toxic precursors on the other (Mercer, 1984).

Marked plant growth-regulating effects have also been observed with DMI fungicides since their introduction to the market. While non-specific signs of phytotoxic activity such as necrosis or leaf drop are only occasionally reported with DMI fungicides, pronounced growth-regulating side effects have been noted on plants (Buchenauer and Grossmann, 1977; Buchenauer and Röhner, 1981). Typically, shorter shoots and internodes (referred to as 'stunting'), smaller dark green leaves and improved stress tolerance of DMI-treated plants have been reported. Accordingly, specific plant growth regulators (PGRs) such as paclobutrazol have been developed from the group of triazoles (Lürssen, 1988).

The positive side effects of fungicidal triazoles have been intensively studied and documented (Buchenauer, 1995) and are an important component of the biological profile of DMI fungicides in various crops, including oilseed rape (prevention of lodging) and cereal seed treatments (increased frost tolerance). Studies on the biochemical mechanisms causing these symptoms in plants have shown that inhibition of the biosynthesis of the plant hormone gibberellin and plant sterols appear to be the most important targets in plants (Sisler et al., 1984; Burden et al., 1989).

DMI fungicides show their strongest effect during the early growth stages and mycelial growth of the fungus. At this stage of development, a constant supply of sterols is required for membrane biosynthesis. The ergosterol reserves in the spore are only sufficient for spore germination and early germ tube growth. As a result of DMI treatment, malformations can be observed in the intracellular membrane system and cell wall of fungi. These effects typically include abnormally swollen and branched hyphae or broken hyphal cell walls (Ragsdale and Sisler, 1973; Burden et al., 1989). Consequently, further growth and reproduction are reduced.

Spectrophotometric tests revealed differences in the way different DMIs bind (Parker et al., 2011). Due to the DMI toxophore consisting of a ring nitrogen bound to the iron atom in the porphyrin of sterol demethylase, typical triazoles (such as triadimenol, tebuconazole, epoxiconazole) produce strong type II difference spectra. As already mentioned, prothioconazole differs slightly in its chemical structure from triazoles, which leads to an altered interaction spectrum.

Three main mechanisms have been identified that lead to gradual development of resistance to DMI fungicides. These mechanisms, which can act simultaneously but are not interrelated, include (Brent and Hollomon, 2007a; Cools and Fraaije, 2008; Leroux and Walker, 2011; Stammler and Semar, 2011; Cools and Fraaije, 2013; Schmitz et al., 2014):

1 increased efflux due to overexpression of efflux pumps,
2 overexpression of the target site (*cyp*51) and, most importantly,
3 target-site alterations.

While resistance to QoI fungicides is mainly due to one of three specific point mutations in the *cytochrome b* gene (Sierotzki and Stammler, 2019), a partly large and variable number of mutations in the *cyp*51 gene lead to different changes in DMI sensitivity.

Several mutations have been described in *Z. tritici* (e.g. L50S, D134G, V136A, Y137F, A379G, I381V, N513K, S524T, etc.), most of which occur in different combinations and lead to different phenotypes (Leroux and Walker, 2011; Cools et al., 2011; Cools and Fraaije, 2013; Huf et al., 2018). *In vitro*, these phenotypes show large variations in sensitivity to different DMI fungicides with resistance factors between 1 and 800 (Leroux and Walker, 2011). Therefore, molecular techniques are not sufficient for shift analyses. Bioassays that survey the EC_{50} (effective concentration at which 50% of fungal growth is inhibited) are imperative to understand the effects of specific mutations and mutation combinations. In the case of *Z. tritici*, many of the *cyp*51 changes discovered before the turn of the millennium, such as L50S, D107V, M145L, N178S, S188N and N513K, are not directly related to sensitivity shifts (Cools and Fraaije, 2013). Most mutations relevant to resistance (ΔY459/G460, Y461S/H, I381V, V136A, A379G and Y459D) were found between 1999 and 2003, while those discovered more recently are still comparatively rare. Although more than 30 *cyp*51-mutations have been already detected a decade ago, only about 11 of them were reported at this time to be involved in reduced DMI susceptibility (Leroux and Walker, 2011).

In addition to bioassays and sequence analyses, complementation studies in which the *Z. tritici cyp*51 is expressed in *Saccharomyces cerevisiae* or molecular modelling help to clarify the effects of mutations on azole sensitivity and the intrinsic function of the protein (Cools et al., 2010; Cools et al., 2011; Mullins et al., 2011). These techniques can be used to analyse the effects of a particular mutation. For example, although I381V in combination with ΔY459/G460 significantly reduces DMI sensitivity, the I381V substitution by itself affects the intrinsic function of the protein, possibly by reducing the volume of the protein's binding pocket (Cools et al., 2010; Mullins et al., 2011). However, incorporation of the ΔY459/G460 deletion leads to an increase in cavity volume, compensating for the reduced cavity volume of I381V when expressed in combination (Mullins et al., 2011). Based on modelling studies, it can be hypothesised whether an alteration affects DMI binding and, if so, whether it affects the size and accessibility of the binding pocket or the localisation of residues that interact with DMIs (Mullins et al., 2011).

Overall, significant differences in sensitivity ranges can be observed between major DMI compounds depending on mutation and phenotype (Leroux and Walker, 2011). Since DMIs exhibit in general cross-resistance, these differences in resistance factors can be explained by the intrinsic properties of the fungicide and by the different impact of the various mutation pattern in

Septoria leaf blotch isolates towards the various DMI fungicides. More details on this subject and on DMI resistance research with *Z. tritici* are given below.

DMI resistance has also been analysed in a number of laboratory mutants. This information adds to the knowledge of sensitivity to pathogens and gives an indication of likely further phenotypes occurring in the field, but at the same time opens the debate on representation, as information on practical relevance is lacking. It appears that the advantage of reduced sensitivity to DMIs is often compensated with a fitness deficit (FRAC, www.frac.info; Karaoglanidis et al., 2001; Köller et al., 1991). Therefore, fitness parameters should generally be determined for all fungicide classes in order to assess the value of resistance-mediating mechanisms as well as their relevance in practice. For example, target-site mutations in *B. cinerea* field isolates have been associated with high levels of *in vitro* resistance to fenhexamid, an SBI class III KRI fungicide. However, these mutations were associated with fitness costs, suggesting that without selection pressure, the spread and persistence of these phenotypes are limited (Billard et al., 2012). Similar results have been found with other pathogen-fungicide class combinations (Karaoglanidis et al., 2001; Pereira et al., 2020a; Wang and Coley-Smith, 1986), while in other studies no clear fitness deficits have been associated with resistant phenotypes (Peever and Milgroom, 1994; Bardas et al., 2008). While DMI-resistant isolates of *M. fructicola* did not suffer from reduced fitness, instability of resistance was observed *in vitro*, which could lead to a return to sensitivity in the absence of DMI pressure in the field (Cox et al., 2007).

In contrast to the major resistance mechanisms in QoIs, which can be easily detected by molecular techniques and are directly related to sensitivity, DMI resistance mechanisms are far more complex. As mentioned above, specific molecular target gene analysis alone does not necessarily reflect the sensitivity status, as other mutations or mutation combinations may be predominant, or mechanisms such as target-site overexpression or efflux transporters may be additionally involved (see Section 6 for more details). Therefore, continuous and mostly annual surveillance programmes have been established by FRAC, www.frac.info, which have shown that annual or regional fluctuations are typical.

6 Examples of resistance

Resistance to DMIs in fungal pathogens affecting monocotyledons has mainly been described in Europe or Australia, while resistance in fungi affecting dicotyledons plays a greater role in the USA. On the other hand, susceptibility changes in wheat powdery mildew in the USA have also been recently investigated (Meyers et al., 2019), and DMI sensitivity changes of *Pseudocercospora fijiensis* are intensively studied in most banana growing regions outside of Europe since more than two decades. Overall, in order to develop effective strategies for resistance management, detailed information on the current sensitivity status of the most important target pathogens and knowledge of the resistance mechanisms relevant under practical conditions are required.

Around seven years after the introduction of DMI fungicides, the first decreases in DMI efficacy were noted, e.g. in *B. graminis* f. sp. *hordei*, *Neocosmospora haematococca*, *Podosphaera xanthii* and *V. inaequalis* (Brent and Hollomon, 2007a). Already in 1985 declining efficacy of DMI against apple scab was reported from an experimental orchard in the USA (Stanis and Jones, 1985) and in Michigan, the main growing area for processing plants, the first population shifts towards DMI resistance were reported in 1991 (Köller et al., 1997). Interestingly, the first sensitivity shift in Indiana, where apples are grown for retail or direct market, was detected much later in 2009 (Chapman et al., 2011).

This was followed by many reports and scientific papers describing reduced DMI sensitivity of different pathogens in various crops, mostly with field isolates but often also with laboratory mutants (Table 3).

DMI resistance of *Clarireedia homoeocarpa*, the causal agent of dollar spot in turfgrass, was first confirmed in 1995 (Golembiewski et al., 1995). Since then, a number of reports have described cross-resistance with DMI compounds and multiple resistance to other fungicide classes (Burpee, 1997; Miller et al., 2002; Jo et al., 2006; Ok et al., 2011). A shift towards lower propiconazole efficacy has been also reported for some *M. fructicola* field populations from Georgia and South Carolina (Zehr et al., 1999; Schnabel et al., 2004). Decreasing DMI sensitivity has been also observed for pathogens being of importance in the postharvest segment, e.g. with the most economically important pathogen in citrus, *Penicillium* spp., causing green mould. In California, poor control of *P. digitatum* was observed with imazalil (Eckert et al., 1994; Holmes and Eckert, 1995; Holmes and Eckert, 1999; Kinay et al., 2007). However, competition studies with imazalil-resistant and -sensitive strains on untreated lemon fruit revealed fitness disadvantages for resistant strains (Holmes and Eckert, 1995; Kinay et al., 2007). Later, slightly reduced postharvest control was observed in natural populations of *Geotrichum candidum* when treated with propiconazole (Yaghmour et al., 2012), and *P. digitatum* isolates with reduced sensitivity to the same DMI were recovered in the laboratory, indicating the likely risk of increasing resistance development and the importance of anti-resistance strategies (McKay et al., 2012).

However, on the global level, the performance of DMI in the field has been successfully maintained. This is partly because many reports represent local findings that may not be entirely relevant to other regions. For example, monitoring of *Z. tritici* for tebuconazole resistance in 2009 showed high diversity

Table 3 List of important plant pathogen species for which DMI sensitivity changes have been first reported, both in field isolates or in laboratory mutants

Pathogen species name	Disease name	Host	First year of report
Blumeria graminis f. sp. *hordei*	Powdery Mildew	Barley	1981
Blumeria graminis f. sp. *tritici*	Powdery Mildew	Wheat	1986
Blumeriella jaapii	Leaf Spot	Cherry	2006
Botrytis cinerea	Grey Mould	Various	1992
Cercospora beticola	Cercospora Leaf Spot	Sugar beet	1989
Cladosporium caryigenum	Scab	Pecan	1997
Clarireedia homoeocarpa	Dollar Spot	Turfgrass	1992
Colletotrichum gloeosporioides	Anthracnose	Mango	2003
Emericella nidulans	Aspergillus Mould	Various	1979
Erysiphe necator	Powdery Mildew	Grapes	1990
Fusarium asiaticum	Head Blight	Wheat	2009
Fusarium graminearum	Head Blight	Wheat	2009
Microdochium nivale	Snow Mold	Cereals	1993
Monilinia fructicola	Brown Rot	Stone Fruit	1989
Neocosmospora haematococca	Foot Rot	Cucurbits	1991
Oculimacula spp.	Eyespot	Wheat	1991
Passalora nattrassii	Leaf Spot	Eggplant	2000
Penicillium digitatum	Green Mold	Citrus	1987
Penicillium italicum	Blue Mold	Various	1982
Phakopsora pachyrhizi	Rust	Soybean	2014
Podosphaera mors-uvae	Powdery Mildew	Blackcurrant	1988
Podosphaera pannosa	Powdery Mildew	Nectarine	2001
Podosphaera xanthii	Powdery Mildew	Cucurbits	1983
Pseudocercospora fijiensis	Black Sigatoka	Banana	1992
Puccinia horiana	White Rust	Chrysanthemum	1992
Puccinia striiformis f. sp. *tritici*	Stripe Rust	Wheat	2000
Pyrenophora teres	Net Blotch	Barley	1985
Pyrenophora tritici-repentis	Tan Spot	Wheat	2005
Rhynchosporium secalis	Leaf Blotch, Scald	Barley	1986
Ustilago avenae	Loose Smut	Oats	1986
Ustilago maydis	Smut	Maize	1981
Venturia inaequalis	Scab	Apple	1985
Venturia nashicola	Japanese Pear Scab	Pear	1998
Zymoseptoria tritici	Septoria Leaf Spot	Wheat	2000

For references, see www.frac.info.

factors (EC_{50} of the least sensitive isolate divided by EC_{50} of the most sensitive isolate) in most European countries, with the exception of Ireland and Sweden. In 2011, Ireland had an adapted population, while the Swedish population remained sensitive (unpublished data Bayer). Molecular biological analyses of the 2009 isolates showed that a different genotype structure in Ireland was responsible for the reduced sensitivity. Since the same genotype prevailed in Sweden as in all other European countries, factors other than genotype must be involved in the differences in sensitivity.

6.1 Case study for demethylation inhibitor sensitivity development and resistance research: Septoria leaf blotch of wheat caused by *Zymoseptoria tritici*

The first reduced sensitivities in field isolates of *Z. tritici* were found in the UK in 2000 (Metcalfe et al., 2000; Mavroedi and Shaw, 2005). Shortly afterwards, alterations in the *cyp*51 gene sequence were associated with changes in sensitivity (Cools et al., 2005a,b; Zhan et al., 2006; Leroux et al., 2007). Leroux et al. (2007) initially categorised these amino acid polymorphisms into different phenotypes as 'Tri R1' to 'Tri R7' depending on the DMI response. Later, this classification was extended to 'Tri R12' (Leroux and Walker, 2011). Mutations at codons 459 to 461 (Y459S/D/N, G460D, ΔY459/G460, Y461S/H) within *cyp*51 were the first alterations associated with reduced sensitivity. These mutations are located in a highly conserved region of the *CYP51* protein near the haem binding site (Cools et al., 2005a,b). By themselves, these mutations resulted in low resistance factors, while additional mutations such as Y131F, V136A, A379G or I381V, located at putative substrate recognition sites, resulted in higher resistance factors (Fraaije et al., 2007; Leroux et al., 2007; Cools and Fraaije, 2008; Stammler et al., 2008).

Since then, a multitude of papers continued to focus on resistance research with *Z. tritici*. Investigations of DMI changes in field dose-response curves related to laboratory sensitivity phenotyping and genotyping assays showed that mutant strains showing low or moderate resistance factors (mutant strain EC_{50} divided by wild-type EC_{50}) are likely to remain sensitive to doses near the maximum permitted field dose, but be insensitive to lower doses, resulting in an observed reduction in field dose-response curvature (Blake et al., 2018).

At the same time, due to the increasing number of *CYP51* haplotypes being meanwhile identified, a new nomenclature has been proposed to describe the distribution and frequency of *CYP51* haplotypes of *Z. tritici* across different European countries based on a collection of isolates sampled in 2016. This new nomenclature consists of a combined letter and number code. The letter describes the number of amino acid alterations in the CYP51 enzyme, whereas the number describes the specific combination of a defined number of amino acid alterations. The E4 haplotype, for example, contains five specific amino acid alterations in CYP51 (L50S, D134G, V136A, I381V, Y461H), whereas the E5 haplotype contains another combination of five amino acid alterations in CYP51 (L50S, V136A, I381V, Y461H, S524T) (Huf et al., 2018). Sequencing of the *CYP51* gene of these isolates revealed at this time 33 different *CYP51* haplotypes, while around 85% of the population was composed of only 9 *CYP51* haplotypes with amino acid alterations at position L50, D134, V136, S188, A379, I381, Y459, G460, Y461, N513 or S524.

Latest resistance research with Septoria leaf blotch described more intensively populations on the local level. Studies with Irish populations considered besides target-site alterations their combination with overexpression and enhanced efflux activity in the reduction of azole sensitivity (Kildea et al., 2019). Although the costs in terms of fitness remain unknown for the constitutive overexpression of *CYP51*, particularly in highly evolved haplotypes, it was concluded that the potential combination of *CYP51* overexpression with evolved *CYP51* haplotypes would undoubtedly cause further reductions in sensitivity and loss of field efficacy.

DMI sensitivity studies with isolates originating from Finland and the Baltic states revealed that the frequencies of *CYP51* mutations D143G, V136A/C, A379G, I381V and S524T were lower than in other European countries, but have increased compared to previous years (Mäe et al., 2020).

3400 isolates sampled between 2012 and 2019 in Denmark and Sweden showed increasing EC_{50} values of epoxiconazole and prothioconazole and decreased field efficacy over time. Interestingly, the field performance of tebuconazole has improved in these countries but remains still at a comparably low level (Heick et al., 2020).

In France, spatio-temporal dynamics of fungicide resistance in *Z. tritici* has been studied using statistical models for the analysis of country-wide trial data generated between 2004 and 2017 and described major differences between the north and south (Garnault et al., 2019). Later on, in a joint work of European researchers, spatio-temporal distribution of DMI fungicide resistance of Septoria leaf blotch based on frequencies of key target-site alterations has been as well investigated throughout Europe, with a focus on the molecular quantification of the S524T allele which was regarded as conferring the highest resistance factor to DMI fungicides (Hellin et al., 2020). Historical leaf DNA samples originating from different European countries revealed that the frequency of all key alterations has been gradually increasing over the past decade. However, although these alterations were described as quickly becoming dominant in Ireland and Great Britain, the authors assumed that scope still exists to delay their selection throughout the wider European population (Hellin et al., 2021). Joint work of European researchers investigated as well potential differences in azole performances against Septoria leaf blotch in 55 field trials carried out between 2015 and 2018, covering ten different European countries and thus a diversity of climatic zones and agricultural practices. During all four seasons, various single azoles were tested and increasing variability in the performances of these azoles against *Z. tritici* was observed across Europe. The frequencies of *CYP51* mutations varied substantially across Europe, with a clear pattern of significantly decreasing frequencies of D134G, V136A and S524T in the local *Z. tritici* populations from west to east. In contrast, no major differences were seen for *CYP51* mutations V136C, A379 and I381V. The azoles showed different levels of cross-resistance, which again depended on specific *CYP51* mutations. Across the four seasons, all single azoles increased the yields between 9% and 11% on average (Jørgensen et al., 2021).

Outside of Europe, rapid and parallel evolution of azole resistance was reported in Australian *Z. tritici* populations. Genetically distinct Tasmanian populations were identified that independently acquired a complex haplotype of resistance to azole fungicides. Whole genome comparisons with the genomes of a global collection of isolates showed very low genome-wide relatedness between this population and other *Z. tritici* populations, ruling out gene flow as the source of the resistance haplotype. Interestingly, the post-azole Australian *Z. tritici* population has acquired the same set of nonsynonymous substitutions in the *CYP51* gene that have been observed in Europe and North America (McDonald et al., 2019).

Particularly for *Z. tritici*, the DMI group can be divided by their cross-resistance patterns. Based on *in vitro* testing, epoxiconazole and prothioconazole are considered to form one group, while tebuconazole, difenoconazole and mefentrifluconazole form a second group with the same and very strong cross-resistance pattern. Due to potential 'pre-selection' by tebuconazole and difenoconazole, leading to high diversity in EC_{50} values for mefentrifluconazole, and expected further adaption with increased use of those three DMIs, for the planning of control strategies cross-resistance patterns ought to be taken into consideration to avoid the build-up of fungicide resistance (Heick et al., 2020) and especially for DMI resistance management.

Interestingly, studies with isolates of *M. fructicola*, *Alternaria alternata*, *C. beticola*, *P. xanthii*, or *Colletotrichum* spp. causing diseases in dicotyledonous crops, clearly demonstrated positive cross-resistance between mefentrifluconazole and other DMIs both *in vitro* and *in planta* experiments (Ishii et al., 2021). Consequently, for any newly developed DMI fungicide and independent from the time of market introduction, it has to be considered that a true (= wild-type) baseline situation for main target pathogens and the particular mode of action of DMIs no longer exists in most agricultural areas with extensive DMI history.

6.2 Impact on demethylation inhibitor sensitivity of cyp51 alterations and overexpression in various pathogens

*Cyp*51 mutations encoding alterations in residues homologous to those in *Z. tritici* have also been found in other fungi. These include Y131F/H in *P. pachyrhizi*, Y134F in *Puccinia triticina* and *P. striiformis* f.sp. *tritici*, Y136F in *B. graminis* f. sp. *tritici*, *B. graminis* f. sp. *hordei* and *E. necator* (all homologous to *Z. tritici* Y137F), and Y132F/H, Y447, G448, F449 in the human pathogen *Candida albicans* (homologous to *Z. tritici* Y137F, Y459D/S/N/C, G460D, Y461H/S) (Délye et al., 1998; Sanglard et al., 1998a; Kelly et al., 1999; Marichal et al., 1999; Wyand and Brown, 2005; Stammler et al., 2009; Schmitz et al., 2014; Tucker et al., 2020; Müller et al., 2021; Cook et al., 2021). In the case of barley powdery mildew, meanwhile, five amino acid substitutions were found by sequencing the *Cyp51B* target gene of Australian isolates and most resistant haplotypes increased in prevalence from 0% in 2009 to 90% in 2011. Yeast strains expressing the *Bgh Cyp51* haplotypes replicated the altered sensitivity to various DMI-fungicides and these results were complemented by *in silico* protein docking studies (Tucker et al., 2020).

In addition to mutations at the target site, resistance to DMIs by overexpression of *cyp*51 has been found in a number of pathogens such as *Alternaria* spp., *Blumeriella jaapii*, *B. cinerea*, *C. beticola*, *Lasiodiplodia theobromae*, *M. fructicola*, *P. digitatum* or *P. pachyrhizi* (Hamamoto et al., 2000; Schnabel and Jones, 2001; Ma et al., 2006; Luo and Schnabel, 2008; Schmitz et al., 2014; Bolton et al., 2016; Li et al., 2020, Zhang et al., 2020a; Zhang et al., 2021). However, evidence suggests that fitness penalties may be present in such DMI-resistant isolates as reported in the case of *L. theobromae*, the cause of stem-end rot in papaya (Li et al., 2020), or of *A. alternata* (He et al., 2019). Much later than the discovery of changes in the *cyp*51 gene, this resistance mechanism was also found in *Z. tritici* (Cools et al., 2012).

While overexpression of the target gene appears to be important for *V. inaequalis* and *M. fructicola* (Schnabel and Jones, 2001; Luo and Schnabel, 2008; Chen et al., 2019), the effects on *Z. tritici* were not only considered at this time to be minor (Cools et al., 2012) but also possibly to become more prominent, as the progressive accumulation of mutations in the *cyp*51 gene can negatively affect the functionality of the protein (Cools et al., 2012). In this context and in relation to *M. fructicola*, resistance research with samples from commercial peach orchards in Brazil provided evidence that DMI resistance can be managed in orchards with high selection pressure on tebuconazole after discontinuing the use of the fungicide for at least three years. By assessing the DMI sensitivity of *M. fructicola* using a mycelial growth assay *in vitro* and by determining the frequency of the G461S mutation in the *cyp51* gene, the work showed that G461S mutation frequency in the state of Sao Paulo was around 80% in the 2012/13 season, but then decreased until it was completely undetectable in 2015/16, apart from a gradual decrease in highly resistant isolates during the same period of time (Pereira et al., 2020a). Prior to these findings, a shift towards lower propiconazole efficacy has also been reported for some *M. fructicola* field populations from Georgia and South Carolina in the USA (Zehr et al., 1999; Schnabel et al., 2004).

In *Fusarium fujikuori*, the plant pathogen causing rice bakanae disease, overexpression of *cyp51a* and *cyp51b* together with the S312T point mutation in CYP51B resulted in DMI resistance, as shown by knockdown of the *cyp51b* gene and complementation between prochloraz-sensitive and prochloraz-resistant strains (Zhang et al., 2020b).

Altered DMI sensitivity has been as well thoroughly investigated in two barley pathogens causing net blotch, *Pyrenophora teres* f. sp. *maculata* and *Pyrenophora teres* f.sp. *teres*, including factors that affect the genotypic composition and the movement of new genotypes in the context of pathogen epidemiology (Ellwood et al., 2019). DMI resistance correlated with point mutations resulting in F489L in *CYP51A* in both species (Mair et al., 2016; Mair et al., 2020) and, as reported for *P. teres* f. sp. *maculata*, with the presence of an insertion element at different positions in the *Cyp51A* upstream region (Mair et al., 2020). Dependent on the DMI, the presence of the mutation, the insertion element or both, resistance factors varied from low to high.

The causal agent of cherry leaf spot disease, *B. jaapii*, is a major disease of sour and sweet cherry. Resistance to DMI fungicides has been reported from orchards in Michigan and has been associated with overexpression of *cyp51* (Ma et al., 2006; Proffer et al., 2006). Overexpression plus mutations of *CYP51* were as well together associated with DMI resistance in *Colletotrichum gloeosporioides* from Chili (Wei et al., 2020). Changes in the expression level of the *P. digitatum cyp51* gene or ATP-binding cassette (ABC) transporters are thought to contribute to increased tolerance to DMIs (Nakaune et al., 1998; Hamamoto et al., 2000; Ghosoph et al., 2007). Reduced sensitivity to DMIs, which may be associated with upregulation of the target gene, has also been observed in leaf spot disease caused by *C. beticola* on sugar beet in North Dakota and Minnesota (Bolton et al., 2012, 2016). Further work in the United States, Serbia, Germany and Japan also described the impact of SNPs in *CYP51* on DMI resistance in *Cercospora* leaf spot, particularly the F144L substitution (Trkulja et al., 2015; Trkulja et al., 2017; Spanner et al., 2021; Kayamori et al., 2021). In addition, a genome-wide association study identified mutations in genes associated with DMI resistance including a 'Regulator of G-protein Signaling' (RGS) protein, an ABC pleiotropic drug resistance transporter, a dual-specificity tyrosine phosphorylation-regulated kinase (DYRK), a gene annotated as a hypothetical protein, as well as a SNP upstream of *CYP51* (Spanner et al., 2021).

Isolates of *Colletotrichum fioriniae* and *Colletotrichum nymphaeae*, containing two paralogous *CYP51* genes that encode sterol 14-demethylase enzymes, were investigated for mediation of differential sensitivity to DMI fungicides. The results suggest species-specific, differential binding of the sterol demethylation inhibitors onto the two CYP51 enzymes. Because some combinations of DMI fungicides tested in this study showed strong synergistic effects against both *C. fioriniae* and *C. nymphaeae* mycelial growth and DMIs appear to target *CYP51* paralogs with preferential affinity, it was suggested that using a mixture of DMIs may improve the efficacy for anthracnose management (Chen et al., 2020). Similar synergistic interactions between DMIs were shown *in vitro* in *Fusarium* spp. in which three *CYP51* paralogs were detected (Liu et al., 2011).

6.3 Impact on demethylation inhibitor sensitivity of efflux transporters

Another mechanism involved in reduced DMI response confers to overexpression of efflux transporters belonging to members of the ATP-binding cassette (ABC) or major facilitator superfamily (MFS). While this mechanism is quite uncommon in phytopathogens, it is known to be a common resistance mechanism in human microbes such as *C. albicans* (Sanglard et al., 1995, 1998b). Examples of phytopathogens with considerable levels of resistance to DMI fungicides due to the upregulation of ABC and MFS transporters are *P. digitatum* and *B. cinerea*. Isolates with this trait appear to be increasing in frequency (Nakaune et al., 1998; Hayashi et al., 2002; Kretschmer et al., 2009). In *Parastagonospora nodorum*, correlations between mutations of MFS transporters and azole sensitivity were observed. Interestingly, the group of isolates lacking *CYP51* resistance mutations, but carrying the more resistant variant of the MFS transporter were highly susceptible, indicating that the MFS transporter mutations depend on *CYP51* mutations to have an effect (Pereira et al., 2020b). Overexpression of efflux pumps leading to a multiple drug resistance (MDR) phenotype has been meanwhile frequently investigated in field isolates of *Z. tritici* (Stergiopoulos et al., 2003; Leroux and Walker, 2011; Stammler and Semar, 2011; Cools et al., 2012), especially as enhanced DMI efflux by upregulated efflux transporters has been previously observed in laboratory strains (Zwiers et al., 2002).

7 Recommended uses for resistance management

The occurrence and spread of resistance are highly dependent on the dosage, number, timing and interval of applications of fungicides that exhibit cross-resistance, factors that are usually modified to manage the development of resistance. Although numerous reports of resistance have been published, the efficacy of DMIs in the field has generally been maintained, at least when newer active ingredients are used in place of older DMIs. This is due to the continuous improvement in the intrinsic properties of newly developed agents. DMIs have different effects on fungal pathogens, and mixtures or alternations of different DMIs can control different phenotypes simultaneously (Cools and Fraaije, 2008). However, extensive use of only one particular DMI mixture is likely to result in the selection of mutations that specifically affect both DMI mixture partners. Preferred mixture partners are non-cross-resistant compounds such as QoIs, inhibitors of succinate dehydrogenase (SDHIs) or multi-site fungicides.

Mixtures of non-cross-resistant fungicides can particularly delay resistance selection. Hobbelen et al. (2011) used mathematical models to predict whether mixtures of a fungicide with a high resistance risk and a fungicide with a low risk delay the selection of *Z. tritici* for resistance to the fungicide with high risk. It was found that the 'effective life' (the number of seasons in which the disease-induced reduction in the integral of canopy green area index could be maintained at < 5% during the yield-generating period) was highest when the full label-recommended dose of the low-risk fungicide was applied, and the dose of the high-risk fungicide was adjusted each season to the level required to maintain effective control. This strategy resulted in a predicted 'effective life span' of ≤ 12 years, compared to 3 to 4 years when using the high-risk fungicide alone. Consequently, mixtures of DMIs with moderate resistance risk with non-cross-resistant compounds are sound tools for resistance management, as the partner fungicides are not expected to specifically counter-select DMI-susceptible or DMI-resistant ('shifted') isolates.

The accumulation of multiple mutations at target sites, as in the case of *Z. tritici*, is likely to lead to fitness deficits. Therefore, reduced selection pressure, which can be achieved by using fungicides with a different mode of action, may lead to a shift back to susceptibility in the pathogen population.

Because DMIs are used in a number of different pathogen-crop settings under varying disease incidence and severity, it is not possible to give general recommendations for use that can be applied in all situations. In general, DMI fungicides should be used preventively rather than curatively to prolong the effectiveness of this unique and highly valuable mode of action as much as possible, using appropriate rates, application intervals and cultivation methods (e.g. crop rotation, less susceptible variety, sowing time, reduction of inoculum source). If DMI shifts have been detected, the number of solo treatments should be reduced and mixtures or alternations with non-cross-resistant fungicides that provide sufficient efficacy against the pathogen population should be increased. The FRAC SBI Working Group provides an annual update on resistance monitoring of the major companies marketing DMI fungicides and develops sensible and practical application recommendations for different crop/pathogen systems (FRAC, www.frac.info).

8 Conclusion

The success of SBI fungicides, and DMI fungicides in particular, is essentially the success of the triazole fungicide class, as all other chemical classes within the SBI mode of action play only a limited role in the market. The largest regional markets for DMIs are Latin America where DMIs are used extensively to control soybean rust in Brazil, followed by Western Europe due to its intensive cereal production.

DMIs are single-site fungicides and inhibit ergosterol biosynthesis by inhibiting sterol C14 demethylase encoded by *cyp*51. Many reports of resistance have been published, but in practice cases of complete loss of efficacy are relatively rare. The reason for this is both the 'shifting-type resistance' characteristic, based on the accumulation of multiple alterations, and the continuous introduction of intrinsically more effective compounds to which fungal populations do not show complete cross-resistance. While overexpression of the target site or efflux transporters has been reported as resistance mechanism, point mutations within the target gene *cyp*51 appear to be most relevant in the field. The most intensively analysed resistance mechanisms are those of cereal pathogens such as *B. graminis* f. sp. *tritici*, *B. graminis* f. sp. *hordei*, *P. graminis*, *P. teres*, *P. tritici-repentis*, *Fusarium* spp. and especially *Z. tritici*. Considerable efforts have been made to understand the effects of certain mutations or combinations of mutations, especially at the molecular level. Although laboratory results (detection of resistance mechanisms, evaluation of EC_{50}) provide valuable information, their relevance needs to be confirmed in the greenhouse or even in the field, as the correlation of laboratory resistance with field performance is less clear than for fungicides that show a 'disruptive resistance characteristic' like QoI fungicides belonging to FRAC Code 11.

Only individual cases are described in which resistance instability was found *in vitro*. However, a partial shift back has been observed when selection pressure decreases, as resistant phenotypes are often associated with fitness deficits. In particular, the constrained evolution of target gene mutations may be associated with fitness losses (Cools et al., 2013). Consequently, in addition to the analysis of resistance mechanisms and their relevance, other aspects such as the stability of resistance and fitness need to be addressed particularly for the characterisation of new upcoming resistance reports.

Sound and comprehensive knowledge of DMI resistance, including the understanding of resistance risk and resistance mechanisms, helps in making and adapting application recommendations for DMI-containing fungicides. It is important to conduct appropriate surveys of potential sensitivity shifts in major plant pathogens, including fungal isolates from different origins (regions, districts, countries), and to follow the development of shifts over several years to identify stable trends. As there can be strong regional differences in susceptibility and year-to-year fluctuations, reports on local populations studied over a period of only one to two years are of limited value in describing the overall DMI susceptibility status of a pathogen. More so, when fungicide mixtures are recommended for managing DMI sensitivity shifts, it is crucial to assess the susceptibility status of all partners in the proposed fungicide mixture to ensure adequate anti-resistance activity.

The DMI susceptibility situation of a particular pathogen in a particular crop can be illustrated by various parameters. These include the strength of resistance measured *in vitro* or *in vivo* (resistance factor), knowledge of the resistance mechanisms and possible fitness differences of the shifted isolates. Another aspect is the description of the sensitivity range of individual isolates via EC_{50} values (diversity factor) and the frequency of highly shifted outliers in certain populations, as the strength of the sensitivity shift determines how well a fungicide application goes. In addition, biological characteristics of the pathogen (life cycle, number of target genes, etc.) can influence the development and spread of resistance, which in turn can be used to make recommendations on fungicide applications.

In summary, strategies for DMI resistance management include various measures. Most importantly, fungicide mixtures or alternations using active ingredients with different modes of action should be considered. In addition, optimised application practices involving the right dosage, timing and number of treatments, as well as other general agronomic practices such as crop rotation, adapted sowing time or less susceptible varieties play an important role in resistance management.

9 References

Anonymous. (2022a). FRAC code list 2022: Fungal control agents sorted by cross-resistance pattern and mode of action. Published Available at: http://www.frac.info.

Anonymous. (2022b). FRAC minutes of the sterol biosynthesis inhibitor (SBI) working group meeting 2022. Published Available at: http://www.frac.info.

Anonymous. (2022c). FRAC Poster® 2022. Published Available at: http://www.frac.info.

Bardas, G. A., Myresiotis, C. K. and Karaoglanidis, G. S. (2008). Stability and fitness of anilinopyrimidine-resistant strains of *Botrytis cinerea*. *Phytopathology* 98(4): 443–50.

Berg, D. and Plempel, M. (1988). *Sterol Biosynthesis Inhibitors: Pharmaceutical and Agrochemical Aspects*. Wiley-VCH Verlag and Horwood Publishing, Cambridge, New York, Basel, Chichester.

Billard, A., Fillinger, S., Leroux, P., Lachaise, H., Beffa, R. and Debieu, D. (2012). Strong resistance to the fungicide fenhexamid entails a fitness cost in *Botrytis cinerea*, as shown by comparisons of isogenic strains. *Pest Manag. Sci.* 68(5): 684–91.

Blake, J. J., Gosling, P., Fraaije, B. A., Burnett, F. J., Knight, S. M., Kildea, S. and Paveley, N. D. (2018). Changes in field dose–response curves for demethylation inhibitor (DMI) and quinone outside inhibitor (QoI) fungicides against *Zymoseptoria tritici*, related to laboratory sensitivity phenotyping and genotyping assays. *Pest Manag. Sci.* 74(2): 302–13.

Bolton, M. D., Birla, K., Rivera-Varas, V., Rudolph, K. D. and Secor, G. A. (2012). Characterization of *CbCyp51* from field isolates of *Cercospora beticola*. *Phytopathology* 102(3): 298–305.

Bolton, M. D., Ebert, M. K., Faino, L., Rivera-Varas, V., de Jonge, R., Van de Peer, Y., Thomma, B. P. H. J. and Secor, G. A. (2016). RNA-sequencing of *Cercospora beticola* DMI-sensitive and -resistant isolates after treatment with tetraconazole identifies common and contrasting pathway induction. *Fungal Genet. Biol.* 92: 1–13.

Brent, K. J. and Hollomon, D. W. (2007a). *Fungicide Resistance in Crop Pathogens: How Can It Be Managed?* (2nd edn.). FRAC Monograph No. 1. British Crop Protection Council.

Brent, K. J. and Hollomon, D. W. (2007b). *Fungicide Resistance: The Assessment of Risk* (2nd edn.). FRAC Monograph No. 2. British Crop Protection Council.

Büchel, K. H. (1986). The history of azole chemistry. In: Green, M. B. and Spilker, D. A. (Eds.), *Fungicide Chemistry: Advances and Practical Applications*. American Chemical Society, Washington, pp. 1–23.

Buchenauer, H. (1995). In: Lyr, H. (Ed.), *Modern Selective Fungicides - Properties, Applications, Mechanisms of Action*. Gustav Fischer Verlag and Wiley-VCH Verlag, Jena, Stuttgart, New York, and Deerfield Beach, pp. 259–90.

Buchenauer, H. and Grossmann, F. (1977). Triadimefon: Mode of action in fungi and plants. *Netherl. J. Plant Path.* 83(Suppl. 1): 93–103.

Buchenauer, H. and Röhner, E. (1981). Effect of triadimefon and triadimenol on growth of various plant species as well as on gibberelin content and sterol metabolism in shoots of barley seedlings. *Pestic. Biochem. Physiol.* 15(1): 58–70.

Burden, R. S., Cooke, D. T. and Carter, G. A. (1989). Inhibitors of sterol biosynthesis and growth in plants and fungi. *Phytochemistry* 28(7): 1791–804.

Burpee, L. L. (1997). Control of dollar spot of creeping bentgrass caused by an isolate of *Sclerotinia homoeocarpa* resistant to benzimidazole and demethylation-inhibitor fungicides. *Plant Dis.* 81(11): 1259–63.

Chapman, K. S., Sundin, G. W. and Beckerman, J. L. (2011). Identification of resistance to multiple fungicides in field populations of *Venturia inaequalis*. *Plant Dis.* 95(8): 921–6.

Chapuis, L., Corio-Costet, M.-R. and Malosse, C. (1996). Sterol composition of the woody plant pathogenic fungus *Eutypa lata*. *Phytochemistry* 42(6): 1599–601.

Chen, S., Hu, M., Schnabel, G., Yang, D., Yan, X. and Yuan, H. (2020). Paralogous *CYP51* genes of *Colletotrichum* spp. mediate differential sensitivity to sterol demethylation inhibitors. *Phytopathology* 110(3): 615–25.

Chen, S., Schnabel, G., Yuan, H. and Luo, C. (2019). LAMP detection of the genetic element 'Mona' associated with DMI resistance in *Monilinia fructicola*. *Pest Manag. Sci.* 75(3): 779–86.

Cook, N. M., Chng, S., Woodman, T. L., Warren, R., Oliver, R. P. and Saunders, D. G. O. (2021). High frequency of fungicide resistance-associated mutations in the wheat yellow rust pathogen *Puccinia striiformis* f. sp. *tritici*. *Pest Manag. Sci.* 77(7): 3358–71.

Cooke, L. R., Locke, T., Lockley, K. D., Phillips, A. N., Sadiq, M. D. S., Coll, R., Black, L., Taggart, P. J. and Mercer, P. C. (2004). The effect of fungicide programmes based on epoxiconazole on the control and DMI sensitivity of *Rhynchosporium secalis* in winter barley. *Crop Prot.* 23(5): 393–406.

Cools, H. J. and Fraaije, B. A. (2008). Are azole fungicides losing ground against *Septoria* wheat disease? Resistance mechanisms in *Mycosphaerella graminicola*. *Pest Manag. Sci.* 64(7): 681–4.

Cools, H. J. and Fraaije, B. A. (2013). Update on mechanisms of azole resistance in *Mycosphaerella graminicola* and implications for future control. *Pest Manag. Sci.* 69(2): 150–5.

Cools, H. J., Fraaije, B. A. and Lucas, J. A. (2005a). Molecular examination of *Septoria tritici* isolates with reduced sensitivities to triazoles. In: *Modern Fungicides and Antifungal Compounds IV: 14th International Reinhardsbrunn Symposium*, Friedrichroda, Thuringia, Germany, April 25–29, 2004, pp. 103–14.

Cools, H. J., Fraaije, B. A. and Lucas, J. A. (2005b). Molecular mechanisms correlated with changes in triazole sensitivity in isolates of *Mycosphaerella graminicola*. In: *Proceedings of the BCPC Congress, Crop Science and Technology*. British Crop Protection Council, Alton, United Kingdom, pp. 267–74.

Cools, H. J., Hawkins, N. J. and Fraaije, B. A. (2013). Constraints on the evolution of azole resistance in plant pathogenic fungi. *Plant Pathol.* 62: 36–42.

Cools, H. J., Mullins, J. G. L., Fraaije, B. A., Parker, J. E., Kelly, D. E., Lucas, J. A. and Kelly, S. L. (2011). Impact of recently emerged *sterol 14α-demethylase* (*CYP51*) variants of *Mycosphaerella graminicola* on azole fungicide sensitivity. *Appl. Environ. Microbiol.* 77(11): 3830–7.

Cools, H. J., Parker, J. E., Kelly, D. E., Lucas, J. A., Fraaije, B. A. and Kelly, S. L. (2010). Heterologous expression of mutated *eburicol 14α-demethylase* (*CYP51*) proteins of *Mycosphaerella graminicola* to assess effects on azole fungicide sensitivity and intrinsic protein function. *Appl. Environ. Microbiol.* 76(9): 2866–72.

Cox, K. D., Bryson, P. K. and Schnabel, G. (2007). Instability of propiconazole resistance and fitness in *Monilinia fructicola*. *Phytopathology* 97(4): 448–53.

Cristani, C. and Gambogi, P. (1993). Laboratory isolation of *Microdochium* (*Fusarium*) *nivale* mutants showing reduced sensitivity to sterol biosynthesis inhibitors. *Riv. Patalogia Veg.* 3: 49–57.

Délye, C., Bousset, L. and Corio-Costet, M. F. (1998). PCR cloning and detection of point mutations in the *eburicol 14a-demethylase* (*CYP51*) gene from *Erysiphe graminis* f. sp. *hordei*, a "recalcitrant" fungus. *Curr. Genet.* 34(5): 399–403.

Eckert, J. W., Sievert, J. R. and Ratnayake, M. (1994). Reduction of imazalil effectiveness against citrus green mold in California packinghouses by resistant biotypes of *Penicillium digitatum*. *Plant Dis.* 78(10): 971–4.

Ellwood, S. R., Piscetek, V., Mair, W. J., Lawrence, J. A., Lopez-Ruiz, F. J. and Rawlinson, C. (2019). Genetic variation of *Pyrenophora teres* f. teres isolates in Western Australia and emergence of a Cyp51A fungicide resistance mutation. *Plant Pathol.* 68(1): 135–42.

Fraaije, B. A., Cools, H. J., Kim, S. H., Motteram, J., Clark, W. S. and Lucas, J. A. (2007). A novel substitution I381V in the *sterol 14α-demethylase* (*CYP51*) of *Mycosphaerella graminicola* is differentially selected by azole fungicides. *Mol. Plant Pathol.* 8(3): 245–54.

Garnault, M., Duplaix, C., Leroux, P., Couleaud, G., Carpentier, F., David, O. and Walker, A. S. (2019). Spatiotemporal dynamics of fungicide resistance in the wheat pathogen *Zymoseptoria tritici* in France. *Pest Manag. Sci.* 75(7): 1794–807.

Ghosoph, J. M., Schmidt, L. S., Margosan, D. A. and Smilanick, J. L. (2007). Imazalil resistance linked to a unique insertion sequence in the PdCYP51 promoter region of *Penicillium digitatum*. *Postharvest Biol. Technol.* 44(1): 9–18.

Golembiewski, R. C., Vargas Jr., J. M., Jones, A. L. and Detweiler, A. R. (1995). Detection of demethylation inhibitor (DMI) resistance in *Sclerotinia homoeocarpa* populations. *Plant Dis.* 79(5): 491–3.

Hamamoto, H., Hasegawa, K., Nakaune, R., Jin Lee, Y. J., Makizumi, Y., Akutsu, K. and Hibi, T. (2000). Tandem repeat of a transcriptional enhancer upstream of the *sterol 14α-demethylase gene* (*CYP51*) in *Penicillium digitatum*. *Appl. Environ. Microbiol.* 66(8): 3421–6.

Hayashi, K., Schoonbeek, H.-J. and de Waard, M. A. (2002). Expression of the ABC transporter *BcatrD* from *Botrytis cinerea* reduces sensitivity to sterol demethylation inhibitor fungicides. *Pestic. Biochem. Physiol.* 73(2): 110–21.

He, M.-H., Wang, Y.-P., Wu, E. J., Shen, L.-L., Yang, L.-N., Wang, T., Shang, L.-P., Zhu, W. and Zhan, J. (2019). Constraining evolution of *Alternaria alternata* resistance to a demethylation inhibitor (DMI) fungicide difenoconazole. *Front. Microbiol.* 10: Article 1609.

Heick, T. M., Matzen, N. and Jørgensen, L. N. (2020). Reduced field efficacy and sensitivity of demethylation inhibitors in the Danish and Swedish *Zymoseptoria tritici* populations. *Eur. J. Plant Pathol.* 157(3): 625–36.

Hellin, P., Duvivier, M., Clinckemaillie, A., Bataille, C., Legrève, A., Heick, T. M., Jørgensen, L. N., Andersson, B., Samils, B., Rodemann, B., Berg, G. and Kildea, S. (2020). Multiplex qPCR assay for simultaneous quantification of CYP51-S524T and SdhC-H152R substitutions in European populations of *Zymoseptoria tritici*. *Plant Pathol.* 69(9): 1666–77.

Hellin, P., Duvivier, M., Heick, T. M., Fraaije, B. A., Bataille, C., Clinckemaillie, A., Legrève, A., Jørgensen, L. N., Andersson, B., Samils, B., Rodemann, B., Berg, G., Hutton, F., Garnault, M., El Jarroudi, M., Couleaud, G. and Kildea, S. (2021). Spatio-temporal distribution of DMI and SDHI fungicide resistance of *Zymoseptoria tritici* throughout Europe based on frequencies of key target-site alterations. *Pest Manag. Sci.* 77(12): 5576–88.

Hobbelen, P. H. F., Paveley, N. D. and van den Bosch, F. (2011). Delaying selection for fungicide insensitivity by mixing fungicides at a low and high risk of resistance development: A modeling analysis. *Phytopathology* 101(10): 1224–33.

Holmes, G. J. and Eckert, J. W. (1995). Relative fitness of imazalil-resistant and -sensitive biotypes of *Penicillium digitatum*. *Plant Dis.* 79(10): 1068–73.

Holmes, G. J. and Eckert, J. W. (1999). Sensitivity of *Penicillium digitatum* and *P. italicum* to postharvest citrus fungicides in California. *Phytopathology* 89(9): 716–21.

Huf, A., Rehfus, A., Lorenz, K. H., Bryson, R., Voegele, R. T. and Stammler, G. (2018). Proposal for a new nomenclature for *CYP51* haplotypes in *Zymoseptoria tritici* and analysis of their distribution in Europe. *Plant Pathol.* 67(8): 1706–12.

Ishii, H., Bryson, P. K., Kayamori, M., Miyamoto, T., Yamaoka, Y. and Schnabel, G. (2021). Cross-resistance to the new fungicide mefentrifluconazole in DMI-resistant fungal pathogens. *Pestic. Biochem. Physiol.* 171: 104737.

Jautelat, M., Elbe, H. L., Benet-Buchholz, J. and Etzel, W. (2004). Chemistry of prothioconazole (JAU 6476). *Pflanz.-Nachr. Bayer* 57: 145–62.

Jo, Y. K., Niver, A. L., Rimelspach, J. W. and Boehm, M. J. (2006). Fungicide sensitivity of *Sclerotinia homoeocarpa* from golf courses in Ohio. *Plant Dis.* 90(6): 807–13.

Jørgensen, L. N., Matzen, N., Heick, T. M., Havis, N., Holdgate, S., Clark, B., Blake, J., Glazek, M., Korbas, M., Danielewicz, J., Maumene, C., Rodemann, B., Weigand, S., Kildea, S.,

Bataille, C., Brauna-Morževska, E., Gulbis, K., Ban, R., Berg, G., Semaskiene, R. and Stammler, G. (2021). Decreasing azole sensitivity of *Z. tritici* in Europe contributes to reduced and varying field efficacy. *J. Plant Dis. Prot.* 128(1): 287–301.

Kang, Z., Huang, L. and Buchenauer, H. (2001). Ultrastructural and cytochemical studies of effects of the fungicide metconazole on *Fusarium culmorum in vitro*. *Z. Pflanzenkrankh. Pflanzenschutz* 108: 419–32.

Karaoglanidis, G. S., Thanassoulopoulos, C. C. and Ioannidis, P. M. (2001). Fitness of *Cercospora beticola* field isolates - Resistant and - Sensitive to demethylation inhibitor fungicides. *Eur. J. Plant Pathol.* 107(3): 337–47.

Kayamori, M., Zakharycheva, A., Saito, H. and Komatsu, K. (2021). Resistance to demethylation inhibitors in *Cercospora beticola*, a pathogen of sugar beet in Japan, and development of unique cross-resistance patterns. *Eur. J. Plant Pathol.* 160(1): 39–52.

Kelly, S. L., Lamb, D. C. and Kelly, D. E. (1999). Y132H substitution in *Candida albicans sterol 14α-demethylase* confers fluconazole resistance by preventing binding to haem. *FEMS Microbiol. Lett.* 180(2): 171–5.

Kildea, S., Heick, T. M., Grant, J., Mehenni-Ciz, J. and Dooley, H. (2019). A combination of target-site alterations, overexpression and enhanced efflux activity contribute to reduced azole sensitivity present in the Irish *Zymoseptoria tritici* population. *Eur. J. Plant Pathol.* 154(3): 529–40.

Kinay, P., Mansour, M. F., Gabler, F. M., Margosan, D. A. and Smilanick, J. L. (2007). Characterization of fungicide-resistant isolates of *Penicillium digitatum* collected in California. *Crop Prot.* 26(4): 647–56.

Köller, W., Smith, F. D. and Reynolds, K. L. (1991). Phenotypic instability of flusilazole sensitivity in *Venturia inaequalis*. *Plant Pathol.* 40(4): 608–11.

Köller, W., Wilcox, W. F., Barnard, J., Jones, A. L. and Braun, P. G. (1997). Detection and quantification of resistance of *Venturia inaequalis* populations to sterol demethylation inhibitors. *Phytopathology* 87(2): 184–90.

Kretschmer, M., Leroch, M., Mosbach, A., Walker, A. S., Fillinger, S., Mernke, D., Schoonbeek, H. J., Pradier, J. M., Leroux, P., de Waard, M. A. and Hahn, M. (2009). Fungicide-driven evolution and molecular basis of multidrug resistance in field populations of the grey mould fungus *Botrytis cinerea*. *PLoS Pathog.* 5(12): e1000696.

Kuck, K. H. (2002). *Modern Fungicides and Antifungal Compounds III* (Eds. Dehne, H. W., Gisi, U., Kuck, K. H., Russel, R. E. and Lyr, H.). Agroconcept, Bonn, pp. 21–8.

Kuck, K. H. and Mehl, A. (2004). Prothioconazole: Sensitivity profile and anti-resistance strategy. *Pflanz.-Nachr. Bayer* 57: 225–35.

Kuck, K. H. and Scheinpflug, H. (1986). Biology of sterol-biosynthesis inhibiting fungicides. In: Bowers, W. S., Ebing, W., Fukuto, T. R., Martin, D., Wegler, R. and Yamamoto, I. (Eds.), *Chemistry of Plant Protection*. Springer-Verlag, Berlin Heidelberg, pp. 65–96.

Kuck, K. H., Scheinpflug, H. and Pontzen, R. (1995). In: Lyr, H. (Ed.), *Modern Selective Fungicides - Properties, Applications, Mechanisms of Action*. Gustav Fischer Verlag and Wiley-VCH Verlag, Jena, Stuttgart, New York, and Deerfield Beach, pp. 205–58.

Kuck, K. H., Stenzel, K. and Vors, J.-P. (2012). Sterol biosynthesis inhibitors. In: Krämer, W., Schirmer, U., Jeschke, P. and Witschel, M. (Eds.), *Modern Crop Protection Compounds* (Vol. 2). Wiley-VCH Verlag GmbH & Co. KGaA, Weinheim, Germany.

Leroux, P., Albertini, C., Cautier, A., Gredt, M. and Walker, A. S. (2007). Mutations in the *CYP51* gene correlated with changes in sensitivity to sterol 14α-demethylation inhibitors in field isolates of *Mycosphaerella graminicola*. *Pest Manag. Sci.* 63(7): 688–98.

Leroux, P. and Marchegay, P. (1991). Caractérisation des souches de *Pseudocercosporella herpotrichoides* agent du Piétin-verse des céréales, resistantes au prochloraze, isolées en France sur blé tendre d'hiver. *Agronomie* 11(9): 767–76.

Leroux, P. and Walker, A. S. (2011). Multiple mechanisms account for resistance to sterol 14alpha-demethylation inhibitors in field isolates of *Mycosphaerella graminicola*. *Pest Manag. Sci.* 67(1): 44–59.

Li, Y., Tsuji, S. S., Hu, M., Câmara, M. P. S., Michereff, S. J., Schnabel, G. and Chen, F. (2020). Characterization of difenoconazole resistance in *Lasiodiplodia theobromae* from papaya in Brazil. *Pest Manag. Sci.* 76(4): 1344–52.

Liu, X., Yu, F., Schnabel, G., Wu, J., Wang, Z. and Ma, Z. (2011). Paralogous *cyp51* genes in *Fusarium graminearum* mediate differential sensitivity to sterol demethylation inhibitors. *Fungal Genet. Biol.* 48(2): 113–23.

Loeffler, R. S. T., Butters, J. A. and Hollomon, D. W. (1992). The sterol composition of powdery mildews. *Phytochemistry* 31(5): 1561–63.

Luo, C. X. and Schnabel, G. (2008). The cytochrome P450 *lanosterol 14α-demethylase* gene is a demethylation inhibitor fungicide resistance determinant in *Monilinia fructicola* field isolates from Georgia. *Appl. Environ. Microbiol.* 74(2): 359–66.

Lürssen, D. (1988). *Sterol Biosynthesis Inhibitors: Pharmaceutical and Agrochemical Aspects*. Wiley-VCH Verlag and Horwood Publishing, Cambridge, New York, Basel, Chichester.

Ma, Z., Proffer, T. J., Jacobs, J. L. and Sundin, G. W. (2006). Overexpression of the 14alpha-demethylase target gene (*CYP51*) mediates fungicide resistance in *Blumeriella jaapii*. *Appl. Environ. Microbiol.* 72(4): 2581–5.

Mäe, A., Fillinger, S., Sooväli, P. and Heick, T. M. (2020). Fungicide sensitivity shifting of *Zymoseptoria tritici* in the Finnish-Baltic region and a novel insertion in the MFS1 promoter. *Front. Plant Sci.* 11: Article 385.

Mair, W. J., Deng, W., Mullins, J. G., West, S., Wang, P., Besharat, N., Ellwood, S. R., Oliver, R. P. and Lopez-Ruiz, F. J. (2016). Demethylase inhibitor fungicide resistance in *Pyrenophora teres* f. sp. *teres* associated with target site modification and inducible overexpression of *Cyp51*. *Front. Microbiol.* 7: 1279.

Mair, W. J., Thomas, G. J., Dodhia, K., Hills, A. L., Jayasena, K. W., Ellwood, S. R., Oliver, R. P. and Lopez-Ruiz, F. J. (2020). Parallel evolution of multiple mechanisms for demethylase inhibitor fungicide resistance in the barley pathogen *Pyrenophora teres* f. sp. *maculata*. *Fungal Genet. Biol.* 145: 103475.

Marichal, P., Koymans, L., Willemsens, S., Bellens, D., Verhasselt, P., Luyten, W., Borgers, M., Ramaekers, F. C. S., Odds, F. C. and Vanden Bossche, H. (1999). Contribution of mutations in the cytochrome P450 *14α-demethylase* (*Erg11p*, *Cyp51p*) to azole resistance in *Candida albicans*. *Microbiology (Reading)* 145(10): 2701–13.

Mauler-Machnik, A., Rosslenbroich, H. J., Dutzmann, S., Applegate, J. and Jautelat, S. (2002*). Brighton Crop Protection Conference, Pests and Diseases*, pp. 389–94.

Mavroedi, V. I. and Shaw, M. W. (2005). Sensitivity distributions and cross-resistance patterns of *Mycosphaerella graminicola* to fluquinconazole, prochloraz and azoxystrobin over a period of 9 years. *Crop Prot.* 24(3): 259–66.

McDonald, M. C., Renkin, M., Spackman, M., Orchard, B., Croll, D., Solomon, P. S. and Milgate, A. (2019). Rapid parallel evolution of azole fungicide resistance in Australian populations of the wheat pathogen *Zymoseptoria tritici*. *Appl. Environ. Microbiol.* 85(4): e01908-18.

McKay, A. H., Förster, H. and Adaskaveg, J. E. (2012). Toxicity and resistance potential of selected fungicides to *Galactomyces* and *Penicillium* spp. causing postharvest fruit decays of citrus and other crops. *Plant Dis.* 96(1): 87–96.

Mehl, A., Schmitz, H., Stenzel, K. and Bloomberg, J. (2019). DMI fungicides (FRAC Code 3): Sensitivity status of key target pathogens, field versus laboratory resistance, and resistance mechanisms. In: Stevenson, K. L., McGrath, M. T. and Wyenandt, C. A. (Eds.), *Fungicide Resistance in North America* (2nd edn.). APS Press, St. Paul, MN, pp. 51–68.

Mercer, E. I. (1984). The biosynthesis of ergosterol. *Pestic. Sci.* 15(2): 133–55.

Metcalfe, R. J., Shaw, M. W. and Russell, P. E. (2000). The effect of dose and mobility on the strength of selection for DMI fungicide resistance in inoculated field experiments. *Plant Pathol.* 49(5): 546–57.

Meyers, E., Arellano, C. and Cowger, C. (2019). Sensitivity of the U.S. *Blumeria graminis* f. sp. *tritici* population to demethylation inhibitor fungicides. *Plant Dis.* 103(12): 3108–16.

Miller, G. L., Stevenson, K. L. and Burpee, L. L. (2002). Sensitivity of *Sclerotinia homoeocarpa* isolates to propiconazole and impact on control of dollar spot. *Plant Dis.* 86(11): 1240–6.

Müller, M. A., Stammler, G. and May De Mio, L. L. (2021). Multiple resistance to DMI, QoI and SDHI fungicides in field isolates of *Phakopsora pachyrhizi*. *Crop Prot.* 145: e105618.

Mullins, J. G., Parker, J. E., Cools, H. J., Togwa, R. C., Lucas, J. A., Fraaije, B. A., Kelly, D. E. and Kelly, S. L. (2011). Molecular modelling of the emergence of azole resistance in *Mycosphaerella graminicola*. *PLoS ONE* 6(6): e20973.

Nakaune, R., Adachi, K., Nawata, O., Tomiyama, M., Akutsu, K. and Hibi, T. (1998). A novel ATP-binding cassette transporter involved in multidrug resistance in the phytopathogenic fungus *Penicillium digitatum*. *Appl. Environ. Microbiol.* 64(10): 3983–8.

Ok, C. H., Popko Jr., J. T., Campbell-Nelson, K. and Jung, G. (2011). *In vitro* assessment of *Sclerotinia homoeocarpa* resistance to fungicides and plant growth regulators. *Plant Dis.* 95(1): 51–6.

Olaya, G. and Geddens, R. (2019). The methyl benzimidazole carbamate fungicides (FRAC code 1). In: Stevenson, K. L., McGrath, M. T. and Wyenandt, C. A. (Eds.), *Fungicide Resistance in North America* (2nd edn.). APS Press, St. Paul, MN, pp. 29–40.

Parker, J. E., Warrilow, A. G. S., Cools, H. J., Martel, C. M., Nes, W. D., Fraaije, B. A., Lucas, J. A., Kelly, D. E. and Kelly, S. L. (2011). Mechanism of binding of prothioconazole to *Mycosphaerella graminicola CYP51* differs from that of other azole antifungals. *Appl. Environ. Microbiol.* 77(4): 1460–5.

Paul, P. A., Lipps, P. E., Hershman, D. E., McMullen, M. P., Draper, M. A. and Madden, L. V. (2008). Efficacy of triazole-based fungicides for *Fusarium* head blight and deoxynivalenol control in wheat: A multivariate meta-analysis. *Phytopathology* 98(9): 999–1011.

Peever, T. L. and Milgroom, M. G. (1994). Lack of correlation between fitness and resistance to sterol biosynthesis-inhibiting fungicides in *Pyrenophora teres*. *Phytopathology* 84(5): 515–9.

Pereira, W. V., Morales, R. G. F., Bauer, A. I. G., Kudlawiec, K. and May-De-Mio, L. L. (2020a). Discontinuance of tebuconazole in the field restores sensitivity of *Monilinia fructicola* in stone fruit orchards. *Plant Pathol.* 69(1): 68–76.

Pereira, D., McDonald, B. A. and Croll, D. (2020b). The genetic architecture of emerging fungicide resistance in populations of a global wheat pathogen. *Genome Biol. Evol.* 12(12): 2231–44.

Pirgozliev, S. R., Edwards, S. G., Hare, M. C. and Jenkinson, P. (2002). Effect of dose rate of azoxystrobin and metconazole on the development of *Fusarium* head blight and the accumulation of deoxynivalenol (DON) in wheat grain. *Eur. J. Plant Pathol.* 108(5): 469–78.

Proffer, T. J., Berardi, R., Ma, Z., Nugent, J. E., Ehret, G. R., McManus, P. S., Jones, A. L. and Sudin, G. W. (2006). Occurrence, distribution, and polymerase chain reaction-based detection of resistance to sterol demethylation inhibitor fungicides in populations of *Blumeriella jaapii* in Michigan. *Phytopathology* 96(7): 709–17.

Ragsdale, N. N. (1975). Specific effects of triarimol on sterols biosynthesis in *Ustilago maydis*. *Biochim. Biophys. Acta* 380(1): 81–96.

Ragsdale, N. N. and Sisler, H. D. (1973). Mode of action of triarimol in *Ustilago maydis*. *Pestic. Biochem. Physiol.* 3(1): 20–9.

Sanglard, D., Ischer, F., Calabrese, D., de Micheli, M. and Bille, J. (1998a). Multiple resistance mechanisms to azole antifungals in yeast clinical isolates. *Drug Resist. Updat.* 1(4): 255–65.

Sanglard, D., Ischer, F., Koymans, L. and Bille, J. (1998b). Amino acid substitutions in the cytochrome P-450 *lanosterol 14α-demethylase* (*CYP51A1*) from azole-resistant *Candida albicans* clinical isolates contribute to resistance to azole antifungal agents. *Antimicrob. Agents Chemother.* 42(2): 241–53.

Sanglard, D., Kuchler, K., Ischer, F., Pagani, J. L., Monod, M. and Bille, J. (1995). Mechanisms of resistance to azole antifungal agents in *Candida albicans* isolates from AIDS patients involve specific multidrug transporters. *Antimicrob. Agents Chemother.* 39(11): 2378–86.

Schmitz, H. K., Medeiros, C. A., Craig, I. R. and Stammler, G. (2014). Sensitivity of *Phakopsora pachyrhizi* towards quinone-outside-inhibitors and demethylation-inhibitors, and corresponding resistance mechanisms. *Pest Manag. Sci.* 70(3): 378–88.

Schnabel, G., Bryson, P. K., Bridges, W. C. and Brannen, P. M. (2004). Reduced sensitivity in *Monilinia fructicola* to propiconazole in Georgia and implications for disease management. *Plant Dis.* 88(9): 1000–4.

Schnabel, G. and Jones, A. L. (2001). The *14α-demethylase* (*CYP51A1*) gene is overexpressed in *Venturia inaequalis* strains resistant to myclobutanil. *Phytopathology* 91(1): 102–10.

Shephard, M. C. (1985). Fungicide behavior in the plant: Systemicity. In: *British Crop Protection Council Monograph.*

Sheridan, J. E., Grbavac, N. and Sheridan, M. H. (1985). Triadimenol insensitivity in *Pyrenophora teres*. *Trans. Br. Mycol. Soc.* 85(2): 338–41.

Shirane, N., Takenaka, H., Ueda, K., Hashimoto, Y., Katoh, K. and Ishii, H. (1996). Sterol analysis of DMI-resistant and -sensitive strains of *Venturia inaequalis*. *Phytochemistry* 41(5): 1301–08.

Sierotzki, H. and Stammler, G. (2019). Resistance of plant pathogens to QoI fungicides (FRAC code 11). In: Stevenson, K. L., McGrath, M. T. and Wyenandt, C. A. (Eds.),

Fungicide Resistance in North America (2nd edn.). APS Press, St. Paul, MN, pp. 97–113.

Sisler, H. D. and Ragsdale, N. N. (1977). Fungitoxicity and growth regulation involving aspects of lipid biosynthesis. *Neth. J. Pl. Pathol.* 83(S1): 81–91.

Sisler, H. D., Ragsdale, N. N. and Waterfield, W. F. (1984). Biochemical aspects of the fungitoxic and growth regulatory action of fenarimol and other pyrimidin-5-ylmethanols. *Pestic. Sci.* 15(2): 167–76.

Spanner, R., Taliadoros, D., Richards, J., Rivera-Varas, V., Neubauer, J., Natwick, M., Hamilton, O., Vaghefi, N., Pethybridge, S., Secor, G. A., Friesen, T. L., Stukenbrock, E. H. and Bolton, M. D. (2021). Genome-wide association and selective sweep studies reveal the complex genetic architecture of DMI fungicide resistance in *Cercospora beticola*. *Genome Biol. Evol.* 13(9): evab209. doi: 10.1093/gbe/evab209.

Stammler, G., Carstensen, M., Koch, A., Semar, M., Strobel, D. and Schlehuber, S. (2008). Frequency of different *CYP51*-haplotypes of *Mycosphaerella graminicola* and their impact on epoxiconazole-sensitivity and -field efficacy. *Pest Manag. Sci.* 27(11): 1448–56.

Stammler, G., Cordero, J., Koch, A., Semar, M. and Schlehuber, S. (2009). Role of the Y134F mutation in *cyp51* and overexpression of *cyp51* in the sensitivity response of *Puccinia triticina* to epoxiconazole. *Crop Prot.* 28(10): 891–7.

Stammler, G. and Semar, M. (2011). Sensitivity of *Mycosphaerella graminicola* (anamorph: *Septoria tritici*) to DMI fungicides across Europe and impact on field performance. *EPPO Bull.* 41(2): 149–55.

Stanis, V. F. and Jones, A. L. (1985). Reduced sensitivity to sterol-inhibiting fungicides in field isolates of *Venturia inaequalis*. *Phytopathology* 75(10): 1098–101.

Stergiopoulos, I., van Nistelrooy, J. G. M., Kema, G. H. J. and De Waard, M. A. (2003). Multiple mechanisms account for variation in base-line sensitivity to azole fungicides in field isolates of *Mycosphaerella graminicola*. *Pest Manag. Sci.* 59(12): 1333–43.

Tomlin, C. D. (2011). The e-pesticide manual, 5.2 British Crop Protection Council (Ed.).

Trkulja, N., Milosavljević, A., Atanisavljevic, R., Mitrović, M., Jović, J., Toševski, I. and Bošković, J. (2015). Occurrence of *Cercospora beticola* populations resistant to benzimidazoles and demethylation-inhibiting fungicides in Serbia and their impact on disease management. *Crop Prot.* 75: 80–7.

Trkulja, N. R., Milosavljević, A. G., Mitrović, M. S., Jović, J. B., Toševski, I. T., Khan, M. F. R. and Secor, G. A. (2017). Molecular and experimental evidence of multi-resistance of *Cercospora beticola* field populations to MBC, DMI and QoI fungicides. *Eur. J. Plant Pathol.* 149(4): 895–910.

Tucker, M. A., Lopez-Ruiz, F., Cools, H. J., Mullins, J. G. L., Jayasenad, K. and Oliver, R. P. (2020). Analysis of mutations in West Australian populations of *Blumeria graminis* f. sp. *hordei CYP51* conferring resistance to DMI fungicides. *Pest Manag. Sci.* 76(4): 1265–72.

Wang, Z.-N. and Coley-Smith, J. R. (1986). Studies on some characteristics of dicarboximide-resistant isolates of *Botrytis cinerea* from protected lettuce. *Plant Pathol.* 35(4): 544–50.

Warner, S. A., Eierman, D. F., Sovocool, G. W. and Domnas, A. J. (1982). Cycloartenol-derived sterol biosynthesis in the *Peronosporales*. *Proc. Natl. Acad. Sci. U. S. A.* 79(12): 3769–72.

Wei, L. L., Chen, W. C., Zhao, W. C., Wang, J., Wang, B. R., Li, F. J., Wei, M. D., Guo, J., Chen, C. J., Zheng, J. Q. and Wang, K. (2020). Mutations and overexpression of *CYP51* associated with DMI-resistance in *Colletotrichum gloeosporioides* from chili. *Plant Dis.* 104(3): 668–76.

Wyand, R. A. and Brown, J. K. M. (2005). Sequence variation in the *CYP51* gene *of Blumeria graminis* associated with resistance to sterol demethylase inhibiting fungicides. *Fungal Genet. Biol.* 42(8): 726–35.

Yaghmour, M. A., Bostock, R. M., Adaskaveg, J. E. and Michailides, T. J. (2012). Propiconazole sensitivity in populations of *Geotrichum candidum*, the cause of sour rot of peach and nectarine, in California. *Plant Dis.* 96(5): 752-8.

Zehr, E. I., Luszcz, L. A., Olien, W. C., Newall, W. C. and Toler, J. E. (1999). Reduced sensitivity in *Monilinia fructicola* to propiconazole following prolonged exposure in peach orchards. *Plant Dis.* 83(10): 913–6.

Zhan, J., Stefanato, F. L. and McDonald, B. A. (2006). Selection for increased cyproconazole tolerance in *Mycosphaerella graminicola* through local adaptation and in response to host resistance. *Mol. Plant Pathol.* 7(4): 259–68.

Zhang, C., Imran, M., Xiao, L., Hu, Z., Li, G., Zhang, F. and Liu, X. (2021). Difenoconazole resistance shift in *Botrytis cinerea* from tomato in China associated with inducible expression of *CYP51*. *Plant Dis.* 105(2): 400–7.

Zhang, Y., Mao, C. X., Zhai, X. Y., Jamieson, P. A. and Zhang, C. Q. (2021). Mutation in *cyp51b* and overexpression of *cyp51a* and *cyp51b* confer multiple resistant to DMIs fungicide prochloraz in *Fusarium fujikuroi*. *Pest Manag. Sci.* 77(2): 824–33.

Zhang, Y., Zhou, Q., Tian, P., Li, Y., Duan, G., Li, D., Zhan, J. and Chen, F. (2020). Induced expression of *CYP51* associated with difenoconazole resistance in the pathogenic *Alternaria* sect. on potato in China. *Pest Manag. Sci.* 76(5): 1751–60.

Ziogas, B. N. and Malandrakis, A. A. (2015). Sterol biosynthesis inhibitors: C14 demethylation (DMIs). In: Ishii, H. and Hollomon, D. W. (Eds.), *Fungicide Resistance in Plant Pathogens - Principles and a Guide to Practical Management*. Springer, Japan, Tokyo, pp. 199–216.

Zwiers, L. H., Stergiopoulos, I., van Nistelrooy, J. G. M. and de Waard, M. A. (2002). ABC transporters and azole susceptibility in laboratory strains of the wheat pathogen *Mycosphaerella graminicola*. *Antimicrob. Agents Chemother.* 46(12): 3900–6.

Chapter 11

Quinone outside inhibitor fungicide resistance: selection patterns and the current situation

Stefano F. F. Torriani and Helge Sierotzki, Syngenta Crop Protection AG, Switzerland

1 Introduction

Fungicides with mitochondrial respiration-inhibiting mechanisms, grouped in the mode of action 'C,' are the most abundant among the classes listed by Fungicide Resistance Action Committee (FRAC) (www.frac.info). This class of fungicides is divided into eight subgroups (C1 to C8) depending on the specific enzymatic target and cross-resistance behavior:

- Inhibitors of complex I NADH oxido-reductase (C1 – group 39);
- Inhibitors of complex II succinate dehydrogenase (C2 – group 7);
- Inhibitors of complex III cytochrome bc1 (ubiquinol oxidase) at Qo site (C3 – group 11 and 11A);
- Inhibitors of complex III cytochrome bc1 at Qi site (C4 – group 21);
- Inhibitors of complex III cytochrome bc1 at Qo site stigmatellin-binding subsite (C8 – group 45);
- Inhibitors of complex V oxidative phosphorylation (C5 – group 29);
- ATP synthase (C6 – group 30); and
- Proposed uncouplers of oxidative phosphorylation and ATP transport (C7 – group 38).

http://dx.doi.org/10.19103/AS.2022.0116.11

Of the eight 'C-subgroups,' the fungicide binding at the complex III Qo site of the mitochondrial encoded gene cytochrome b (code 11, 11A and 45) are the most abundant with two new active ingredients included in the past decade (mandestrobin and metyltetraprole). The inner pocket of the cytochrome b enzyme is targeted by another group of compounds (group 21). Two fungicides have recently been included in this group - fenpicoxamid and florylpicoxamid - with a broader spectrum of plant pathogen control compared with the first members of this group - cyazofamid and amisulbrom - which showed oomycete control. This class of fungicides will not be discussed in this chapter.

Fungicide binding to the complex III Qo site are classified by FRAC into different groups according to their resistance patterns. Three different resistance mechanisms lead to different cross-resistance profiles for group 45, group 11, and newly introduced group 11A (Fig. 1). The chemistry, synthesis, diversity, structure–activity relationship, biokinetic, transport distribution, and influence on plant physiology of quinone outside inhibitors (QoI) fungicides have been summarized and updated in Gewehr and Sauter (2019).

All QoI fungicides bind into the ubiquinol oxidizing outer pocket. The most frequent binding mode of the compounds, belonging to group 11, is characterized by strobilurin A and myxothiazole binding, which is mediated by the interaction with Glu 272 of the EF helix. Other important interactions are Gly 143 and Phe 129. The newly introduced group 11A fungicide (metyltetraprole) also expresses a strong binding to Glu 272 but is not or little influenced by Gly 143 and Phe 129, respectively (Arakawa et al., 2018, Fig. 2). Group 45 (ametoctradin) was initially characterized by a similar binding mode to

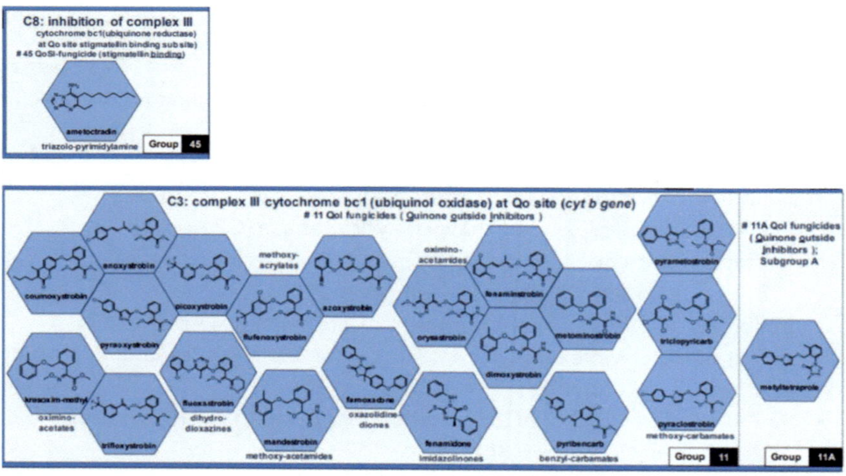

Figure 1 FRAC classification of group 11, group 11A, and group 45. Source: FRAC Mode of Action Poster, 2021.

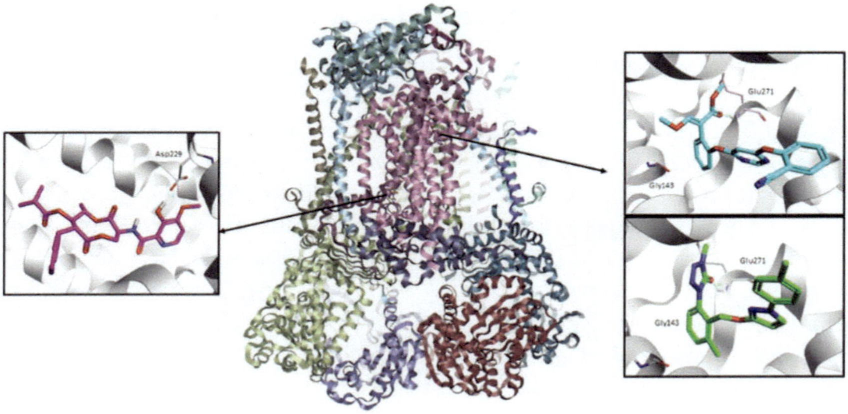

Figure 2 Model of cytochrome b (different domains are highlighted in different colors). The zoomed-in pictures indicate the proposed binding for azoxystrobin (top right) and metyltetraprole (bottom right) in the Qo site and as a comparison (not discussed in the chapter) florylpicoxamid in the Qi site (left) (Stefano Rendine, Syngenta internal).

Stigmatellin into the Qo pocket, which is sufficiently different in order not to be affected by variations at the Gly 143 position (Zhu et al., 2015). However, later studies revealed additional binding properties to the quinone inside inhibitor (QiI) site (Dreinert et al., 2018). The most recent modeling information has inferred the Qi site as important for resistance development in the field (see later).

The relative relevance of QoI fungicides is projected to remain between their current 19% and 15% of the total fungicide market over the next 10 years. They are used in many crops, including cereals, soybean, grapes, vines, peanuts, turf, maize, oil seed rape, potatoes, cotton, and tea, as well as many fruits and vegetables. The 13 most sold QoI fungicides accounted for more than US$3 billion sales in 2020, with azoxystrobin, trifloxystrobin, pyraclostrobin, picoxystrobin and kresoxim-methyl, the leading fungicides (Agbio Crops, December 2021). Despite QoI resistance being observed in a range of pathogens, this group of fungicides is still included in most spray programs as a mixture partner, likely due to its very broad spectrum and good crop tolerance.

Registration application in the European Union (EU) for the novel fungicide compound with the ISO common name metyltetraprole was submitted in 2018 (https://www.basf.com/global/en/media/news-releases/2018/06/p-18-240.html). Dependent on regulatory approval, market introduction is expected in the EU from 2022 onwards.

2 The history of resistance and resistance mechanisms

Following the introduction of QoIs in the market in 1996, the first example of resistance was detected in 1998 in *Podosphaera xanthii* (Vielba-Fernandez et al., 2020). In 1999, resistance was observed in *Blumeria graminis* f. sp. *tritici* (powdery mildew on wheat), in *Mycosphaerella fijiensis* (black sigatoka on banana), and in *Plasmopara viticola* (downy mildew of grapes). The spread of resistance in fungal plant pathogen populations up to 2015 has been summarized by Sierotzki (2015). Today, around 40 species have developed resistance to QoI (www.FRAC.info). Among the most prominent examples of rapid resistance development are soybean rust caused by *Phakopsora pachyrhizi* (Klosowski et al., 2016) and Cercospora leaf spot of sugar beet caused by *Cercospora beticola* (Birla et al., 2012; Bolton et al., 2013).

The mechanisms leading to resistance to group 11 fungicides have been summarized and reviewed (Sierotzki, 2015). Only a few mutations in cytochrome b have been reported for most of the resistance cases observed to date. The G143A mutation was first described in 1996 (Brasseur et al., 1996) as having a strong effect on the binding of QoI (myxothiazol) in combination with a minimal impact on fitness. This mutation is responsible for most of the resistance cases associated with relevant field performance reduction (Table 1).

Initially, it was claimed that the larger alanine residue clashes sterically with the phenolic linker between the pharmacophore and the hydrophobic tail (Link et al., 2003). However, more recent studies (Arakawa et al., 2018) have revealed that the pharmacophore part (known as part C in Gewehr and Sauter, 2019) is mainly involved in resulting in the resistant phenotype of G143A-possessing strains. Artificial mutagenesis highlighted the possibility of selecting for serine at position 143 as shown for *Pyricularia grisea* (Ruan, 2022) and *Cercospora beticola* (Malandrakis et al., 2011), which led to a resistant phenotype and comparable resistance factor as for alanine. Despite G143S having only a weak fitness disadvantage, this mutation has not been detected in field isolates. The impact of the G143A mutation on the field performance of the QoI fungicides is dependent on the frequency of the mutation in the local pathogen population. The second most frequent mutation is at codon 129 of cytochrome b, leading to a change of phenylalanine to leucine (F129L). This mutation is claimed to have a smaller resistance factor than G143A and therefore does not significantly affect the field performance of QoI-containing products.

Variations in resistance factors associated with F129L do, however, depend on the pathogen, the QoI fungicide considered (Patel et al., 2011) and the frequency of the mutation in the population. Besides the two most common mutations, another variant at codon 137 (G137R) was observed in plant pathogens collected from the field (Sierotzki et al., 2007). Recently, other mutations (L299F or N256S + L299F) have been reported in *Puccinia horiana*

Table 1 List of QoI-resistant and -sensitive plant pathogen species monitored

Species name	Common name	Host	Geography	Sensitivity/resistance mechanisms	Intron b2 after position G143	Comments
Alternaria alternata	Alternaria blight	Pistachio	US	G143A	No	Abundant
Alternaria alternata	Early blight	Potato, Tomato	EU	G143A	No	Abundant
Alternaria arborescens	Alternaria blight	Pistachio	US	G143A	No	Abundant
Alternaria brassicicola		Various	EU	Sensitive	No	
Alternaria brassicae		Various	EU	Sensitive	No	
Alternaria dauci		Carrot	EU	G143A	No	Moderate
Alternaria helianthi		Sunflower	EU	Sensitive		
Alternaria mali	Alternaria blotch	Apple	US	G143A	No	
Alternaria solani	Early blight	Potato, Tomato	US, EU	F129L	Yes	Abundant
Alternaria tenuissima	Alternaria blight	Pistachio	US	G143A	No	Abundant
Alternaria tomatophila		Tomato	US, EU	G143A	No	Abundant
Blumeria graminis f.sp. *hordei*	Powdery mildew	Barley	EU	G143A	No	Abundant
Blumeria graminis f.sp. *tritici*	Powdery mildew	Wheat	EU	G143A	No	Abundant
Bipolaris maydis	Northern leaf spot	Corn	EU	Sensitive		
Botrytis cinerea	Gray mold	Strawberries	EU	G143A	10% with intron	Abundant
Bremia lactucae	Downy mildew	Lettuce	EU	Sensitive		
Cercospora sojina	Frogeye spot	Soybean	US, BR	G143A	No	Abundant

(*Continued*)

Table 1 (*Continued*)

Species name	Common name	Host	Geography	Sensitivity/resistance mechanisms	Intron b2 after position G143	Comments
Cercospora spp.		Soybean	BR	G143A	No	Abundant
Cercospora zeae-maydis	Corn gray leaf spot	Corn	US	G143A	No	
Cercospora beticola	Cercospora leaf spot	Sugar beet	US, EU, JA	G143A	No	Abundant
Colletotrichum lentis[a]	Anthracnose	Lentil	US	G143A	No	
Cochliobolus carbonum		Various		Sensitive	Yes	
Colletotrichum acutatum	Anthracnose	Citrus	US	G143A	No	
Colletotrichum graminicola[b]	Anthracnose	Turf grass	US	G143A	No	Abundant
Corynespora cassiicola	Leaf spot, target spot	Cucumber	Japan	G143A	No	Abundant
Didymella bryoniae	Gummy stem blight	Cucurbit	US, EU	G143A	No	
Didymella rabiei[c]	Ascochyta blight	Chickpeas	US	G143A	No	
Erysiphe necator	Powdery mildew	Grape	US, EU	G143A	No	Abundant
Erysiphe heraclei	Powdery mildew	Carrot	EU	G143A	No	
Exserohilum turcicum	Northern leaf blight	Corn	EU	Sensitive	Yes	
Fusicladium effusum[d]	Pecan scab	Pecan	USA	G137S	Yes	
Colletotrichum gloeosporioides	Anthracnose	Strawberries	US, Japan	G143A	No	Abundant

Guignardia bidwellii	Black rot	Grape		Sensitive	Yes	
Guignardia citricarpa	Citrus black spot	Citrus		Sensitive	Yes	
Microdochium nivale	Snow mold, ear blight	Wheat	EU, Japan	G143A	No	Abundant
Microdochium majus	Snow mold, ear blight	Wheat	EU, Japan	G143A	No	Abundant
Monilinia species	Brown rot	Stone fruit	EU	Sensitive	Yes	
Mycosphaerella fijiensis	Black sigatoka	Banana	Lat. America, PH	G143A	No	Abundant
Zymoseptoria tritici	Septoria leaf blotch	Wheat	US, EU	G143A, F129L	No	Abundant
Mycosphaerella musicola	Yellow sigatoka	Banana	S. America, Australia	G143A	No	
Mycovellosiella nattrassii[e]	Leaf mold	Eggplant	Japan	G143A	No	
Neofabrea alba, perenanns	Bull's eye rot	Apple	EU	Sensitive		
Passalora fulva[f]	Leaf mold	Tomato	Japan	F129L	Yes	
Peronospora destructor	Downy mildew	Onion	EU	Sensitive	No	
Phaeosphaeria nodorum	S. nodorum leaf blotch	Wheat	EU	G143A	No	
Phakopsora pachyrhizi	Soybean rust	Soybean	BR	F129L	Yes	Abundant
Phytophthora infestans	Late blight	Potato	EU	Sensitive	No	
Plasmopara viticola	Downy mildew	Grape	EU	G143A, F129L	No	Abundant
Plasmopara halstedii	Downy mildew	Sunflower	EU	Sensitive		
Plenodomus lingam, biglobosus	Phoma	OSR	EU	Sensitive		

(*Continued*)

Table 1 (*Continued*)

Species name	Common name	Host	Geography	Sensitivity/resistance mechanisms	Intron b2 after position G143	Comments
Podosphaera leucotricha	Powder mildew	Apple	EU	Sensitive	Yes	
Pseudoperonospora cubensis	Downy mildew	Cucurbits	EU, Asia	G143A	No	Abundant
Puccinia horiana[g]	White rust	Chrysanthemum	Japan	L275F, L299F or N256S + L299F	Yes	
Puccinia hordei	Dwarf rust	Barley	EU	Sensitive	Yes	Non-target-site reduced sensitivity observed
Puccinia triticina	Wheat rust	Wheat	EU	Sensitive	Yes	
Puccinia striiformis	Yellow rust	Wheat	EU	Sensitive	Yes	
Puccinia sorghi[h]	Rust	Millet		Sensitive	Yes	
Pyrenopeziza brassicae	Light leave spot	OSR	EU	Sensitive		
Pyrenophora teres	Net blotch	Barley	EU	F129L	Yes	Abundant
Pyrenophora tritici-repentis	Tan spot	Wheat	EU	G143A, F129L, G137R	No	Abundant
Pyricularia grisea	Gray leaf spot	Turf grass Rice	US JP, VN, IT	G143A, F129L G143A, I300F	No No	
Pythium aphanidermatum	Pythium blight	Turf grass	US	F129L	Yes	Abundant
Pythium irregulare G1	Pythium blight	Various	EU	G143A	No	Limited
Pythium ultimum	Pythium blight	Various	EU			Natural insensitivity observed in some strains

Ramularia areola, pseudoglycines	Gray mildew	Cotton	Brazil	G143A	No	Abundant
Ramularia collo-cygni	Ramularia	Barley	EU	G143A	No	Abundant
Rhynchosporium commune	Scald	Barley	EU	G143A	No	Limited
Rhizoctonia solani AG1.1A	Sheath blight	Rice	US	F129L	No	Limited
Sclerotonia sclerotiorum	White mold	Various	EU	Sensitive	No	
Sphaerotheca fuliginea	Powdery mildew	Cucurbits	EU, Asia	G143A	No	Abundant
Stemphylium botryosum		Spinach	US	G143A	No	Abundant
Stemphylium vesicarium	Brown spot	Asparagus, Pear	EU	G143A	No	Abundant
Venturia inaequalis	Scab	Apple	EU	G143A	No	Abundant
Venturia pirina	Scab	Pear	US	G143A	No	

References see www.frac.info and [a]Hoffmeister et al. (2022); [b]Sierotzki (2015); [c]Delgado et al. (2013); [d]Standish et al. (2016); [e]Ishii et al. (2007); [f]Watanabe et al. (2017); [g]Matsuzaki et al. (2021); [h]Grasso et al. (2006).

to confer resistance to QoI (Matsuzaki et al., 2021). This species contains the b2 intron after the G143 position in the cytochrome b gene. In the same species, mutation L275F has previously been monitored (Matsuura, 2019) with moderate resistance factors. Other mutations have been generated in lab studies, such as M296V (Miao et al., 2020), but these mutations have not yet been selected in the field and do not represent a challenge for QoI performances. The reason has not been fully elucidated but is most likely associated with the fitness penalty. Besides target site mutations, another resistance mechanism known as alternative oxidase (AOX) was reported to reduce sensitivity to QoI. Despite the importance of the AOX in homeostasis and abiotic and biotic stress mitigation (Vanlerberghe, 2013), all current available studies suggest that AOX can induce reduced sensitivity in *in vitro* studies on artificial media. This mechanism is not conferring field-relevant resistance (Esser et al., 2014).

It is commonly known that the presence of an intron type II, directly or within a few nucleotides after codon 143, is likely to prevent the evolution of the G143A-mediated strong resistance to QoI. Introns type II are self-splicing elements and need to form a particular hairpin loop structure, including the triplet coding for the G at position 143 in the cytochrome b gene to be correctly spliced out. If a mutation occurs in the triplet, splicing does not take place and a non-functional transcript is generated (Grasso et al., 2006). The intron at this position has been reported in a number of plant pathogens, most of which evolved mutation F129L associated with lower resistance factors. Bypass mechanisms have been reported (Vallieres et al., 2011), but not yet in relevant agronomic plant pathogens. This example, however, shows that resistance can emerge due to other mutations or mechanisms, as described for the soybean rust pathogen, *Phakopsora pachyrhizi* (Klosowski et al., 2916). Since the intron type II is described as a mobile element, it is principally possible that it moves and would allow the evolution of stronger resistance due to the G143A mutation.

Another proposed bypass mechanism is related to the whole gene transfer between closely related species, as happened for a virulence gene between *Stagonospora nodorum* and *Pyrenophora tritici-repentis* (Friesen et al., 2006). As a matter of fact, a gene transfer has been detected in a single isolate of *Pyrenophora teres*. Co-infection of *Pyrenophora tritici-repentis* and *Pyrenophora teres* on barley seem to have enabled the partial replacement of the *Pyrenophora teres* cytb gene by the G143A-containing gene fragment from *Pyrenophora tritici-repentis* (Hoffmeister et al., 2022). Subsequent intensive population monitoring did not reveal any further individuals of *Pyrenophora teres* with the G143A mutation.

Resistance to the recently announced QoI group 11A (metyltetraprole) has not been described from field isolates. A recent mechanistic study using the yeast model revealed the interaction of different mutations with metyltetraprole.

The G143A mutation does not affect the binding and is therefore not considered a resistance allele for QoI group 11A; however, other mutations have been proposed for further resistance risk studies, such as different amino acids at position 129 and 143 or mutations at position 132 and 275 (Serot et al., 2022). However, in-depth cross-resistance studies using G143A possessing strains of *Zymoseptoria tritici* and F129L possessing *Pyrenophora teres* (Matsuzaki and Iwahashi, 2019; Suemoto et al., 2019) showed no or only minor impact of these mutations on sensitivity to metyltetraprole.

The resistance risk for group 11A has been estimated by FRAC to be high, which implies that resistance is expected to occur in high-risk pathogens under disease-conducive agronomic conditions and if the density of the usage exceeds a certain percentage of the whole cropping area in a region (Debarre et al., 2009). Specific anti-resistance recommendations have been published by FRAC (www.frac.info).

The fungicide group 45 [Quinone outside inhibitor stigmatellin-binding type (QoSI)], including ametoctradin, has been reported to bind to the Qo site in a slightly different manner compared to the QoI fungicides. The binding was more like stigmatellin and was therefore classified accordingly in the FRAC code list (Fehr et al., 2016). QoSI group is non-cross-resistant to the QoI groups 11 and 11A. More recent results from a functional binding study indicate an unusual double binding of ametoctradin to both the Qo and Qi sites (Dreinert et al., 2018). In the Qo site, mutations at codon 147 and 278 in the cytochrome b were reported to influence sensitivity. The mutation S34L in the Qi site has been detected in field isolates of *Plasmopara viticola* (Zito et al., 2020; Fontaine et al., 2019). Mutation analysis of the mitochondrial complex III Qi site revealed that several mutations could affect the activity of ametoctradin and that the S34L mutation found in field isolates might be associated with a fitness penalty (Mounkoro et al., 2019). Despite the putative binding of group 45 to the Qi site of cytochrome b, no cross-resistance has been detected to Qi-site inhibitors of group 21 (Zito et al., 2020). The use of fungicides belonging to group 45 seems to have increased the frequency of the AOX-related mechanism in *Plasmopara viticola* populations. As described earlier, the practical relevance of this mechanism to reduce field performance remains unclear (Fontaine et al., 2019).

The following case studies have been selected to describe different QoI resistance evolutionary patterns and dynamics in relevant agronomic pathogens worldwide.

3 Case study 1: *Phakopsora pachyrhizi* (soybean – Brazil)

In the last two decades, soybean production has increased significantly in Brazil. Totaling more than 110 million tons, Brazil is catching up with the US as one of the largest producers worldwide. Soybean is currently the most prevalent crop in the country with more than 34 million ha of growing area. The total soybean cultivated surface is comparable to the area cultivated by the other seven major crops put together (Cattelan and Dall'Agnol, 2018). Soybean rust is not an endemic disease in Brazil, with the first report of *Phakopsora pachyrhizi* in the country in 2002 (Rossi, 2003). This pathogen found optimal climatic and agronomic conditions in Brazil for rapid spread, becoming within a few years the most relevant soybean disease.

Besides agronomic practices such as early sowing of resistant cultivars and eradication of secondary hosts, the most effective disease control strategy is currently related to the use of fungicides. Fungicides belonging to QoI group 11, often in association with demethylation inhibitors (DMIs), have historically been the backbone of disease and resistance management. Today, fungicide spray programs mainly include products containing QoI, DMI, succinate dehydrogenase inhibitors (SDHI), and multisites.

The first report of QoI adaptation was in the 2012–2013 season following around 10 years of selection. Considering the high local agronomic risk, the first report of QoI adaptation was surprisingly late. However, once selected, in just two seasons the Brazilian QoI-sensitive population of *Phakopsora pachyrhizi* was completely replaced by less sensitive isolates harboring mutation cytb-F129L (Fig. 3a). The rapid geographic distribution and quick selection of the QoI resistance allele can be explained by the high agronomic risk characterizing soybean cultivation in Brazil. A DMI sensitivity shift was observed in Brazil a few years earlier, between 2007 and 2010. The decreased sensitivity of *Phakopsora pachyrhizi* to DMI might have led to a higher selection of QoI. The intensive use of QoI in Brazilian soybean field contributed to the evolution of QoI resistance based on mutation cytb-G143A in other soybean pathogens such as different *Cercospora* species, *Colletotrichum truncatum*, and *Corynespora cassiicola* (Mello et al., 2018, 2022). The adaptation mechanisms described for QoI and DMI fungicides for *Phakopsora pachyrhizi* do not confer complete resistance and products belonging to these fungicide classes maintain acceptable efficacy in the field.

In the last few years, several initiatives to optimize fungicide use enabled the adoption and implementation of anti-resistance guidelines. Diversification of spray programs including fungicides with different mechanisms of action such as SDHIs and multisites reduced the selection to single components. Recently, in some locations, higher frequencies of QoI wild-type populations were observed (Fig. 3b). The frequency of less sensitive strains could be negatively influenced by any potential fitness cost associated with the mechanism of resistance. Whether mutation F129L alone or in combination with other respiration inhibitor resistance alleles is associated with fitness cost is of great interest but has not yet been fully elucidated.

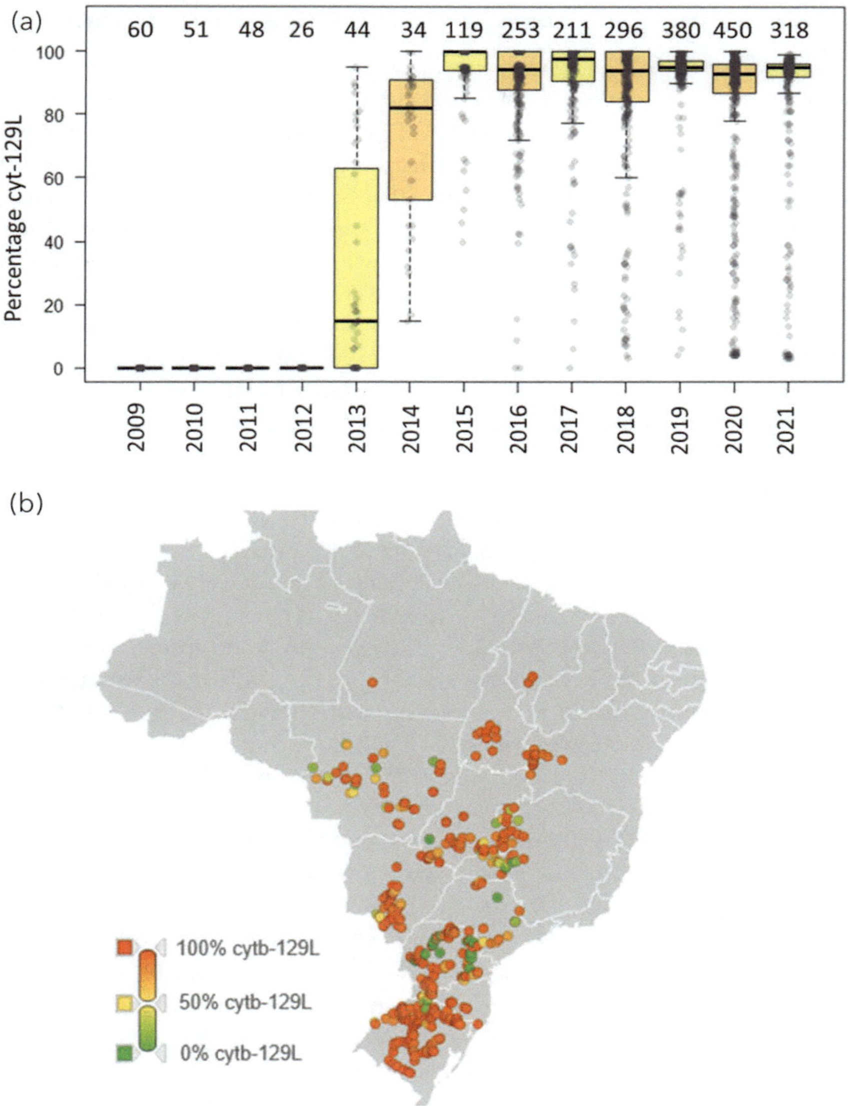

Figure 3 (a) Evolution of cytb-F129L percentage monitored for *Phakopsora pachyrhizi* populations collected from 2008–2009 to 2021–2022. A total of 2293 populations from all relevant soybean-producing areas of Brazil were monitored. A quick selection of cytb-F129L was observed between the 2012–2013 and 2014–2015 seasons. A greater number of populations with a lower percentage of cytb-F129L were observed in the last two seasons. (b) Geographic distribution and percentage of cytb-F129L of 768 *Phakopsora pachyrhizi* populations collected between 2019–2020 and 2020–2021. Despite an overall high frequency of cytb-F129L, a limited number of fully sensitive populations to QoI were observed. Quantification of cytb-F129L was carried out by pyrosequencing (Syngenta internal data).

4 Case study 2: *Cercospora beticola* (sugar beet – Europe)

Cercospora beticola, the causal agent of Cercospora leaf spot, is the most serious disease in sugar beet worldwide. This disease severely impacts root development and is associated with a significant reduction in harvest and extractable sucrose. Disease management relies on good agronomic practices such as planting tolerant varieties and adopting crop rotation. However, the most effective control of the disease is achieved if the aforementioned measures are integrated with timely fungicide applications (Kaiser et al., 2010; Karaoglanidis and Ioannidis, 2010). Among the fungicide classes routinely used to control *Cercospora beticola* are methyl benzimidazole carbamates, DMI, QoI, and multisites. Since their introduction into the market, QoI fungicides have represented a critical component of disease management, due to their broad spectrum, preventative, and curative characteristics.

In 2010, QoI resistance was monitored in Italy following a decade of intense use. The resistance mechanism is related to the mutation cytb-G143A associated with high resistance factors. In 2012, a total of 189 strains were collected from Italy ($n = 72$), France ($n = 49$), and Austria ($n = 68$). *In vitro* fungicide sensitivity tests showed a frequency of 86% resistance in Italy, 10% in France, and 9% in Austria. Resistance to QoI rapidly increased and spread throughout different countries. Molecular quantification of cytb-G143A by pyrosequencing reported low to moderate frequencies in most European countries between 2013 and 2015 (Fig. 4b). Between 2016 and 2018, most European countries were characterized by *Cercospora beticola* populations with moderate to high frequency of cytb-G143A, with only Poland showing low frequency (Fig. 4c). Recently, cytb-G143A increased to a very high frequency in most European countries with only Romania, Poland, and Russia showing moderate frequency. To date, no fitness costs have been associated with mutation cytb-G143A in *Cercospora beticola* and no recovery in sensitivity has been observed. In 2021, all monitored populations showed at least 40% of cytb-G143A mutation (Fig. 4a).

Currently, the limited amount of effective fungicide classes makes disease management of Cercospora leaf spot problematic and new solutions are required. Until new fungicide classes are registered, integrated pest management strategies need to be implemented. The use of Cercospora leaf spot-resistant crop varieties, biological control agents, and appropriate hygienic practices such as crop rotation and field hygiene obtained by the removal of diseased plant debris are all potential complementary approaches to fungicide applications.

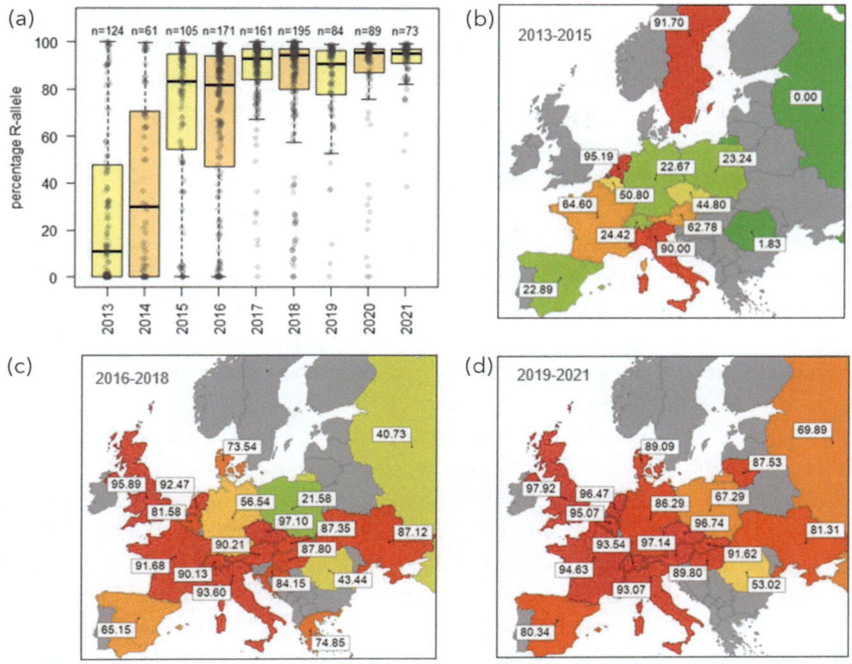

Figure 4 (a) Evolution of cytb-G143A percentage monitored for *Cercospora beticola* populations collected from the 2013-2021 season. A total of 1063 populations from all relevant sugar beet-producing areas of Europe were monitored. Quick selection of cytb-G143A was observed until 2015. (b) Geographic distribution is expressed as the mean country percentage of cytb-G143A from 2013 to 2015. (c) Geographic distribution is expressed as the mean country percentage of cytb-G143A from 2016 to 2018. (d) Geographic distribution is expressed as the mean country percentage of cytb-G143A from 2019 to 2021. Quantification of cytb-G143A was carried out by pyrosequencing (Syngenta internal data).

5 Case study 3: *Plasmopara viticola* (grapes – France)

Downy mildew in grapes caused by oomycete pathogen *Plasmopara viticola* is one of the most severe diseases in vineyards. Many of the most famous grape cultivars are highly susceptible to this obligate parasite, making disease management complex. Fungicide treatments are an essential control strategy to ensure a good quality yield. Since *Plasmopara viticola* can complete several infection cycles during a single growing season, regular fungicide treatments are necessary. The first report to control *Plasmopara viticola* with a product showing fungicidal activity was from 1882, when Pierre Marie Alexis Millardet set the basis for the development of the 'Bordeaux mixture' (Millardet, 1885).

Massi et al. (2021) described the history of chemicals used to control *Plasmopara viticola* and its associated resistance evolution. Several fungicide classes can be used to control *Plasmopara viticola*, including QoI, QoSI, and QiI respiration inhibitors. All three fungicide classes target the complex III of the mitochondrial respiration encoded by the mitochondrial cytb gene.

Plasmopara viticola is a pathogen showing high risk to evolve fungicide resistance, due to its high reproductive frequency (Calo et al., 2013) and complex lifecycle, including cycles of sexual and asexual reproduction. Resistance mechanisms related to target site mutations in cytb and activation of alternative respiration (AOX) have been described for all three fungicide classes inhibiting respiration. In contrast to target site mutations, which are specific to QoI, QoSI, and QiI, AOX-related resistance mechanisms have been proposed to affect all complex III inhibitors.

A few years following the introduction of QoI into the market, resistance based primarily on cytb-G143A was reported (Heaney et al., 2000; Gullino et al., 2004; Sierotzki et al., 2005). Mutation cytb-F129L was also described in *Plasmopara viticola*. As for other pathogens such as *Z. tritici* and *Pyrenophora tritici-repentis,* for which multiple QoI resistance alleles are described, cytb-G143A is the dominating mutation due to the high resistance factors and absence to date of relevant fitness costs. QoSI target site resistance has been described by the mutation cytb-34L in lab and since 2015 in field populations (Fountaine et al., 2019; Mounkoro et al., 2019). QiI target site resistance is explained by amino acid change cytb-201S and insertion/deletion variants involving cytb residues 203 and 204 (e.g. E203- DE-V204) (Cherrad et al., 2018). Multiple resistance to QoI and QoSI has been observed in strains showing the co-presence of mutation cytb-G143A and cytb-34L (Fontaine et al., 2019). To our knowledge and intern monitoring programs, there are to date no strains combining QiI- (cytb-201S and indels 203-204) and QoI- (cytb-G143A) related mutations.

Syngenta monitoring data based on quantification of cytb-G143A showed a selection of this mutation from 2010 to 2015 in France (Syngenta internal data). Interestingly, the frequency of cytb-143 decreased in the following years. During the period between 2019 and 2021, overall lower frequencies of cytb-G143A were observed in French populations compared to the 2010–2012 period (Fig. 5a). Monitoring data based on quantification of the single cytb mutations affecting the sensitivity to QoI and QoSI suggests a negative correlation between cytb-G143A (QoI) and cytb-203-204 indel (QoSI) (Fig. 5b). High variability from site to site is monitored for QoI- and QoSI-related molecular profiles (Fig. 5c and d).

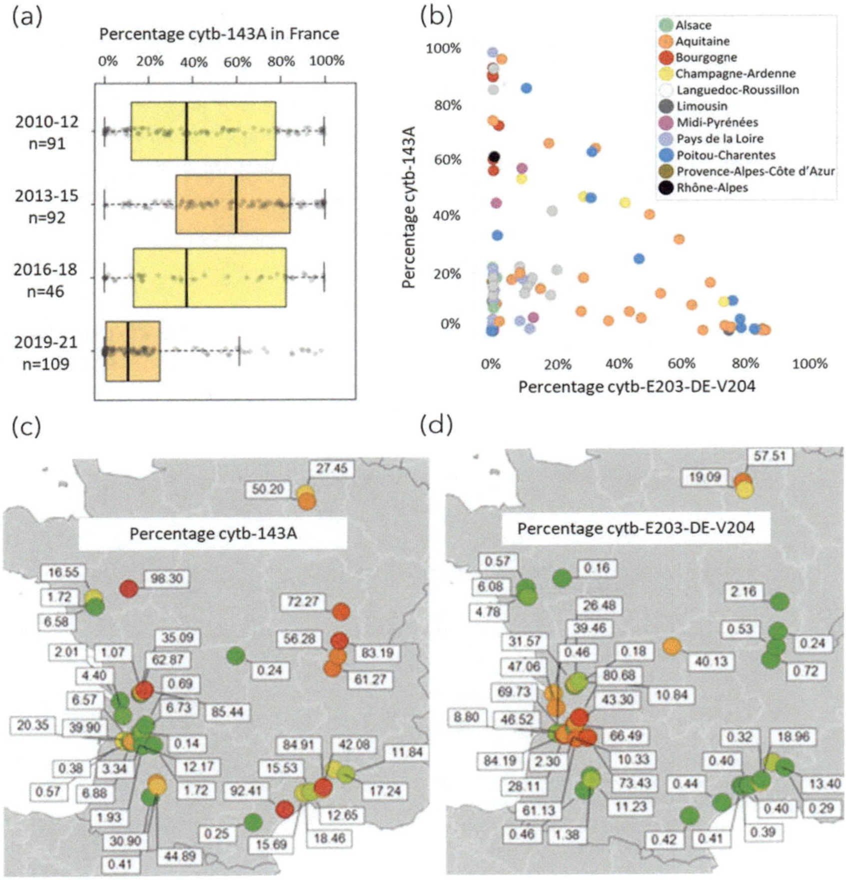

Figure 5 (a) Evolution of cytb-G143A percentage in *Plasmopara viticola* populations collected in France from 2010 to 2021. A total of 338 populations from major wine-producing areas of France were monitored. Mutation cytb-G143A increased in frequency from 2010 to 2015. A trend toward higher sensitivity to QoI has been monitored in the last few years. (b) A negative correlation was observed between cytb-G143A and cytb-E203-DE-V204 affecting QoI and QoSI, respectively. (c) Geographic distribution of cytb-G143A percentage. (d) Geographic distribution of cytb-E203-DE-V204 percentage. In panels b, c, and d, populations from 2020 and 2021 were considered.

6 Current global sensitivity: QoI groups 11, 11A and 45

6.1 QoI group 11

Overall information regarding QoI sensitivity has been obtained from 70 pathogens. Around 25 fungal species have been recorded as sensitive toward QoI code 11 fungicides. Twelve pathogens contain an intron b2 directly after the amino acid position G143, so are unlikely to develop the G143A mutation. Half acquired another mutation whilst the remainder are wild-type (sensitive). Eleven pathogens were reported to be resistant due to a mutation other than G143A (e.g. F129L) or had another mutation in addition to the G143A (Table 1). To our knowledge, no cases of a pathogen species are reported to possess two different cytb mutations in one individual. Species with resistance to QoI show significant differences in frequency and distribution. There are pathogen species with high frequency and widespread resistance in areas of intensive use of QoI fungicides, whilst in other areas resistance is still rather rare.

In 2020, the FRAC QoI working group reported data for 26 pathogen/crop systems in 33 different countries worldwide, resulting in more than 410 data points (Fig. 6). The reported cases are based on year/pathogen/country

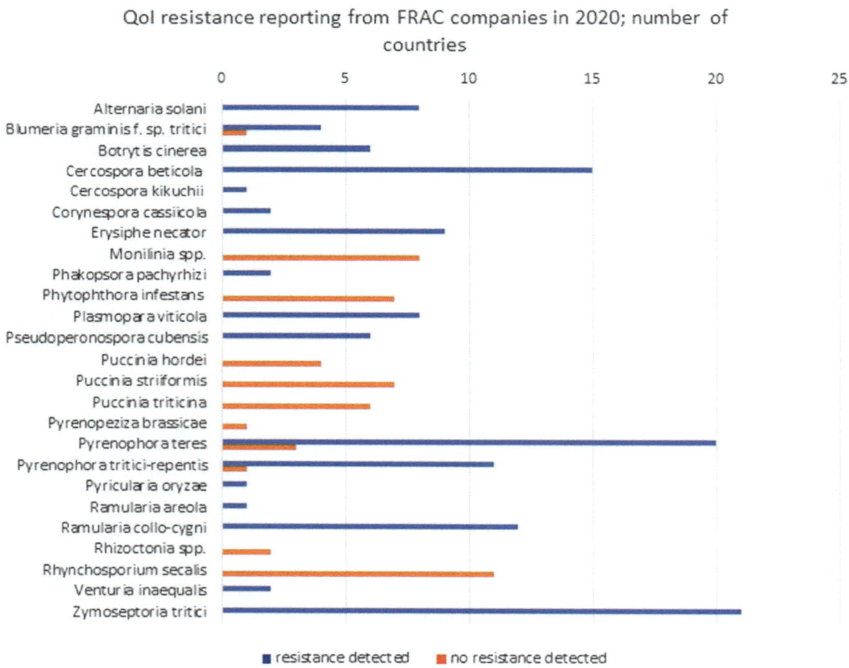

Figure 6 Reported QoI sensitivity monitoring in 2020: Number of countries for which sensitivity monitoring has been performed and reported to the FRAC QoI working group. No resistance indicates that the EC50 values of all tested isolates/samples/populations are within the range of baseline sensitivity or that all tested samples were showing wild-type cytb genotypes. Resistance detected means that either phenotypically resistant isolates, samples, populations, or resistance alleles were detected in cytb. The detailed frequency of resistance for each country can range from few detected isolates (very low frequency) to high frequency and is accessible at www.frac.info.

level. For example, the 15 cases reported for *Cercospora beticola* reflect the information shared by all members of the FRAC QoI working group, performing monitoring across 15 countries, in which resistance has been detected from a single occurrence to very high frequency levels. It is possible that for some pathogens both resistant and sensitive cases are reported, meaning that in a given year in a certain country, resistance has been detected, whereas in another country no resistant isolates were found (e.g. *B. graminis* f. sp. *tritici*).

In some pathogen/host combinations, the frequency of the QoI resistance reached high levels in the main growing areas of the respective crop, for example in *Z. tritici* in Northern Western Europe or *Plasmopara viticola* in several regions. However, for such cases, there are still significant regions with low or even no resistance, such as *Z. tritici* in Eastern Europe. Additionally, it is important to state that even in cases with strong resistance mechanisms such as G143A, some residual efficacy of QoI remain detectable and in some cases contribute significantly to overall disease control (Blake et al., 2018). In the United Kingdom, the frequency of the G143A mutation is very high and widespread in *Z. tritici*; however, the residual activity of QoI (at different degrees comparing individual compounds) was still visible in field trials (Blake, 2018).

As mentioned in the overview, a notable number of pathogens have not yet developed resistance toward the QoI fungicides during all the years and in all countries monitored to date, despite spray programs containing such products and significant efforts made in monitoring the populations. In 7 out of 26 species reported in 2020, no resistance was detected. Most of the species without resistance fall into the intron-containing group. In this group of pathogens, the aforementioned molecular mechanism is most likely to prevent individuals from acquiring the G143A mutation. A prominent example is cereal rust fungi. Other species are also reported to contain an intron (Table 1), such as *Pyrenophora teres*. Interestingly, few plant pathogenic species have been described as polymorphic with respect to intron presence, such as *Botrytis cinerea*. In rare isolates of *Puccinia hordei,* reduced sensitive isolates were reported, but no molecular mechanism could have been associated with it to date, nor could these isolates have gained relevance in the population and had no impact on the performance of the QoI-containing fungicides (FRAC). The reasons why no resistance has been detected in *Phytophthora infestans* or in *Sclerotinia sclerotiorum* are unknown. In certain species such as *Rhynchosporium secalis,* resistant mutants are occasionally found; however, the G143A genotype does not gain higher frequency for reasons not yet fully understood. It could be due to heteroplasmic situations in the fungal cells (Lesemann et al., 2006; Mosquera et al., 2019), but in most cases, no significant heteroplasmy in mitochondria in plant pathogenic fungi has been detected. The other reason could be that the G143A, despite general observations in most pathogens, does have a fitness penalty in certain species. Studies in *Plasmopara viticola* suggest a weak impact of the mutation on competitiveness (Sierotzki et al., 2007).

6.2 QoI group 11A

Studies on the sensitivity situation toward group 11A fungicide metyltetraprole reveal that in species such as *Z. tritici, Pyrenophora teres, Ramularia collo-cygni, Pyrenophora tritici-repentis, Parastagonospora nodorum, Microdochium majus, Colletotrichum gloeosporioides, Pyricularia oryzae, Venturia inaequalis, Alternaria solani, Passalora fulva*, and *Pestalotiopsis longiseta* (Matsuzaki et al., 2020), all tested isolates are sensitive, which includes wild-type genotypes as well as G143A and F129L cytb mutants. In addition, the AOX and the multidrug resistance affect the activity of the group 11A fungicide only to a minor extent. It is not clear if strains combining the different mechanisms have been explicitly tested, but earlier in the case of code 11 fungicides, it has been hypothesized that such minor mechanisms could enable individuals to survive long enough to develop stronger resistance mechanisms.

No resistance has been reported in FRAC to date, indicating that no selection took place, either from QoI group 11 usage or from any other pressures.

6.3 QoI group 45

Group 45 fungicides bind to the QoI-binding pocket (Fehr et al., 2016) but show a different cross-resistance profile by being not impacted by mutations G143A or F129L affecting the fungicides from group 11. However, in 2016, resistance to ametoctradin was reported in *Plasmopara viticola* in French vineyards at low levels of frequency (Fontaine et al., 2019). Resistance was mediated by the S34L mutation in the cytb gene. The stability of the resistance is not yet clarified and requires further monitoring and specific experimental confirmation. No other resistance reports are currently available.

7 Conclusion and future trends

Fungicides of groups 11, 11A, and 45 remain important components of spray programs in crop protection worldwide, even after 25 years of use in agriculture in the case of group 11, and are forecast to remain important in the foreseeable future. Despite resistance developments, they are valuable for use as mixing partners or even as solo compound products in spray schedules. New developments of compounds with differential binding properties appear to be possible, enriching the respiration-inhibiting fungicide family (Esser et al., 2018). Additionally, there is ongoing research to access the respiration chain, including cytochrome b, for medical purposes (Duvenage et al., 2019).

The important point is to continue to monitor the sensitivity situation alongside biological and mechanistic studies to gain as much information as possible for use recommendations to ensure these further maintain and prolong the effectiveness of QoI fungicides.

Currently, the use guidelines agreed within FRAC member companies include recommendations for groups 11 and 11A as well as group 45 fungicides. Manufacturers' use recommendations, such as rates and intervals, need to be followed to ensure effective disease management, although additional measurements are required to minimize or mitigate resistance development risk. Limitation of the number of applications within a spray program, whether used on its own or as a mixing partner (either ready mixtures or as tank mix), is a key point to mitigate resistance evolution. The recommendation reflects cross-resistance between group 11 and group 11A fungicides. Mixing partners need to be sufficiently active against the target pathogen, including aspects such as duration of activity and distribution behavior within the plant. In addition to the general recommendations, there are currently ten different crop-specific recommendations published on the FRAC webpage for QoI fungicides (www.frac.info), including vines, pome fruit, cucurbits, vegetable crops, potato, tomato, cereals, soybeans, sugar beet, and rice, as well as for other multispray crops.

Use recommendations for group 45 fungicide ametoctradin are also available via the FRAC website and include usage in a preventative manner on all crops and for use on grapes as well as restrictions on the number of applications and in mixtures.

Novel respiration-inhibiting fungicides as tools offer new perspectives for the current QoI. The cytochrome b gene is highly conserved, and it is likely that multiple resistance alleles will be associated with fitness cost or are incompatible. The experience gained to date is useful for establishing next-generation recommendations, helping to keep this mode of action a valuable tool in crop protection. New technologies to aid with early diagnosis of disease epidemics and resistance developments will help to further protect respiration inhibitor fungicides.

8 Where to look for further information

A standard introduction with general information related to the QoI fungicide class can be accessed visiting the QoI working group page of the Fungicide Resistance Action Committee FRAC web page:

- https://www.frac.info/frac-teams/working-groups/qol-fungicides.

Specific recommendations for QoI fungicide class use in different crops including cereals, vines, pome fruit, potato and tomato, soybean, sugar beets, cucurbits, rice, Banana, greenhouse grown non-cucurbit vegetables, and multispray crops are accessible here:

- https://www.frac.info/frac-teams/working-groups/qol-fungicides/recommendations-for-qoi.

Latest monitoring results from season 2022 published in 2023 can be accessed here:

- https://www.frac.info/docs/default-source/working-groups/qol-fungicides/qoi-meeting-minutes/minutes-of-the-2023-qoi-wg-meeting-and-recommendations-for-2022-on-19th-of-january-2023.pdf?sfvrsn=ca4c4e9a_2.

9 References

Ag Bioinvestor. (2022). https://agbioinvestor.com/agbiocrop-agrochemical-market-intelligence/.

Arakawa, A., Kasai, Y., Yamazaki, K. and Iwahashi, F. (2018). Features of interactions responsible for antifungal activity against resistant type cytochrome bc1: A data-driven analysis based on the binding free energy at the atomic level. *PLoS ONE* 13(11): e0207673.

Birla, K., Rivera-Varas, V., Secor, G. A., Khan, M. F. R. and Bolton, M. D. (2012). Characterization of cytochrome b from European field isolates of Cercospora beticola with quinone outside inhibitor resistance. *European Journal of Plant Pathology* 134(3): 475–488. https://doi.org/10.1007/s10658-012-0029-y.

Blake, J. J., Gosling, P., Fraaije, B. A., Burnett, F. J., Knight, S. M., Kildea, S. and Paveley, N. D. (2018). Changes in field dose-response curves for demethylation inhibitor (DMI) and quinone outside inhibitor (QoI) fungicides against *Zymoseptoria tritici*, related to laboratory sensitivity phenotyping and genotyping assays. *Pest Management Science* 74(2): 302–313.

Brasseur, G., Saribas, A. S. and Daldal, F. (1996). A compilation of mutations located in the cytochrome b subunit of the bacterial and mitochondrial BC1 complex. *Biochimica et Biophysica Acta* 1275(1–2): 61–69.

Calo, S., Billmyre, R. B. and Heitman, J. (2013). Generators of phenotypic diversity in the evolution of pathogenic microorganisms. *PLoS Pathogens* 9(3): e1003181. https://doi.org/10.1371/journal.ppat.1003181.

Cattelan, A. J. and Dall'Agnol, A. (2018). The rapid soybean growth in Brazil. *OCL* 25(1): D102.

Cherrad, S., Hernandez, C., Steva, H. and Vacher, S. (2018). *Résistance de Plasmopara viticola aux Inhibiteurs du complexe III: Un point sur la caractérisation phénotypique et génotypique des souches. 12ème Conférence Internationale sur les maladies des plantes*. Tours, France: AFPP.

Debarre, F., Lenormand, T. and Gandon, S. (2009). Evolutionary epidemiology of drug-resistance in space. *PLoS Computational Biology* 5(4): e1000337.

Dreinert, A., Wolf, A., Mentzel, T., Meunier, B. and Fehr, M. (2018). The cytochrome BC1 complex inhibitor Ametoctradin has an unusual binding mode. *Biochimica et Biophysica Acta. Bioenergetics* 1859(8): 567–576.

Duvenage, L., Munro, C. A. and Gourlay, C. W. (2019). The potential of respiration inhibition as a new approach to combat human fungal pathogens. *Current Genetics* 65(6): 1347–1353.

Esser, L., Yu, C. A. and Di Xia, D. (2014). Structural basis of resistance to anti-cytochrome BC1 inhibitors: Implication for drug improvement. *Current Pharmaceutical Design* 20(5): 704–724.

Fehr, M., Wolf, A. and Stammler, G. (2016). Binding of the respiratory chain inhibitor ametoctradin to the mitochondrial BC1 complex. *Pest Management Science* 72(3): 591–602.

Fontaine, S., Remuson, F., Caddoux, L. and Barrès, B. (2019). Investigation of the sensitivity of *Plasmopara viticola* to amisulbrom and ametoctradin in French vineyards using bioassays and molecular tools. *Pest Management Science* 75(8): 2115–2123.

Friesen, T. L., Stukenbrock, E. H., Liu, Z., Meinhardt, S., Ling, H., Faris, J. D., Rasmussen, J. B., Solomon, P. S., McDonald, B. A. and Oliver, R. P. (2006). Emergence of a new disease as a result of interspecific virulence gene transfer. *Nature Genetics* 38(8): 953–956.

Gewehr, M. and Sauter, H. (2019). Strobilurins and other complex. In: Jeschke, P., Witschel, M., Krämer, W. and Schirmer, U. (Eds.), *Inhibitors in Modern Crop Protection Compounds* (3rd edn., Vol. III). Wiley-VCH Press.

Grasso, V., Palermo, S., Sierotzki, H., Garibaldi, A. and Gisi, U. (2006). Cytochrome b gene structure and consequences for resistance to Qo inhibitor fungicides in plant pathogens. *Pest Management Science* 62(6): 465–472.

Gullino, M. L., Gilardi, G., Tinivella, F. and Garibaldi, A. (2004). Observations on the behaviour of different populations of Plasmopara viticola resistant to QoI fungicides in Italian vineyards. *Phytopathology Mediterrean* 43: 341–350.

Heaney, S. P., Hall, A. A., Davis, S. A. and Olaya, G. (2000). Resistance to fungicides in the QoI-STAR cross resistance group: Current perspectives. In: *Proceedings of the Brighton Crop Protection Conference - Pests and Diseases* (Vol. 2, pp. 755–762).

Hoffmeister, M., Martens, G., Yarnell, L., Blain, N., Wieja, B., Riediger, N. and Stammler, G. (2022). Method for QoI sensitivity monitoring in Colletotrichum lentis by a species-specific Pyrosequencing assay. *Canadian Journal of Plant Pathology* 44(1): 128–135.

Hoffmeister, M., Mehl, A., Hinson, A., Siepe, I., Taufferner, T., Stammler, G. and Felsenstein, F. (2022). Acquired QoI resistance in Pyrenophora teres through an interspecifc partial gene transfer by Pyrenophora triticirepentis? *Journal of Plant Diseases and Protection* 129(5): 1073–1086.

Ishii, H., Yano, K., Date, H., Furuta, A., Sagehashi, Y., Yamaguchi, T., Sugiyama, T., Nishimura, K. and Hasama, W. (2007). Molecular characterization and diagnosis of QoI resistance in cucumber and eggplant fungal pathogens. *Phytopathology* 97(11): 1458–1466.

Kaiser, U., Kluth, C. and Märländer, B. (2010). Variety-specific epidemiology of Cercospora beticola Sacc. and consequences for threshold-based timing of fungicide application in sugar beet. *Journal of Phytopathology* 158(4): 296–306.

Karaoglanidis, G. S. and Ioannidis, P. M. (2010). Fungicide resistance of Cercospora beticola in Europe. In: Lartey, R. T., Weiland, J. J., Panella, L., Crous, P. W. and Windels, C. E. (Eds.), *Cercospora Leaf Spot of Sugar Beet and Related Species*. St. Paul: The American Phytopathological Society.

Klosowski, A. C., May De Mio, L. L., Miessner, S., Rodrigues, R. and Stammler, G. (2016). Detection of the F129L mutation in the cytochrome b gene in Phakopsora pachyrhizi. *Pest Management Science* 72(6): 1211–1215. https://doi.org/10.1002/ps.4099.

Lesemann, S. S., Schimpke, S., Dunemann, F. and Deising, H. B. (2006). Mitochondrial heteroplasmy for the cytochrome b gene Controls the level of strobilurin resistance in the apple powdery mildew fungus Podosphaera leucotricha (Ell. & Ev.) E.S. Salmon. *Journal of Plant Diseases and Protection* 113(6): 259–266.

Link, Th. A., Iwata, M., Björkman, J., van der Spoel, D., Stocker, A. and Ixata, S. (2003). Molecular modelling of inhibitors at the Qi and Qo Sites in cytochrome BC1 complex. In: Voss, G. and Ramos, G. (Eds.), *Chemistry of Crop Protection*. Weinheim, Germany: Wiley-VCH Press.

Malandrakis, A. A., Markoglou, A. N., Nikou, D. C., Vontas, J. G. and Ziogas, B. N. (2011). Molecular diagnostic for detecting the cytochrome b G143S – QoI resistance mutation in *Cercospora beticola*. *Pesticide Biochemistry and Physiology* 100(1): 87–92.

Massi, F., Torriani, S. F. F., Borghi, L. and Toffolatti, S. L. (2021). Fungicide resistance evolution and detection in plant pathogens: Plasmopara viticola as a case study. *Microorganisms* 9(1): 119. https://doi.org/10.3390/microorganisms9010119.

Matsuura, S. (2019). Does QoI (strobilurin) resistance in isolates of Puccinia horiana, the causal agent of chrysanthemum white rust, occur in western Japan? *Journal of Plant Diseases and Protection* 126(5): 469–473.

Matsuzaki, Y., Harada, T. and Iwahashi, F. (2021). New cytochrome b haplotypes, harboring L299F or N256S + L299F substitutions, were found in azoxystrobin-resistant Puccinia horiana, the causal agent of chrysanthemum white rust. *European Journal of Plant Pathology* 160(4): 963–972.

Matsuzaki, Y. and Iwahashi, F. (2019). Biological profile of metyltetraprole, a new QoI-fungicide: discovery of the tetrazolinone chemical class and in vitro/in vivo analysis of activity against QoI-resistant fungal strains. In: Deising, H. B., Fraaije, B., Mehl, A., Oerke, E. C., Sierotzki, H. and Stammler, G. (Eds.), *Modern Fungicides and Antifungal Compounds IX. Proceedings of the 19th International Reinhardsbrunn Symposium.* DPG Spectrum Phytomedizin.

Matsuzaki, Y., Kiguchi, S., Suemoto, H. and Iwahashi, F. (2020). Antifungal activity of metyltetraprole against the existing QoI-resistant isolates of various plant pathogenic fungi: Metyltetraprole against QoI-R isolates. *Pest Management Science* 76(5): 1743–1750.

Mello, F. E., Lopes-Caitar, V. S., Xavier, S. A., Mehl, A., Marcelino-Guimaraes, F. C., Verreet, J. A., Balbi-Peña, M. I. and Godoy, C. V. (2018). Identification of QoI mutation in soybean pathogens in Brazil. In: *Conference: International Congress of Plant Pathology*, Boston, USA.

Mello, F. E., Lopes-Caitar, V. S., Xavier-Valencio, S. A., da Silva, H. P., Franzenburg, S., Mehl, A., Verreet, J., Balbi-Peña, M. I., Marcelino-Guimaraes, F. C. and Godoy, C. V. (2022). Resistance of Corynespora cassiicola from soybean to QoI and MBC fungicides in Brazil. *Plant Pathology* 71(2): 373–385. https://doi.org/10.1111/ppa.13474.

Miao, J., Zhao, G., Wang, B., Du, Y., Li, Z., Gao, X., Zhang, C. and Liu, X. (2020). Three point-mutations in cytochrome b confer resistance to trifloxystrobin in Magnaporthe oryzae. *Pest Management Science* 76(12): 4258–4267.

Millardet, A. (1885). Traitement du mildiou par la melange du sulfate de cuivre et de chaux. *Agriculturists Prat* 49: 707–710.

Mosquera, S., Chen, L. H., Aegerter, B., Miyao, E., Salvucci, A., Chang, T. C., Epstein, L. and Stergiopoulos, I. (2019). Cloning of the cytochrome b gene from the tomato powdery mildew fungus Leveillula taurica reveals high levels of allelic variation and heteroplasmy for the G143A mutation. *Frontiers in Microbiology* 10: 663.

Mounkoro, P., Michel, T., Benhachemi, R., Surpateanu, G., Iorga, B. I., Fisher, N. and Meunier, B. (2019). Mitochondrial complex III Qi-site inhibitor resistance mutations found in laboratory selected mutants and field isolates. *Pest Management Science* 75(8): 2107–2114.

Patel, J. S., Meinhardt, S. W., Sierotzki, H., Stammler, G., Gudmestad, N. C. and Adhikari, T. B. (2011). A two-step molecular detection method for Pyrenophora tritici-repentis isolates insensitive to QoI fungicides. *Plant Disease* 95(12): 1558–1564.

Ruan, H., Tian, P., Shi, N., Du, Y. and Chen, F. (2022). Characterization of pyraclostrobin-resistant *Magnaporthe oryzae*. Journal of Phytopathology 170(4): 233–241.

Serot, C., Michel, Th., Lucas, V. and Meunier, B. (2022). *Study of Metyltetraprole, an Unusual Agrofungicide Targeting the Q o -Site of Cytochrome BC 1 Complex. ffhal-03582522.*

Sierotzki, H. (2015). Respiration inhibitors: Complex III. In: *Fungicide Resistance in Plant Pathogens*. Japan: Springer, pp. 119–143. https://doi.org/10.1007/978-4-431-55642-8_9.

Sierotzki, H., Kraus, N., Assemat, P., Stanger, C., Cleere, S. and Windass, J. (2005). Evolution of resistance to QoI fungicides in Plasmopara viticola populations in Europe. In: Dehne, H. W., Gisi, U., Kuck, K. H., Russell, P. E. and Lyr, H. (Eds.), *Modern Fungicides and Antifungal Compounds* (Vol. IV). Alton, UK: BCPC, pp. 73–80.

Sierotzki, H., Kraus, N., Pepin, S., Fernandes, N. and Gisi, U. (2007). Dynamics of QoI resistance in *Plasmopara viticola*. In: Dehne, H. W., Deising, H. B., Gisi, U., Kuck, K. H. P. E. and Lyr, R. H. (Eds.), *Modern Fungicides and Antifungal Compounds*, p. V.

Suemoto, H., Matsuzaki, Y. and Iwahashi, F. (2019). Metyltetraprole, a novel putative complex III inhibitor, targets known QoI-resistant strains of Zymoseptoria tritici and Pyrenophora teres. *Pest Management Science* 75(4): 1181–1189.

Vallieres, C., Trouillard, M., Dujardin, G. and Meunier, B. (2011). Deleterious effect of the Qo inhibitor compound resistance-conferring mutation G143A in the intron-containing cytochrome b gene and mechanisms for bypassing it. *Applied and Environmental Microbiology* 77(6): 2088–2093.

Vanlerberghe, G. C. (2013). Alternative oxidase: a mitochondrial respiratory pathway to maintain metabolic and signaling homeostasis during abiotic and biotic stress in plants. *International Journal of Molecular Sciences* 14(4): 6805–6847.

Vielba-Fernandez, A., Polonio, Á., Ruiz-Jiménez, L., de Vicente, A., Pérez-García, A. and Fernández-Ortuño, D. (2020). Fungicide resistance in powdery mildew fungi. *Microorganisms* 8(9): 1431.

Watanabe, H., Horinouchi, H., Muramoto, Y. and Ishii, H. (2017). Occurrence of azoxystrobin-resistant isolates in *Passalora fulva*, the pathogen of tomato leaf mould disease. *Plant Pathology* 66(9): 1472–1479.

Zhu, X., Zhang, M., Liu, J., Ge, J. and Yang, G. (2015). Ametoctradin is a potent Qo Site Inhibitor of the mitochondrial respiration complex III. *Journal of Agricultural and Food Chemistry* 63(13): 3377–3386. https://doi.org/10.1021/acs.jafc.5b00228.

Zito, R., Meyer, L., Aumont, C., Fehr, M. and Stammler, G. (2020). State of knowledge on molecular mechanisms leading to a reduced sensitivity of *P. viticola* towards ametoctradin and complex III inhibitors. In: Deising, H. B., Fraaije, B., Mehl, A., Oerke, E. C., Sierotzki, H. and Stammler, G. (Eds.), *Modern Fungicides and Antifungal Compounds* (Vol. IX). Braunschweig: Deutsche Phytomedizinische Gesellschaft, pp. 117–118.

Chapter 12

Understanding resistance to succinate dehydrogenase inhibitor fungicides

Wesley Mair, Centre for Crop and Disease Management, Curtin University, Australia

1 Introduction

The succinate dehydrogenase inhibitors (SDHIs), also known as the carboxamide or group 7 compounds, represent the fastest growing class of fungicides (Sierotzki and Scalliet, 2013). Introduced in 1969, the compounds released in the first decade, such as carboxin (1969), benodanil (1974), fenfuram (1974), and oxycarboxin (1975), were effective mainly to control *Basidiomycetes* such as *Ustilago* spp., *Tilletia* spp., and *Puccinia* spp. (Glättli et al., 2011). The following two decades saw the release of mepronil (1981), flutolanil (1986), thifluzamide (1997), and furametpyr (1997), all mainly targeting *Rhizoctonia* spp. on various crops (Stammler et al., 2015). Beginning with the introduction of boscalid in 2003, the spectrum of activity of this class has substantially widened, encompassing now also ascomycete and deuteromycete pathogens on a wide variety of hosts, including canola, cereals, fruits, grapes, vegetables, and turf (Glättli et al., 2011; Rehfus, 2018; Stammler et al., 2015). The number of SDHI fungicides has also increased substantially, with 24 compounds currently registered in this class (FRAC, 2022). Resistance to SDHIs emerged rapidly following the introduction of this class of fungicides (Abiko et al., 1977; Avenot and Michailides, 2007), and has now been reported in at least 30 species (Tables 1–3).

http://dx.doi.org/10.19103/AS.2022.0116.12

Table 1 SdhB: position number based on alignment to reference sequence from *Pyrenophora teres* f. sp. *teres* (NCBI accession number EFQ89326)

Amino acid substitution(s) in archetype	Homologous position in other species	Reference
C147	C137R in SEPTTR	Fraaije et al. (2012)
Y162	Y137C in VENTIN	FRAC (2016)
S228	S217L in RAMUCC	Piotrowska et al. (2017)
	S218F in SEPTTR	Scalliet et al. (2012)
P230	P220L/T in SEPTTR	Scalliet et al. (2012)
	P225F/H/L/T in BOTRCI	Kleemann and Mehl (2019); Leroux et al. (2010); Stammler et al. (2008)
	P225L in PLEOAL	FRAC (2015)
	P226L in SCLESC	Wang et al. (2022)
S231	S221P/T in SEPTTR	Fraaije et al. (2012)
N235	N224I in RAMUCC	Piotrowska et al. (2017)
	N225H/I/T in SEPTTR	FRAC (2017); Rehfus (2018); Scalliet et al. (2012)
	N230I in BOTRCI	Leroux et al. (2010)
R275	R264P in RAMUCC	Rehfus et al. (2019a)
	R265P/L in SEPTTR	FRAC (2018); Fraaije et al. (2012); Scalliet et al. (2012)
H277Y	H229Y in XANTOZ	Zhang et al. (2010)
	H242R/Y in UNCINE	Cherrad et al. (2017); Graf et al. (2017)
	H243Y in PENIDI	Xu et al. (2022)
	H243R/Y in PODOXA	Miyamoto et al. (2020)
	H245L/Y in PYRIOR	Guo et al. (2016); Liu et al. (2022b)
	H246Y in RHIZCE	Sun et al. (2017)
	H249Y in ASPEFL	Masiello et al. (2020)
	H249L/N/Y in EUROOR	Shima et al. (2009)
	H249Y in RHIZSO	Miao et al. (2021)
	H257L in USTIMA	Broomfield and Hargreaves (1992)
	H260R in BLUMJA	Outwater et al. (2019)
	H266R/Y in RAMUCC	Rehfus et al. (2019b)
	H267R/Y in SCLEHO	Lee et al. (2021)
	H267F/L/N/Q/R/Y in SEPTTR	Fraaije et al. (2012); Scalliet et al. (2012); Rehfus (2018); Skinner et al. (1998)
	H270R/Y in ASPEFU	Fraaije et al. (2020)
	H272L/R/V/Y in BOTRCI	De Miccolis Angelini et al. (2010); Leroux et al. (2010); De Miccolis Angelini et al. (2014)
	H272Y/R in BOTREL	FRAC (2015)

(*Continued*)

Table 1 (*Continued*)

Amino acid substitution(s) in archetype	Homologous position in other species	Reference
	H272Y/R in PLEOAL	FRAC (2015)
	H272R/Y in PENIEX	Malandrakis et al. (2017)
	H273Y in SCLESC	FRAC (2015)
	H277L/R/Y in ALTEAL	Avenot et al. (2008); Sierotzki et al. (2011)
	H277R/Y in DIDYBR	Avenot et al. (2012)
	H277R/Y in MSPHTA	Pearce et al. (2019)
	H277Y in STGSCI	Li et al. (2019)
	H278R/Y in ALTESO	Mallik et al. (2013)
	H278L/R/Y in CORYCA	Zhu et al. (2022); Miyamoto et al. (2010)
T278	T253I in VENTIN	Huf et al. (2017)
	T267I in RAMUCC	Rehfus et al. (2019b)
	T268A/I in SEPTTR	FRAC (2018); Fraaije et al. (2012)
I279	I244V in UNCINE	Stergiopoulos et al. (2022)
	I268V in RAMUCC	Rehfus et al. (2019b)
	I269V in SEPTTR	Fraaije et al. (2012); Scalliet et al. (2012)
	I274V in BOTRCI	Malandrakis et al. (2022)
	I279V in MSPHTA	Pearce et al. (2019)
	I279V in STGSCI	Li et al. (2019)
	I280V in ALTESO	Einspanier et al. (2022)
	I280V in CORYCA	Zhu et al. (2019)
N281	N271K in SEPTTR	Scalliet et al. (2012)
K288	K283N in BOTRCI	Liu et al. (2022b)

Table 2 SdhC: position number based on alignment to reference sequence from *Pyrenophora teres* f. sp. *teres* (NCBI accession number EFQ89104)

Amino acid substitution(s) in archetype	Homologous position in other species	Reference
K49E	NA	FRAC (2016); Torriani et al. (2015)
L60	F48L in RHIZSO	Zhao et al. (2019)
T68	T77I in GIBBZE SdhC2	Chen et al. (2021a)
	T79I/N in SEPTTR	Scalliet et al. (2012); FRAC (2016)
	P80H in BOTRCI	Liu et al. (2022b)
	P80L in SCLEHO	Lee et al. (2021)
W69	W80S in SEPTTR	FRAC (2017)
S72	S83G in SEPTTR	Scalliet et al. (2012)

(*Continued*)

Table 2 (*Continued*)

Amino acid substitution(s) in archetype	Homologous position in other species	Reference
S73P	I61F in RHIZSO	Zhao et al. (2022)
	S73P in CORYCA	Miyamoto et al. (2010)
	S73P in MSPHTA	Pearce et al. (2019)
	A78V in GIBBZE SdhC1	Shao et al. (2022)
	A83V in GIBBZE SdhC2	Sun et al. (2020)
	A84I/V in SEPTTR	Scalliet et al. (2012)
	G85A in BOTRCI	Li et al. (2022)
	I85V in PUCCSI	Cook et al. (2021)
	A86V in PODOXA	Miyamoto et al. (2020)
	I86F in PHAKPA	Simões et al. (2018)
L74	L85P in SEPTTR	Fraaije et al. (2012); Scalliet et al. (2012)
N75S	N75S in ALTESO	Einspanier et al. (2022)
	N75S in CORYCA	FRAC (2020)
	N86A/K/S in SEPTTR	Rehfus et al. (2018)
	N87S in RAMUCC	FRAC (2018)
R76	R86C/H in GIBBZE SdhC2	Chen et al. (2021a); Sun et al. (2020)
	R87C in SEPTTR	Scalliet et al. (2012)
I77	V88D in SEPTTR	Scalliet et al. (2012)
T78	T90I in EUROOR	Shima et al. (2009)
G79R	G79R in ALTEAL	Yang et al. (2015)
	G79R in MSPHTA	Pearce et al. (2019)
	G90R in SEPTTR	Stammler et al. (2011)
	G91R in ASPEFL	Masiello et al. (2020)
	G91R in RAMUCC	FRAC (2018)
	G91R in SCLEHO	Popko et al. (2018)
V81	I93V in BOTRCI	Li et al. (2022)
H134R	H134R in ALTEAL	Avenot et al. (2009)
	H134Q/R in ALTESO	Mallik et al. (2013); Metz et al. (2019)
	H134R in CORYCA	Zhu et al. (2022)
	H134Q/R in MSPHTA	Pearce et al. (2019)
	H139Y in RHIZCE	Sun et al. (2017)
	H145R in SEPTTR	Scalliet et al. (2012)
	H146L/Q/R in RAMUCC	FRAC (2018); Piotrowska et al. (2017)
	H146R in SCLESC	FRAC (2015)

(*Continued*)

Table 2 (*Continued*)

Amino acid substitution(s) in archetype	Homologous position in other species	Reference
S135R	S135R in ALTEAL	Sierotzki et al. (2011)
	S135R in MSPHTA	Pearce et al. (2019)
G138	G150R in SCLEHO	Popko et al. (2018)
	G151R in PODOXA	Miyamoto et al. (2020)
R140	R151M/S/T in SEPTTR	FRAC (2018)
	R152M in RAMUCC	Rehfus et al. (2019a)
H141	H151R in VENTIN	Huf et al. (2017)
	H152R in SEPTTR	Stammler et al. (2011)
	H153R in RAMUCC	FRAC (2018)
F150	I161S in SEPTTR	Kirikyali et al. (2017)
V155	V166M in SEPTTR	Rehfus et al. (2018)
G159	G169D/S in UNCINE	Graf (2017)
	G171D in RAMUCC	FRAC (2018)
	G172D in PODOXA	Miyamoto et al. (2020)

Table 3 SdhD: position number based on alignment to reference sequence from *Pyrenophora teres* f. sp. *teres* (NCBI accession number EFQ94661)

Amino acid substitution(s) in archetype	Homologous position in other species	Reference
L9	V9A in BOTRCI	Liu et al. (2022a)
R63	R47W in SEPTTR	Dooley et al. (2016)
I66	I50F in SEPTTR	FRAC (2018)
S118	S89P in CORYCA	Miyamoto et al. (2010)
	S121P in PODOXA	Miyamoto et al. (2020)
D124E/N	D95E in CORYCA	Zhu et al. (2019)
	D112E in MSPHTA	Pearce et al. (2019)
	D123E in ALTEAL	FRAC (2015)
	D123E in ALTESO	Mallik et al. (2013)
L130	M114V in SEPTTR	FRAC (2018)
H134R	H105R in in CORYCA	Zhu et al. (2019)
	H116D/Y in RHIZCE	Sun et al. (2017)
	H122R in MSPHTA	Pearce et al. (2019)
	H132R in BOTRCI	Leroux et al. (2010)
	H132R in SCLESC	Stammler et al. (2011)

(*Continued*)

Table 3 (*Continued*)

Amino acid substitution(s) in archetype	Homologous position in other species	Reference
	H133P/Q/R/T in ALTEAL	Malandrakis et al. (2018); Adaskaveg et al. (2019); Avenot et al. (2009)
	H133R in ALTESO	Mallik et al. (2013)
	H137R in PODOXA	Miyamoto et al. (2020)
H136	H118Y in RHIZCE	Sun et al. (2017)
G138V	G109V in CORYCA	FRAC (2018)
I143	I127V in SEPTTR	Fraaije et al. (2012)
D145G	D124E in EUROOR	Shima et al. (2009)
	D129E/G/S/T in SEPTTR	Fraaije et al. (2012); Scalliet et al. (2012)
F147	L129I in RHIZCE	Sun et al. (2017)
E178K	NA	Rehfus et al. (2016)

2 Mode of action

2.1 Succinate dehydrogenase (succinate-ubiquinone oxidoreductase, Complex II)

The succinate dehydrogenase (Sdh) enzyme complex (also known as succinate-ubiquinone oxidoreductase or respiratory Complex II) couples the oxidation of succinate to fumarate with the reduction of ubiquinone to ubiquinol (Horsefield et al., 2006). Sdh is a key component linking the tricarboxylic acid (TCA, also known as the citric acid or Krebs) cycle with the synthesis of adenosine triphosphate (ATP) by oxidative phosphorylation, as part of the mitochondrial electron transport chain (Iverson et al., 2012). The Sdh enzyme complex is a membrane-bound heterotetramer consisting of four nuclear-encoded subunits (Fig. 1a); it is the only enzyme of the TCA cycle that is membrane-bound, and the only component of the mitochondrial electron transport chain fully encoded in the nucleus (Lemire and Oyedotun, 2002). The soluble catalytic dimer domain is composed of the subunits SdhA and SdhB and faces the mitochondrial matrix, while the hydrophobic transmembrane domain, composed of subunits SdhC and SdhD, anchors the Sdh complex to the inner mitochondrial membrane (Scalliet et al., 2011; Maklashina et al., 2010). SdhA and SdhB are highly conserved, while SdhC and SdhD are highly divergent across species (Ackrell, 2000). SdhA is a flavoprotein, with a covalently bound electron acceptor, flavin adenine dinucleotide (FAD); and a dicarboxylate-binding site,

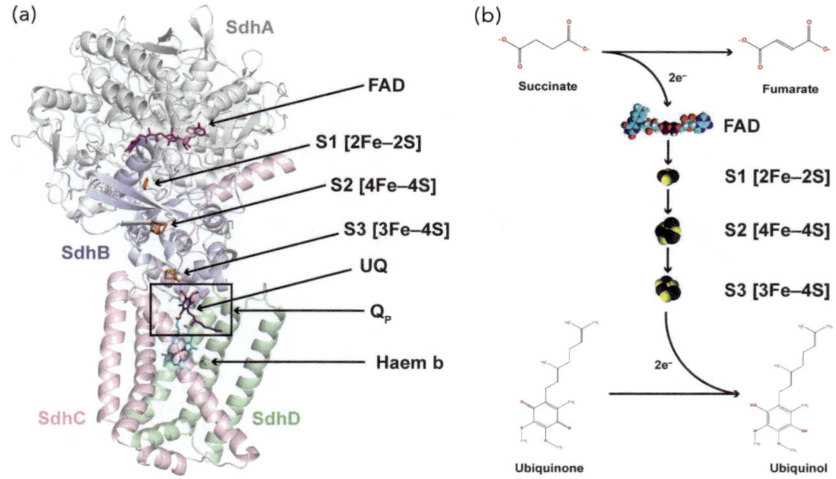

Figure 1 (a) Homology model of the succinate dehydrogenase (Sdh) enzyme complex from *Pyrenophora teres* with bound ubiquinone substrate. Grey: SdhA (flavoprotein); blue: SdhB (iron–sulphur protein); pink: SdhC (cytochrome b560); green: SdhD (cytochrome b small unit); magenta: flavin adenine dinucleotide (FAD); orange: three iron–sulphur clusters, S1 [2Fe–2S], S2 [4Fe–4S], and S3 [3Fe–4S]; purple: ubiquinone (UQ); cyan: haem b; boxed: proximal quinone-binding site (Q$_P$). SDH domains were modelled using RoseTTAFold (Baek et al., 2021) aligned to PDB 2WQY, and visualised in PyMol 1.7.6.3 (Schrödinger Inc., New York, USA). (b) Schematic model of the oxidation of succinate to fumarate coupled with the reduction of ubiquinone to ubiquinol. At SdhA, succinate binds and is oxidised to fumarate, the next substrate of the TCA cycle. Electrons are accepted by the flavin adenine dinucleotide (FAD) of SdhA, and sequentially transferred (downwards arrows) through the iron–sulphur clusters of SdhB, with S1 acting as the electron acceptor from FAD and S3 the electron donor to ubiquinone. At the quinone-binding (Q$_P$) site, at the interface of SdhB, SdhC, and SdhD, ubiquinone accepts electrons and is reduced to ubiquinol. Chemical structures drawn in JChemPaint (Krause et al., 2000) and ACD ChemSketch 2022.1.2 (Advanced Chemistry Development, Toronto, Canada).

where succinate binds and is oxidised to fumarate, the next substrate of the TCA cycle (Fig. 1b) (Maklashina et al., 2010; Iverson et al., 2012). SdhB is an iron–sulphur protein with three iron–sulphur clusters, S1 [2Fe–2S], S2 [4Fe–4S], and S3 [3Fe–4S]; electrons are sequentially transferred through the iron–sulphur clusters with S1 acting as the electron acceptor from FAD and S3 the electron donor to ubiquinone (Scalliet et al., 2012; Hägerhäll, 1997). SdhC, the cytochrome b560, and SdhD, the cytochrome b small unit, together interface with the SdhB subunit to form the proximal quinone-binding (Q$_P$) site (Fig. 2), a hydrophobic pocket proximal to the S3 cluster, where ubiquinone accepts electrons and is reduced to ubiquinol (Hägerhäll, 1997; Yankovskaya et al., 2003; Szeto et al., 2007); certain residues of the Q$_P$ site are highly conserved across species (Horsefield et al., 2006). At the distal quinone-binding (Q$_D$) site

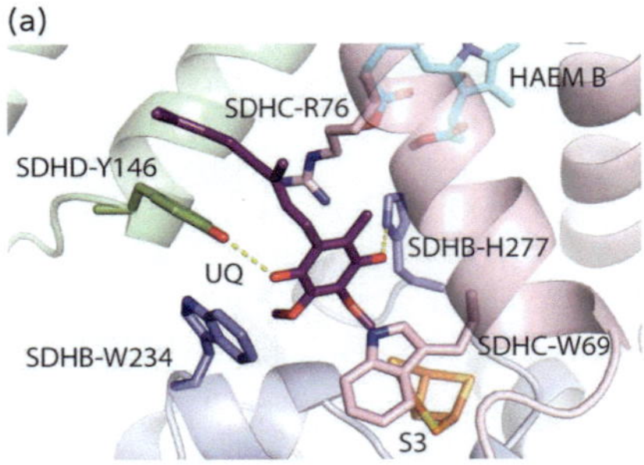

Figure 2 Homology model of the quinone-binding (Q_P) site of the *Pyrenophora teres* Sdh complex, with predicted binding modes of ubiquinone substrate (a) and bixafen SDHI fungicide (b). Blue: SdhB (iron-sulphur protein); pink: SdhC (cytochrome b560); green: SdhD (cytochrome b small unit); orange: iron-sulphur cluster S3 [3Fe-4S]; cyan: haem b; purple: ubiquinone (UQ); magenta: bixafen (BIX); yellow dotted lines: predicted hydrogen bonds (H-bonds). Key residues of the polar cavity (SdhB-W234, SdhB-H277, SdhC-R76, and SdhD-Y146) and the hydrophobic pocket (SdhC-W69) of the Q_P site are also shown. The SDHI compound bixafen binds more strongly in the Q_P site than the UQ substrate. The amide bond moiety of the bixafen is the key feature for binding at the Q_P site, forming two predicted H-bonds with SdhC-R76 and one with SdhB-H277. UQ is predicted to form one H-bond each with SdhB-H277 and SdhD-Y146. SDH domains were modelled using RoseTTAFold (Baek et al., 2021) aligned to PDB 2WQY, docking was performed using the software gnina (McNutt et al., 2021), and visualised in PyMol 1.7.6.3 (Schrödinger Inc., New York, USA).

formed by SdhC and SdhD, electrons are transferred to Complex III of the mitochondrial electron transport chain via the quinone pool (Hägerhäll, 1997; Lemire and Oyedotun, 2002; Horsefield et al., 2006). Between the Q_P and Q_D sites, at the interface of SdhC and SdhD, is situated a haem b group (Maklashina et al., 2010; Sierotzki and Scalliet, 2013); residues SdhC-*H134* and SdhD-*H134* (amino acid numbering based on archetype sequence from *Pyrenophora teres*, as per Mair et al. (2016)) coordinate the central iron of the haem (Stammler et al., 2015). The function of the haem b group is unclear, as it is not required for Sdh assembly or ubiquinone reduction (Oyedotun et al., 2007); though it may have a role in maintaining structural stability and high catalysis rate of the enzyme (Glättli et al., 2011).

2.2 Succinate dehydrogenase inhibitor (carboxamide, group 7) fungicides

The SDHI fungicides are a structurally diverse class, currently consisting of 24 different compounds classified into 12 chemical groups (Fig. 3) (FRAC, 2022). The SDHIs act as competitive inhibitors with ubiquinone (Zhu et al., 2014), binding more deeply and strongly than ubiquinone at the Q_P site (Fig. 2) (Glättli et al., 2011), and blocking reoxidation of the S3 iron-sulphur cluster and

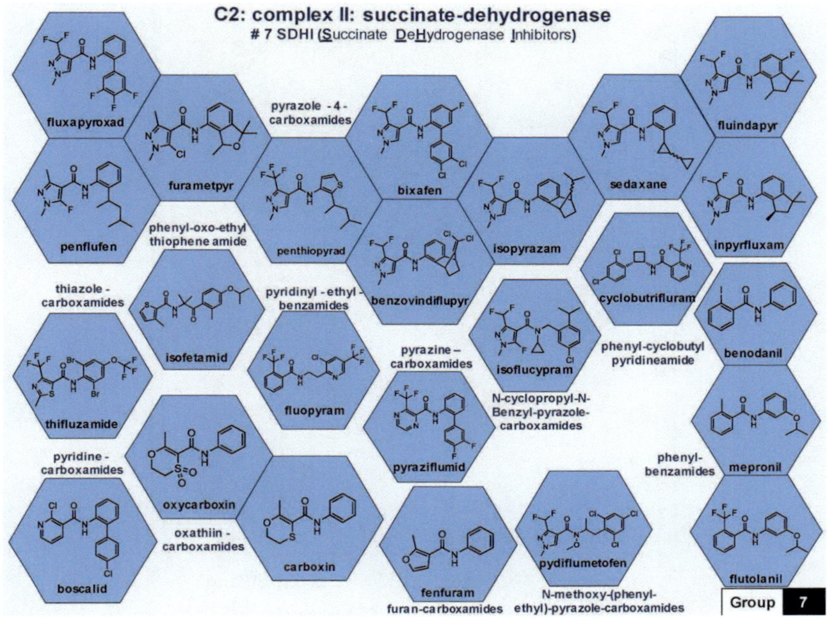

Figure 3 The succinate dehydrogenase inhibitor (SDHIs) fungicides. Also known as complex II (C2) inhibitors, carboxamides, or Group 7 fungicides, this class currently consists of 24 different compounds classified into 12 chemical groups: oxathiin-carboxamides, phenyl-benzamides, furan-carboxamides, pyrazole-4-carboxamides, thiazole-carboxamides, pyridine-carboxamides, pyridinyl-ethyl-benzamides, *N*-methoxy-(phenylethyl)-pyrazole-carboxamides, phenyl-cyclobutyl-pyridine amides, *N*-cyclopropyl -*N*-benzyl-pyrazole-carboxamides, pyrazine-carboxamides, and phenyl-oxo-ethylthiophene amides. Adapted from: FRAC Poster 2022, ©2022 Fungicide Resistance Action Committee (https://www.frac.info/).

transfer of the electrons to ubiquinone (Stammler et al., 2015; Skinner et al., 1998). The SDHI thereby both blocks the TCA cycle and inhibits the synthesis of ATP (Zhang et al., 2007; Li et al., 2019).

The essential feature of all SDHIs is the central amide bond (Fig. 4) (Sierotzki and Scalliet, 2013; Glättli et al., 2011; Scalliet et al., 2012), which acts as a hydrogen bond acceptor with amino acid residues at the polar cavity of the Q_p site, SdhB-*W234* and SdhD-*Y146*; these are strictly conserved residues also necessary for ubiquinone binding (Sierotzki and Scalliet, 2013). Another shared feature of SDHIs is an aromatic ring in the 'aniline block' moiety, located on the amine side of the amide bond (Glättli et al., 2011; Stammler et al., 2015), which stabilises the SDHI at the hydrophobic pocket via π–π or hydrophobic interactions with residues of SdhB and SdhC, including the conserved residue SdhB-*P230* (Glättli et al., 2011; Stammler et al., 2015). The aniline block consists of a linker moiety, generally the aromatic group; the linker may additionally

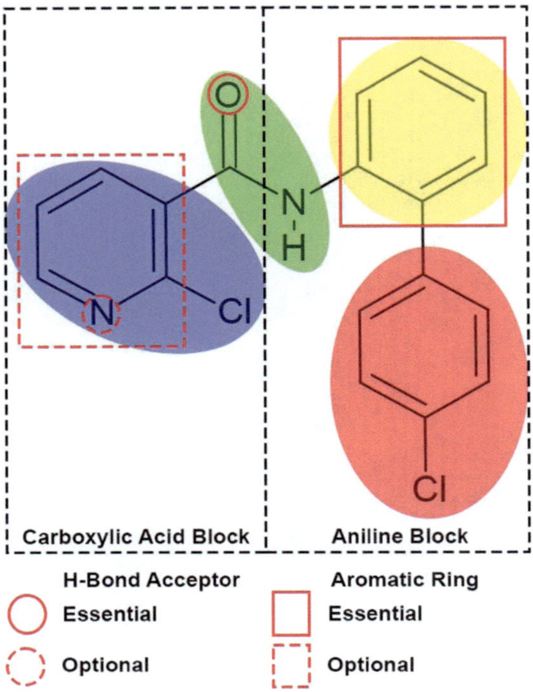

Figure 4 Chemical structure of boscalid with features of the SDHI class highlighted. The essential feature of all SDHIs is the central amide bond (green), which acts as a hydrogen bond acceptor (solid red circle) with the polar cavity of the Q_P site. Another shared feature of SDHIs is an aromatic ring (solid red square) within the 'aniline block' (left-hand side), on the amine side of the amide bond, which stabilises the SDHI at the hydrophobic pocket. The aniline block consists of a 'linker' moiety (yellow), generally the aromatic group; the linker may additionally carry a 'hydrophobic rest' moiety (pink), which interacts with the 'hydrophobic groove' on the surface of the protein. In the 'carboxylic acid block' (right-hand side), on the carbonyl side of the amide bond, the acidic core or head moiety (blue) consists of a substituted or non-substituted five- or six-member aromatic ring (broken red square). The heteroatom substituent (broken red circle) found in the core ring of most SDHIs also acts as a hydrogen bond acceptor. Chemical structure drawn in ACD ChemSketch 2022.1.2 (Advanced Chemistry Development, Toronto, Canada).

carry a hydrophobic rest moiety, which interacts with the 'hydrophobic groove' formed by residues of SdhC and SdhD on the surface of the protein (Sierotzki and Scalliet, 2013). The acidic core or head moiety, also known as the 'carboxylic acid block', consists of a substituted or non-substituted five- or six-member aromatic ring on the carbonyl side of the amide bond (Sierotzki and Scalliet, 2013); it binds deepest in the Q_P site at the polar cavity via π–π interactions, and the heteroatom substituent found in the core ring of most SDHIs acts as a hydrogen bond acceptor with the conserved residue SdhB-*H277* located at the deepest part of the cavity (Stammler et al., 2015; Glättli et al., 2011;

Sierotzki and Scalliet, 2013). The aniline block shows greater variation, while the carboxylic acid block is more conserved and is the basis of classification of the SDHI compounds (Stammler et al., 2015; Sierotzki and Scalliet, 2013). The core, linker, and hydrophobic rest moieties all affect the spectrum and activity of an SDHI (Sierotzki and Scalliet, 2013). The 'first-generation' SDHIs, beginning with carboxin introduced in 1969, have a relatively narrow spectrum, effective mainly against Basidiomycete fungi (Walker et al., 2011; Rehfus, 2018). The 'second-generation' SDHIs, beginning with boscalid introduced in 2003, have a considerably broader spectrum, including also the *Ascomycetes* and *Deuteromycetes* (Glättli et al., 2011; Rehfus, 2018).

2.3 Other modes of action

There is also evidence that certain SDHI compounds have other anti-fungal properties, in addition to the inhibition of Sdh enzyme activity. In *Sclerotinia sclerotiorum*, treatment with the pyrazine-carboxamide pyraziflumid or the *N*-methoxy-(phenylethyl)-pyrazole-carboxamide pydiflumetofen were found to cause electrolyte leakage, indicating cell membrane damage, and decreased production of oxalic acid, an important virulence determinant (Hou et al., 2018b; Huang et al., 2019). However, the pyridine-carboxamide boscalid, while highly effective at controlling *S. sclerotiorum*, had no effect on cell membrane permeability (Hu et al., 2018). Similarly, treatment of *Colletotrichum siamense* and *Colletotrichum nymphaeae* with the pyrazole-4-carboxamide benzovindiflupyr resulted in electrolyte leakage, in addition to inhibitory activity against the Sdh enzyme (Liang et al., 2022). By contrast, boscalid, while inhibiting Sdh, had no effect on cell membrane permeability in *Colletotrichum* spp. (Liang et al., 2022). Hou et al. (2018a) observed that treatment of *Bipolaris maydis* with benzovindiflupyr resulted in increased cell membrane permeability; these findings were confirmed by transmission electron microscopy showing damage to the cell membrane, cell wall, and organelles.

3 Resistance mechanisms

3.1 Target site

3.1.1 Target site mutations

The earliest report of field resistance to first-generation SDHIs occurred within two years of the introduction of oxycarboxin to control *Puccinia horiana* on chrysanthemum (Abiko et al., 1977). The development of resistance to the first of the second-generation SDHIs, boscalid, also occurred within two years of the introduction of this compound to control *Alternaria alternata* on pistachio

(Avenot and Michailides, 2007). SDHI fungicide resistance has now been found in at least 30 species (Table 4), associated with at least 46 mutated amino acid positions across the SdhB, SdhC, and SdhD subunits (Tables 1–3). As SdhA does not form part of the ubiquinone-binding active site, SDHI resistance is not associated with any mutations in this subunit (Piotrowska et al., 2017).

Amino acid substitutions at the target site entail a trade-off between a lower SDHI-binding affinity that reduces inhibition of Sdh and lower ubiquinone-binding affinity that decreases enzyme efficiency; some mutations can result in

Table 4 Abbreviations of species names

Abbreviation (EPPO code)	Name of pathogen
ALTEAL	*Alternaria alternata*
ALTESO	*Alternaria solani*
ASPEFL	*Aspergillus flavus*
ASPEFU	*Aspergillus fumigatus*
BLUMJA	*Blumeriella jaapii*
BOTRCI	*Botrytis cinerea*
BOTREL	*Botrytis elliptica*
CORYCA	*Corynespora cassiicola*
DIDYBR	*Stagonosporopsis cucurbitacearum*
EUROOR	*Aspergillus oryzae*
GIBBZE	*Fusarium graminearum*
MSPHTA	*Didymella tanaceti*
PENIDI	*Penicillium digitatum*
PENIEX	*Penicillium expansum*
PHAKPA	*Phakopsora pachyrhizi*
PLEOAL	*Stemphylium vesicarium*
PODOXA	*Podosphaera xanthii*
PUCCSI	*Puccinia striiformis f.* sp. *tritici*
PYRIOR	*Pyricularia oryzae*
RAMUCC	*Ramularia collo-cygni*
RHIZCE	*Rhizoctonia cerealis*
RHIZSO	*Rhizoctonia solani*
SCLEHO	*Clarireedia homoeocarpa*
SCLESC	*Sclerotinia sclerotiorum*
SEPTTR	*Zymoseptoria tritici*
STGSCI	*Stagonosporopsis citrulli*
UNCINE	*Erysiphe necator*
USTIMA	*Ustilago maydis*
VENTIN	*Venturia inaequalis*
XANTOZ	*Xanthomonas oryzae*

over 90% reduction in catalytic activity (Scalliet et al., 2012; Shima et al., 2009). Mutations may also result in lower enzyme stability or increased production of superoxide radicals (Szeto et al., 2007; Guo and Lemire, 2003; Maklashina et al., 2001). As the conserved residues SdhB-*W234* and SdhD-*Y146* are essential for ubiquinone binding, they have never been found to be mutated in field isolates or laboratory-generated mutants (Sierotzki and Scalliet, 2013; Scalliet et al., 2012).

Many mutations have been found within the binding site, where they exert a direct influence on binding affinity of the SDHI compound. For instance, the B-*H277* residue, within the polar cavity of the Q_p site and in direct proximity with the core moiety of the SDHI, is frequently mutated to a tyrosine or other amino acid, removing the H-bond interaction with the heteroatom (Shi et al., 2021b; Scalliet et al., 2012). This residue has been found to be mutated in most species for which SDHI resistance has been reported (Stammler et al., 2015; Sierotzki and Scalliet, 2013). Likewise, the B-*P230* residue, interacting hydrophobically with the linker moiety, is frequently found substituted to a bulkier or more polar residue that reduces SDHI binding affinity (Glättli et al., 2011). Both of these residues are within close proximity to the S3 [3Fe-4S] iron–sulphur cluster, and hence substitutions at these positions may also affect the transfer of electrons to ubiquinone (Glättli et al., 2011; Stammler et al., 2015). The residues C-*H134* and D-*H134* are also frequently found to be mutated (Glättli et al., 2011; Stammler et al., 2015; Sierotzki and Scalliet, 2013). These mutations may exert an indirect affect by altering the binding mode of haem b, as the propionate side chains of the haem form part of the Q_p site (Sierotzki and Scalliet, 2013; Piotrowska et al., 2017); substitutions may also result in the loss of a coordinating partner for the haem b iron (Rehfus, 2018). Many other mutations have been found far away from the binding site, and therefore must indirectly affect fungicide binding by causing long-range conformational rearrangements (Glättli et al., 2011; Stammler et al., 2015; Sierotzki and Scalliet, 2013). However, there does not appear to be any clear relationship between the proximity of a mutation to the binding site and the degree of resistance conferred (Stammler et al., 2015). Different amino acid substitutions at the same position can have very different effects on SDHI sensitivity; and depending on the species, the same mutation at homologous positions can have different effects (Scalliet et al., 2012).

A particular amino acid substitution may also have very different effects depending on the specific SDHI compound. Some mutations confer cross-resistance across many or all SDHIs, while for other mutations negative cross-resistance, where a mutation reduces sensitivity to one compound but increases sensitivity to another, has been reported (Scalliet et al., 2012). For example, the B-*H277* residue has been found to be most frequently mutated to a tyrosine in many species (Sierotzki and Scalliet, 2013). This mutation results in reduced sensitivity to most SDHI compounds but increased sensitivity to the pyridinyl-ethyl-benzamide fluopyram (Shi et al., 2021b; Scalliet et al., 2012). The substitution of tyrosine for histidine removes the hydrogen bond formed with the heteroatom in the core ring of most SDHIs, resulting in lower binding affinity (Shi et al., 2021b; Scalliet et al., 2012). However, fluopyram lacks a H-bond acceptor heteroatom in the core ring (Scalliet et al., 2012), and instead fluopyram is hypothesised to form H-bonds with other residues in the binding pocket via an altered binding mode, or between the trifluoromethyl group on the core ring and the mutated tyrosine residue (Shi et al., 2021b; Fraaije et al., 2012; Zhu et al., 2022), resulting in increased binding affinity for this compound. On the other hand, a substitution to leucine at the same position causes cross-resistance to all SDHIs, including fluopyram (Scalliet et al., 2012).

Two or more simultaneous mutations in the Sdh complex have been observed in some species, including *Alternaria alternata* (Avenot et al., 2014; Landschoot et al., 2017), *Alternaria solani* (Landschoot et al., 2017), *Corynespora cassiicola* (Zhu et al., 2022; Sun et al., 2022), *Didymella tanaceti* (Pearce et al., 2019), *Rhizoctonia cerealis* (Sun et al., 2017), and *Zymoseptoria tritici* (Fraaije et al., 2012). In *A. alternata* and *A. solani*, double mutants had considerably higher resistance to SDHIs compared to the equivalent single mutants (Avenot et al., 2014; Landschoot et al., 2017). In *C. cassiicola*, Sun et al. (2022) found that some double mutations had a synergistic effect, some additive effects, and some were antagonistic, depending on the combination of mutations and the fungicides tested. Zhu et al. (2022) found that there was no consistent shift in sensitivity in *C. cassiicola* double mutants compared to the equivalent single mutants. In *D. tanaceti*, double mutants with both C-H134Q (*H134*) and C-S135R (*S135*) showed similar SDHI sensitivity to C-S135R single mutants (Pearce et al., 2019). Most double mutants of *Z. tritici* showed greater sensitivity to most SDHIs than the equivalent single mutants (Fraaije et al., 2012).

3.1.2 Paralogs

Resistance may also be conferred by paralogs of the *Sdh* complex genes. Yamashita and Fraaije (2018) observed resistance to the SDHIs isofetamid and fluopyram, but not bixafen, or to chemically unrelated fungicides, in field strains of *Z. tritici* that had no mutations in the *SdhB*, *SdhC*, or *SdhD* subunits. Subsequently, Steinhauer et al. (2019) determined that resistance was conferred by a dispensable paralog of *SdhC*, *ZtSDHC3*, present as a standing variation in 20–30% of the European *Z. tritici* population. The residue I78 in the alternative-SdhC, homologous to A84I/V (*S73*) in the *Z. tritici* SdhC, conferred specific resistance to the stretched heterocycle amide SDHIs such as isofetamid and fluopyram. Among isolates with the SdhC paralog, there was also a wide range of resistance levels to isofetamid and fluopyram, correlated with gene overexpression due to transposon insertions in the *ZtSDHC3* promoter, as well as alternative splicing of *ZtSDHC3* (Steinhauer et al., 2019). The alternative-SdhC competes with SdhC for formation of the Sdh complex (Steinhauer et al., 2019). Similarly, multiple paralogs of *SdhC* have also been reported in *Fusarium graminearum* (Chen et al., 2021a). $SdhC_1$ and $SdhC_2$, which also compete to form the Sdh complex, harbour different mutations, and may have different roles in SDHI sensitivity: $SdhC_1$ has lower expression levels in certain SDHI-resistant mutants (Chen et al., 2021a).

3.1.3 Heterokaryosis

In binucleate or multinucleate fungi, the presence of multiple, genetically distinct nuclei may permit changes in the proportion of nuclei carrying resistance mutations, in response to fungicide selective pressure (De Miccolis Angelini et al., 2015). In the multinucleate fungus *Rhizoctonia solani*, the level of resistance to thifluzamide was correlated with the proportion of nuclei carrying the mutation B-H249Y (*H277*) (Miao et al., 2021). Sensitive and highly resistant isolates were homokaryotic, carrying only the H249 or Y249 allele, respectively; a range of intermediate sensitivity levels were correlated with varying proportions of nuclei with the mutant allele. Experimental evolution both *in vitro* and *in planta* showed how the proportion of the Y249 allele in heterokaryons increased under repeated fungicide selection, and heterokaryons could be converted to highly resistant Y249 homokaryons. Similarly, in the multinucleate fungus *S. sclerotiorum*, isolates homokaryotic for the mutation B-P226L (*P230*) had higher resistance levels to boscalid than heterokaryons; heterokaryons could be converted to highly resistant L226 homokaryons under repeated fungicide selection (Wang et al., 2022). In the binucleate, dikaryotic fungus *Rhizoctonia cerealis*, resistance to thifluzamide was higher in strains homokaryotic for the mutation H246Y (*H277*) in *SdhB* than strains heterokaryotic for this mutation, whereas there was no difference in thifluzamide sensitivity between strains homokaryotic and heterokaryotic for mutation H139Y (*H134*) in *SdhC* (Sun et al., 2017).

3.1.4 Other resistance mechanisms

Target site alterations not resulting in amino acid substitutions have also been associated with SDHI resistance. Chen et al. (2021a) observed laboratory-generated mutants of *F. graminearum* resistant to boscalid, pydiflumetofen, thifluzamide, and flubeneteram, where no amino acid substitutions were observed in *SdhB*, *SdhD*, or either paralog of *SdhC*. Resistance was correlated with an A-to-T transversion at –57 in the 5' UTR region of $SdhC_1$; the role of the 5′ UTR mutation in conferring the SDHI-resistant phenotype was validated by construction of gene-replacement mutants (Chen et al., 2021a). The presence of the 5′ UTR significantly decreased expression of *SdhC₁* under fungicide induction. The competition between *SdhC₁* and *SdhC₂* in formation of the Sdh complex, and the differing affinities of these paralogs to SDHIs, may explain the observed differences in SDHI sensitivity in 5′ UTR mutants with lower expression of *SdhC₁* (Chen et al., 2021a). Popko et al. (2018) observed several field isolates of *Sclerotinia homoeocarpa* with resistance to penthiopyrad, isofetamid, fluxapyroxad, and boscalid, where no amino acid substitutions were observed in *SdhB*, *SdhC*, or *SdhD*; however, resistance was correlated with the presence of the silent mutation CTT to CTC at codon 181 of *SdhB*. The authors hypothesised that the synonymous mutation in the coding sequence may result in altered protein folding, as has previously been described by Kimchi-Sarfaty et al. (2007) for alternative codons in the human *MDR1* gene.

3.2 Non-target site

3.2.1 Efflux

Active efflux, or phase III detoxification (Sang et al., 2018), reduces the intracellular concentration of toxic compounds (Del Sorbo et al., 2000). In fungi, efflux is mediated mainly by the ATP-binding cassette (ABC) transporters and the major facilitator superfamily (MFS) transporters (Costa et al., 2014), which together account for almost half of known transporters in microorganisms (Pao et al., 1998). The ABC transporters are primary transporters (Yan, 2015) with conserved nucleotide-binding domains, which couple translocation of substrates across the membrane to hydrolysis of ATP (Thomas and Tampé, 2020). The MFS are secondary active transporters (Yan, 2015) with a characteristic structural fold of 12–14 transmembrane helices, which translocate substrates using the transmembrane electrochemical gradient (Yan, 2013). The ability of transporters to efflux structurally diverse substrates (Del Sorbo et al., 2000) means they may be associated with multi-drug resistance (MDR) phenotypes, where resistance is observed to multiple unrelated chemical classes (Costa et al., 2014; Thomas and Tampé, 2020).

Both ABC and MFS transporters have been implicated in SDHI resistance. In MDR field strains of *Botrytis cinerea* showing resistance to penthiopyrad, resistance was associated with overexpression of the ABC transporter *BcatrB* (MDR1 strains), overexpression of the MFS transporter *BcmfsM2* (MDR2 strains), or overexpression of both *BcatrB* and *BcmfsM2* (MDR3 strains) (Kretschmer et al., 2009; Leroux and Walker, 2013). Overexpression of *BcatrB* was associated with gain-of-function mutations in the transcription factor *Mrr1*, while overexpression of *BcmfsM2* was associated with a long-terminal-repeat (LTR) retrotransposon insertion in the *BcmfsM2* promoter (Kretschmer et al., 2009). In MDR field strains of *Z. tritici*, resistance to boscalid and bixafen was associated with constitutive overexpression of the MFS transporter *MgMFS1* (Omrane et al., 2015). Overexpression of this gene in field strains was associated with three to four different insertion events in the promoter of *MgMFS1* (Omrane et al., 2017). The gene for the ABC transporter *ABCt-2* was also found to be constitutively overexpressed in a laboratory-generated *Z. tritici* isolate resistant to fluxapyroxad, fluopyram, and carboxin (Gutiérrez-Alonso et al., 2017). Three ABC transporters (*FvABCG1*, *FvABCG2*, and *FvABCG3*) and one MFS transporter (*g12362*) were overexpressed in a MDR *Fusarium virguliforme* field strain resistant to fluopyram (Sang et al., 2022). Both ABC and MFS transporters were also found to be overexpressed in *R. solani* laboratory-generated isolates resistant to thifluzamide (Zhao et al., 2022). In MDR field strains of *S. homoeocarpa*, constitutive overexpression of two ABC transporters, *ShatrD* and *ShPDR1*, was associated with resistance to boscalid (Sang et al., 2015); overexpression was correlated with a gain-of-function mutation in the transcription factor *ShXDR1* (Sang et al., 2018). In *Monilinia fructicola*, resistance to boscalid was not associated with point mutations in any of the *Sdh* genes, but was correlated with resistance to tolnaftate, a known substrate for multidrug transporters; therefore, the authors hypothesized that transporters may be responsible for SDHI resistance in this pathogen (Chen et al., 2013).

3.2.2 Detoxification

Detoxification of xenobiotic compounds by phase I and phase II metabolic enzymes may also contribute to reduced sensitivity to SDHI fungicides. Phase I enzymes mainly catalyse hydrolysis, oxidation, or reduction reactions (Iyanagi, 2007); the products are then typically either exported by phase III efflux pumps or further metabolised by phase II enzymes (Westhouse and Car, 2013). Phase II enzymes mainly catalyse glucuronidation, sulfation, or acetylation reactions (Iyanagi, 2007), converting the xenobiotic to a non-toxic metabolite (Chen, 2012). Four cytochrome P450 (phase I) genes and two glutathione *S*-transferases (phase II) genes were highly overexpressed in a *F. virguliforme* field strain resistant to fluopyram (Sang et al., 2022); alongside overexpression of ABC and MFS transporter (phase III) genes, the overexpression of these metabolic enzymes was identified as a potential mechanism of SDHI fungicide resistance. A MDR field strain of *S. homoeocarpa* showing resistance to boscalid was associated with constitutive overexpression of three phase I cytochrome P450 genes, *CYP561*, *CYP65*, and *CYP68*, as well as two ABC transporters; overexpression was caused by a gain-of-function mutation in the transcription factor *ShXDR1* (Sang et al., 2018).

3.2.3 Melanisation

Melanin has been identified as reducing susceptibility of the fungal cell to toxic compounds, binding certain toxic compounds and thereby preventing them from reaching their target sites (Nosanchuk and Casadevall, 2003). Increased melanisation has been associated with reduced sensitivity to SDHIs in *Z. tritici* (Krishnan et al., 2018). Upregulation of genes in the melanin biosynthetic cluster was correlated with either the presence of SNPs or the absence of transposable element insertions, in the promoter of the transcription factor *Zmr1* (Krishnan et al., 2018). More heavily melanised strains had lower *in vitro* sensitivity to the SDHI bixafen.

4 Fitness cost and resistance risk assessment

Fitness is defined as the ability of an individual, genotype, or population to survive and reproduce in a given environment (Orr, 2009; Elena and Lenski, 2003; Pringle and Taylor, 2002). As fungicide resistance mutations may be associated with fitness penalties, evaluating the fitness of resistant isolates is a key part in the assessment of the risk of fungicide resistance developing in the field (Liu et al., 2022b; Xu et al., 2022; Sun et al., 2017; Zhao et al., 2019). Such risk assessments assist in the development of anti-resistance strategies (Lalève et al., 2014a).

Fitness may be assessed either in terms of predicted fitness, that is the measurement of one or more traits of an individual isolate, or by realised fitness, that is the competitiveness between wild type and resistant isolates (Fan et al., 2015; Kim and Xiao, 2011). Predicted fitness may be a composite of a number of individual components from any of stages in the life cycle of the fungus, each of which contributes to overall fitness; usually this encompasses growth rate, fecundity, and viability (Pringle and Taylor, 2002; Orr, 2009; Lalève et al., 2014a). Realised fitness is determined by direct competition between isolates or genotypes, usually by measuring relative growth of each competitor following one or more rounds of co-inoculation (Elena and Lenski, 2003).

4.1 Alternaria alternata

In *A. alternata*, the fitness of boscalid-resistant field isolates was measured in terms of *in vitro* hyphal growth rate, conidia production, conidia germination rate, and virulence on pistachio leaves (Avenot and Michailides, 2007). No significant differences between the resistant isolates and the wild type were observed for any of the fitness parameters. A subsequent study by Avenot et al. (2009) assessed the sensitivity to oxidative stress in boscalid-resistant field isolates with the mutations B-H277R (*H277*), B-H277Y (*H277*), C-H134R (*H134*),

D-D123E (*D124*), or D-H133R (*H134*). All resistant isolates were hypersensitive to oxidative stress, compared to the wild type.

Fan et al. (2015) assessed the fitness of boscalid-resistant field isolates carrying the mutations B-H277Y (*H277*), C-H134R (*H134*), and D-D123E (*D124*). Predicted fitness was measured in terms of *in vitro* hyphal growth rate, conidia production on various growth media, conidia germination rate, sensitivity to oxidative stress, sensitivity to osmotic stress, and virulence on peach fruit. For B-H277Y and C-H134R, realised fitness was also measured in terms of *in planta* competition, by five successive transfers to peach fruit of spore mixtures of B-H277Y or C-H134R isolates in 1:1 ratio with wild type; the rate of germ tube formation was measured on boscalid-amended media. D-D123E isolates had a higher mycelial growth rate but lower sporulation rate and greater sensitivity to oxidative stress, compared to the wild type. B-H277Y isolates had lower sensitivity to oxidative stress and higher virulence on peach fruit, compared to the wild type. C-H134R isolates did not significantly differ from the wild type for any of the measured parameters. The competition experiment indicated B-H277Y or C-H134R isolates competed effectively with wild-type strains.

Landschoot et al. (2017) measured the *in vitro* hyphal growth rates and conidia germination rates of boscalid-resistant field isolates with the mutations B-H277Y (*H277*), C-H134R (*H134*), D-D123E (*D124*), or D-H133R (*H134*). There was no significant difference between isolates for any of these parameters, and SDHI resistance was not associated with any fitness cost in *A. alternata*.

Lichtemberg et al. (2019) assessed the fitness of SDHI-resistant field isolates with the mutations B-H277Y (*H277*), or C-H134R (*H134*). After the 1st and 12th successive transfers on media, predicted fitness was measured in terms of hyphal growth rate, spore production, spore germination, and pathogenicity on pistachio leaves. Realised fitness was also measured in terms *in vitro* competition, by five successive transfers to media of spore mixtures of B-H277Y or C-H134R isolates in 1:1 ratio with wild type; the rate of germ tube formation was measured on boscalid-amended media. C-H134R was lower than the wild type across all parameters tested. B-H277Y had markedly lower sporulation than the wild type, but other parameters were not significantly different to the wild type. Both B-H277Y and C-H134R increased in frequency under successive transfers of spore mixtures, indicating they could compete with the wild type *in vitro*.

4.2 Alternaria solani

In *A. solani*, Landschoot et al. (2017) analysed the *in vitro* hyphal growth rates and conidia germination rates of boscalid-resistant field isolates with the following mutations or combinations of mutations: B-H278R (*H277*), B-H278Y (*H277*), C-H134R (*H134*), B-H278R + C-H134R, and B-H278Y + C-H134R. There was no significant difference between isolates for any of these parameters, and SDHI resistance was not associated with any fitness cost in *A. solani*.

Bauske and Gudmestad (2018) assessed the fitness of field isolates resistant to boscalid, with the mutations B-H278R (*H277*), B-H278Y (*H277*), C-H134R (*H134*), D-D123E (*D124*), or D-H133R (*H134*). *In vitro* hyphal growth rates, conidia germination rates, and *in planta* virulence on tomato leaves were measured. There were no significant differences between any of the resistant mutants and the wild types for *in vitro* growth rate and conidia production. Isolates with D-D123E were more aggressive than the wild type on tomato leaf, whereas no difference was observed between isolates carrying other mutations compared to the wild type *in planta*. No mutation was associated with any fitness penalties, and D-D123E isolates appear to be more fit than the wild type.

4.3 *Aspergillus oryzae*

Shima et al. (2009) measured the succinate-cytochrome c reductase activity of mitochondrial extracts from laboratory-generated mutants of *Aspergillus oryzae*, with mutations B-H249L (*H277*), B-H249N (*H277*), B-H249Y (*H277*), C-T90I (*T78*), and D-D124E (*D145*). The succinate-cytochrome c reductase activity for *SdhB* mutants was markedly lower (up to 82% lower, for B-H249L), compared to the wild type; *SdhC* or *SdhD* mutants had activities between that of *SdhB* mutants and the wild type.

4.4 *Botrytis cinerea*

The first study of fitness costs associated with SDHI resistance in *B. cinerea* assessed laboratory-generated mutants resistant to boscalid for *in vitro* hyphal growth rate, conidia production and germination rate, and *in planta* virulence on aubergine leaf (Zhang et al., 2007). Resistant isolates had similar growth rates and conidia germination rates than wild types, but had lower sporulation rates and were less virulent on aubergine.

Kim and Xiao (2011) conducted the first fitness study of field isolates with SDHI resistance. Predicted fitness was measured *in vitro* by hyphal growth rate at different temperatures, conidia production, conidia germination rate, sensitivity to osmotic stress, and *in planta* virulence on apple fruit. Realised fitness was measured by *in planta* competition, by four successive transfers of spore mixtures of mutant isolates in various ratios with wild type to apple fruit, after which resistance was assessed by growth of isolates on boscalid-amended media. The boscalid-resistant isolates had lower conidia production than the wild types but otherwise displayed no penalties with respect to predicted fitness or virulence. However, no boscalid-resistant isolates were observed in the competition assay after four disease cycles, indicating that boscalid resistance was associated with decreased competitive ability.

Veloukas et al. (2014) analysed the fitness of B-P225F (*P230*), B-N230I (*N235*), or B-H272L/R/Y (*H277*) field isolates also carrying the G143A mutation in *cytochrome b*. Fitness parameters assessed were hyphal growth rate, conidia production, conidia germination, sclerotia production, and virulence on bean leaves, at several temperatures. Competition was also assessed by four successive transfers of spore mixtures to apple fruit. B-H272R isolates had lower sclerotia germination rates and hyphal growth rates but were otherwise similar to the wild types. Other mutations incurred significant fitness costs. The frequency of all mutations in competition experiments declined in the absence of fungicide selection, indicating decreased competitive ability for the mutants.

Lalève et al. (2014b) assessed the Sdh enzyme activity, respiration rate, and hyphal growth rate on different carbon sources for laboratory-generated mutants with mutations B-P225F/L/T, B-N230I, or B-H272L/R/Y. All mutations, with the exception of B-H272Y, resulted in lower respiration rate and Sdh enzyme activity. A subsequent study by Lalève et al. (2014a) also examined the effect of these mutations on hyphal growth rate at different temperatures; conidia production and germination rates; sclerotia production, viability, and latency period; sensitivity to oxidative stress; mycelial wettability; pathogenicity on tomato and green bean; and *in vitro* competitiveness in spore mixtures with wild types. The B-H272Y was associated with the lowest fitness penalties, whereas the B-P225L and B-H272R mutations were associated with the greatest fitness penalties.

Li et al. (2022) assessed the fitness of laboratory-generated pydiflumetofen-resistant isolates with the mutations B-P225L, or C-G85A (*S73*) + C-I93V (*V81*), in terms of mycelial growth rate at various temperatures, conidia production and germination rate, sclerotia size and germination rate, and virulence on tomato fruit. In contrast to Lalève et al. (2014a), conidia production and germination rate, sclerotia germination rate, and virulence on tomato were similar between mutant and wild-type isolates. Some C-G85A + C-I93V mutants produced larger sclerotia, and some B-P225L had greater mycelial biomass than the wild type.

Liu et al. (2022a) examined the fitness of fluxapyroxad-resistant field isolates with the following mutations or combinations of mutations: B-P225F, B-N230I, B-N230I + D-V9A, B-N230I + B-K283N (*K288*) + D-V9A, C-G37S (*R26.9*), and B-H272R + C-P80H (*T68*). Fitness was measured in terms of hyphal growth rate, conidia production and germination rate, sclerotia production, virulence on tomato and apple fruits, and sensitivity to osmotic, oxidative, and cell wall stress. Fluxapyroxad-resistant mutants suffered fitness penalties with regard to growth rate, conidia and sclerotia production, virulence, and tolerance of osmotic, oxidative, and cell wall stress.

A study by Malandrakis et al. (2022) also analysed the fitness of field isolates carrying B-P225F/H, B-N230I, or B-H272R in addition to G143A in *cyt b*, as well as those with mutation B-I274V (*I279*). Isolates were measured for hyphal growth rate, conidia production, and virulence on tomato fruit. B-I274V was associated with reduced conidia production and virulence, while B-P225F was associated with reduced conidia production, virulence, and growth rate.

4.5 *Corynespora cassiicola*

Shi et al. (2021a) studied the *in vitro* and *in planta* fitness of laboratory-generated mutants of *C. cassiicola* with the following mutations: B-H278R (*H277*), B-H278Y (*H277*), B-I280V (*I279*), D-D95E (*D124*), D-H105R (*H134*), and D-G109V (*G138*). The parameters of fitness measured were hyphal growth rates at different temperatures, hyphal growth rates on different carbon sources, conidia production and germination, pathogenicity on cucumber leaves, and sensitivity to osmotic stresses, cell wall damage, metal ions, and the alternative oxidase inhibitor salicylhydroxamic acid (SHAM). No fitness penalty was found for mutations B-H278R, S-D95E, or D-G109V; whereas B-H278Y, B-I280V, and D-H105R had decreased hyphal growth rates, decreased conidia production, and greater sensitivity to osmotic stress. However, a subsequent study by Shi et al. (2021b) analysed laboratory-generated B-H278Y (*H277*) and B-I280V (*I279*) mutants of *C. cassiicola* for *in vitro* and *in planta* fitness according to mycelial growth rate, conidial production, Sdh enzyme activity, pathogenicity on cucumber leaves, and sensitivity to oxidative stresses, osmotic stresses, cell wall damage, metal ions, and SHAM. No fitness penalty was found; both B-H278Y and B-I280V conferred increased Sdh enzyme activity and lower sensitivity to osmotic stress.

Sun et al. (2022) analysed laboratory-generated single and double mutants of *C. cassiicola* with the following mutations or combinations of mutations: B-H278R (*H277*), B-H278Y (*H277*), B-I280V (*I279*), D-D95E (*D124*), D-H105R (*H134*), D-G109V (*G138*), B-I280V + D-D95E, B-H278R + D-D95E, B-H278Y + D-D95E, B-I280V + D-G109V, B-H278R + D-G109V, B-H278Y + D-G109V, and B-I280V + D-H105R. The measures of fitness were as follows: hyphal growth rate at various temperatures; conidia production; conidia germination rate; Sdh enzyme activity; virulence on cucumber leaf; and sensitivity to osmotic stress, oxidative stress, metal ions, cell wall damage, and SHAM. All double mutants had lower fitness than the wild type; moreover, all double mutants, except for B-I280V+D-95E, had lower fitness than isolates with the corresponding single mutation.

By contrast, Zhu et al. (2022) measured fitness in field isolates of *C. cassiicola* with the following mutations or combinations of mutations: B-H278L (*H277*), B-H278Y (*H277*), B-I280V (*I279*), C-S73P (*S73*), C-N75S (*N75*), C-H134R (*H134*),

D-D95E (*D124*), D-G109V (*G138*), B-I280V + C-S73P, B-I280V + C-N75S, and C-S73P + C-N75S. Fitness was assessed in terms of Sdh enzyme activity, hyphal growth rates at various temperatures, conidia production and germination rate, sensitivity to osmotic stress, sensitivity to oxidative stress, and virulence on cucumber leaves. No fitness cost was associated with any of the mutations or combinations of mutations.

4.6 *Didymella bryoniae*

Mao et al. (2020) evaluated both *in vitro* and *in planta* fitness of *D. bryoniae* laboratory mutants with mutations B-H277R (*H277*) and B-H277Y (*H277*). The traits assessed included hyphal growth rate, biomass, and morphology; sensitivity to osmotic stress; sensitivity to glucose; sensitivity to various metal cations; sensitivity to oxidative stress; sensitivity to cell wall and cell membrane damage; cell membrane permeability; and pathogenicity on watermelon leaves. Overall, most isolates did not significantly differ from the wild type with regard to hyphal growth traits, sensitivity to glucose or metal cations, or pathogenicity on watermelon leaves. Cell membrane permeabilities of resistant isolates were significantly different from wild types, while mutants were also less sensitive than the wild types to cell wall damage. The risk of *D. bryoniae* developing SDHI resistance was assessed as medium to high.

4.7 *Fusarium asiaticum*

In *F. asiaticum*, the fitness of pydiflumetofen-resistant laboratory mutants was assessed with regards to hyphal growth rate, conidia production, mycotoxin production, sensitivity to oxidative stress, and virulence on wheat head (Chen et al., 2021b). No differences were observed with regard to growth rate, conidia production, or virulence. All resistant mutants had greater mycotoxin production than the wild type, and some mutants had lower sensitivity than wild type to oxidative stress. Pydiflumetofen resistance was not associated with any fitness penalty, and the risk of SDHI resistance in *F. asiaticum* was assessed as moderate to high.

4.8 *Fusarium graminearum*

In *F. graminearum*, the fitness of laboratory-generated pydiflumetofen-resistant mutants with $SdhC_2$-A83V (*S73*), $SdhC_2$-R86C (*R76*), or $SdhC_2$-R86H (*R76*) mutations was assessed (Sun et al., 2020). Fitness was assessed in terms of *in vitro* hyphal growth rate and conidia production, and *in planta* virulence on wheat heads. Mutants carrying $SdhC_2$-A83V (*S73*) showed no fitness penalties with regard to growth rate, conidia production, or virulence. $SdhC_2$-R86C mutants had reduced hyphal growth rates, while in C-R86H mutants growth rates were severely reduced. Both $SdhC_2$-R86C (*R76*) and $SdhC_2$-R86H (*R76*) were associated with reduced virulence. The risk of SDHI resistance in *F. graminearum* was assessed as moderate.

Chen et al. (2021a) assessed the fitness of flubeneteram-resistant laboratory mutants with the mutations $SdhC_2$-T77I (*T68*), $SdhC_2$-R86C (*R76*), or an A-to-T transversion at −57 of the 5′ UTR of $SdhC_1$. The following parameters were assessed as measures of fitness: mycelial growth rate on different carbon sources; conidia production; mycotoxin production; Sdh enzyme activity; virulence on wheat head and coleoptile; sensitivities to oxidative stress, osmotic stress, cell wall-damaging agents, and cell membrane-damaging agents. All resistant mutants had reduced growth rate, conidia production, mycotoxin production, and virulence, indicating fitness costs associated with these mutations. *F. graminearum* was assessed as at a moderate risk of developing resistance to SDHIs.

Shao et al. (2022) evaluated the fitness of pydiflumetofen-resistant field isolates carrying the mutation $SdhC_1$-A78V (*S73*), in terms of hyphal growth rate, conidia production, sensitivity to oxidative stress, and virulence on corn silk and on wheat coleoptile. No fitness penalty in the resistant mutants was observed for any of the measured parameters. The risk of SDHI resistance in *F. graminearum* was assessed as high.

4.9 *Magnaporthe oryzae*

Liu et al. (2022b) assessed the *in vitro* and *in planta* fitness of laboratory-generated mutants of *Magnaporthe oryzae* with the mutations B-H245L (*H277*) and B-H245Y (*H277*), with regard to growth rate on different agar media, conidia production, appressorium formation, and pathogenicity on barley; no difference between mutants and wild types was observed. The risk of *M. oryzae* developing SDHI resistance was assessed as medium to high.

4.10 *Penicillium digitatum*

In *Penicillium digitatum*, Xu et al. (2022) assessed the *in vitro* and *in planta* fitness of B-H243Y (*H277*) laboratory mutants with regard to hyphal growth rate on agar, conidia production, virulence on citrus fruit, and sensitivities to osmotic stress, oxidative stress, and SHAM. Overall, B-H243Y mutants showed great variation in growth rate, conidia production, virulence, and sensitivity to exogenous stresses, but there was no general reduction in fitness compared to the wild type across any of the measured traits. *P. digitatum* was assessed as high risk for developing resistance to SDHIs.

4.11 *Penicillium expansum*

In *Penicillium expansum*, the *in vitro* and *in planta* fitness of laboratory-generated B-H272R (*H277*) and B-H272Y (*H277*) mutants was studied by Malandrakis et al. (2017). The measures of fitness were hyphal growth rate on agar, conidia production, conidial germination rate, sensitivity to osmotic stress, production of mycotoxins, and pathogenicity on apple fruit. Overall, mutants were not significantly different than wild types for osmotic sensitivity, conidia production, or pathogenicity on apple, while some mutants had lower hyphal growth and conidia germination rates. There was no significant relationship between SDHI sensitivity and mycotoxin production. *P. expansum* was assessed as high risk for developing resistance to SDHIs.

4.12 *Phakopsora pachyrhizi*

Claus et al. (2022) examined the *in vitro* fitness of field isolates of *Phakopsora pachyrhizi* with the mutation C-I86F (*S73*) with regard to the following traits: sensitivity to osmotic stress in saline medium, sensitivity to oxidative stress in the presence of paraquat, and sensitivity to UV radiation. Compared to the wild types, isolates with SdhC-I86F were hypersensitive to both osmotic and oxidative stress. C-I86F mutants were also hypersensitive to the multisite dithiocarbamate fungicide mancozeb. The authors concluded that this mutation was associated with significant fitness penalties in *P. pachyrhizi*.

4.13 *Ramularia collo-cygni*

Piotrowska et al. (2017) determined the growth rate on agar, as a measure of *in vitro* fitness, for laboratory-generated mutants of *Ramularia collo-cygni* with the following mutations: B-S217L (*S228*), B-N224I (*N235*), C-H146Q (*H134*), or C-H146R (*H134*); no differences between mutants and wild types were observed. The *in planta* fitness of B-N224I and C-H146R mutants were also assessed; no difference in infection of barley was observed compared to wild types. *R. collo-cygni* was assessed as high risk for developing resistance to SDHIs.

4.14 *Rhizoctonia cerealis*

In *R. cerealis*, the fitness of laboratory-generated thifluzamide-resistant mutants was examined with regard to hyphal growth rate, sclerotia production, and virulence on wheat leaves (Zhang et al., 2015). No fitness penalty was observed among the mutants for any of the measured traits; however, the authors assessed the risk of *R. cerealis* developing resistance to SDHIs as low-to-moderate, owing to the difficulty of inducing thifluzamide resistance *in vitro*.

A subsequent study by Sun et al. (2017) looked at laboratory-generated thifluzamide-resistant mutants both homozygous and heterozygous for the mutations B-H246Y (*H277*), C-H139Y (*H134*), D-H116Y (*H134*), D-H118Y (*H136*), and D-L129I (*F147*). Fitness was assessed *in vitro* in terms of mycelial growth rate and *in planta* in terms of virulence on wheat leaves. Although there were differences observed in growth rate and virulence for some mutants compared to the wild type, in general no fitness penalty could be linked to any Sdh mutation.

4.15 *Rhizoctonia solani*

In *R. solani*, the fitness of both thifluzamide-resistant field isolates and laboratory-generated mutants was assessed (Mu et al., 2017). The traits of fitness measured were hyphal growth rates at different temperatures, mycelial biomass, sclerotial biomass and germination rate, and virulence on rice sheath. No differences were observed for mycelial biomass or germination rate of sclerotia; the mycelial growth rate and virulence was reduced in some mutants. Thifluzamide resistance was associated with some fitness penalties in *R. solani*; the risk of SDHI resistance developing was assessed as low to moderate.

Zhao et al. (2019) assessed the fitness of laboratory-generated flutolanil-resistant mutants, including those with the mutation C-F48L (*L60*). The measures of fitness were *in vitro* growth rates at various temperatures, and *in planta* virulence on sugar beet seedlings. Flutolanil-resistant mutants had lower rates of hyphal growth and were less virulent than the wild type on sugar beet, suggesting that SDHI resistance is associated with a slight fitness cost in *R. solani*. A subsequent study by Zhao et al. (2022) looked at the fitness of laboratory-generated thifluzamide-resistant mutants, including isolates overexpressing ABC and MFS drug transporters, and those with the mutation C-I61F (*S73*). Fitness was again assessed in terms of *in vitro* growth rates at different temperatures, and virulence on sugar beet. This study found no fitness cost for thifluzamide resistance; growth rates for most mutants were similar to or slightly higher than the wild types, while virulence on sugar beet was similar to or only slightly lower than wild type.

4.16 *Sclerotinia sclerotiorum*

Wang et al. (2022) measured the *in vitro* fitness parameters of hyphal growth rate, sclerotia production, and infection cushion formation in laboratory-generated mutants of *S. sclerotiorum* both heterokaryotic and homokaryotic for the mutation B-P226L (*P230*). In terms of mycelial growth rate, heterokaryotic mutants were slightly decreased, and homokaryotic mutants were markedly decreased, compared to wild types. Heterokaryotic mutants had decreased production of sclerotia than wild types, while homokaryotic mutants lost the ability to produce sclerotia entirely. Homokaryotic mutants also had lower or no production of infection cushions. The authors concluded that this mutation was associated with significant fitness penalties in *S. sclerotiorum*.

4.17 *Stemphylium solani*

Bi et al. (2022) analysed the *in vitro* and *in planta* fitness of laboratory-generated fluxapyroxad-resistant mutants of *Stemphylium solani*, in terms of mycelial growth rates at different temperatures, and virulence on tomato leaves. Fluxapyroxad resistance affected mycelial growth rate in a temperature-dependent manner: for some mutant isolates the mycelial growth rate was significantly lower than the wild type only at 10°C, for some mutants growth was significantly lower than the wild type only at 30°C, and some mutants had increased growth compared to the wild type at 20°C and 30°C. All fluxapyroxad-resistant mutants were markedly less virulent than the wild type on tomato leaves. The authors concluded that fluxapyroxad resistance was associated with significant fitness penalties in *S. solani*.

4.18 *Zymoseptoria tritici*

Fraaije et al. (2012) examined the *in planta* pathogenicity and spore production of laboratory-generated carboxin-resistant mutants carrying the B-S221P (*S231*), B-H267N (*H277*), B-H267Y (*H277*), or D-D129E (*D145*) mutations on wheat seedlings. No fitness cost was associated with any of the mutations. Scalliet et al. (2012) analysed the *in vitro* and *in planta* fitness of laboratory-generated mutants resistant to boscalid, carboxin, fluopyram, or isopyrazam. The succinate ubiquinone reductase enzyme activity was measured for 26 mutations. Sensitivity to oxidative stress was assessed for isolates carrying the mutations B-R265P (*R275*), B-H267L (*H277*), B-H267Y (*H277*), B-I269V (*I279*), C-S83G (*S72*), C-A84V (*S73*), C-N86K (*N75*), C-H152R (*H141*), or D-D129G (*D145*). Isolates carrying B-H267L (*H277*), C-A84V (*S73*), C-N86K (*N75*), or C-H152R (*H141*) were also assessed *in planta* for virulence on wheat plants. All mutations resulted in decreased enzyme activity compared to the wild-type enzyme. There was no evidence of hypersensitivity to oxidative stress in any of the mutants. *In planta* tests showed no loss of virulence, and some mutants had increased virulence compared to the wild type.

4.19 Conclusion

The impact of a given Sdh mutation on a trait of fitness is dependent on the species in question. For B-*H277Y*, the most common Sdh mutation found across species (Table 1), Avenot et al. (2009) found hypersensitivity to oxidative stress in *A. alternata*, and Lalève et al. (2014a) likewise for the homologous B-H272Y in *B. cinerea*, whereas no evidence of hypersensitivity to oxidative stress was found for the equivalent B-H243Y in *P. digitatum* (Xu et al., 2022), B-H267Y in *Z. tritici* (Fraaije et al., 2012), B-H277Y in *D. bryoniae* (Mao et al., 2020), or B-H278Y in *C. cassiicola* (Zhu et al., 2022). Also apparent are the differences reported between studies for a particular fitness trait for the same mutation in the same species. To take the B-*H277Y* mutation again as the example, B-H278Y (*H277*) in *C. cassiicola* was associated with reduced hyphal growth rates and reduced conidia production by Sun et al. (2022), and reduced hyphal growth rates, reduced conidia production, and hypersensitivity to osmotic stress by Shi et al. (2021a), whereas the same mutation in the same species was not associated with penalties across any of these three traits in studies by Shi et al. (2021b) and Zhu et al. (2022). Another of the most common mutations, C-*H134R*, was reported by Lichtemberg et al. (2019) as being associated with reduced mycelial growth rate in *A. alternata*, but the same mutation in *A. alternata* was not associated with reduced growth in studies by Landschoot et al. (2017) and Fan et al. (2015). It is also important to note that the realised fitness observed in the competition between isolates may not always accord with the fitness predicted from the measurement of individual fitness components of isolates. For instance, Kim and Xiao (2011) tested *B. cinerea* isolates for hyphal growth rates at different temperatures, conidia production, conidia germination rate, sensitivity to osmotic stress, and *in planta* virulence on apple fruit. Although conidia production was the only trait of predicted fitness reduced in boscalid-resistant isolates, no boscalid-resistant isolates were observed in the competition assay after four disease cycles co-inoculated with wild types, indicating that the mutants were unable to effectively compete with the wild-type strains.

5 Case study

Net-form of net blotch (NFNB), caused by the fungus *Pyrenophora teres* f. sp. *teres* (*Ptt*), is the second-most economically important disease of barley in Australia, causing AUS$19 million in losses annually, with estimated potential losses of AUS$117 million annually in the absence of current control measures (Murray and Brennan, 2010). In part due a lack of resistant or highly resistant barley cultivars (GRDC, 2022), fungicides form a key part of NFNB management. The first SDHI registered for the control of net blotch in Australia was fluxapyroxad, introduced in the 2016 growing season (Platz et al., 2017), followed by bixafen

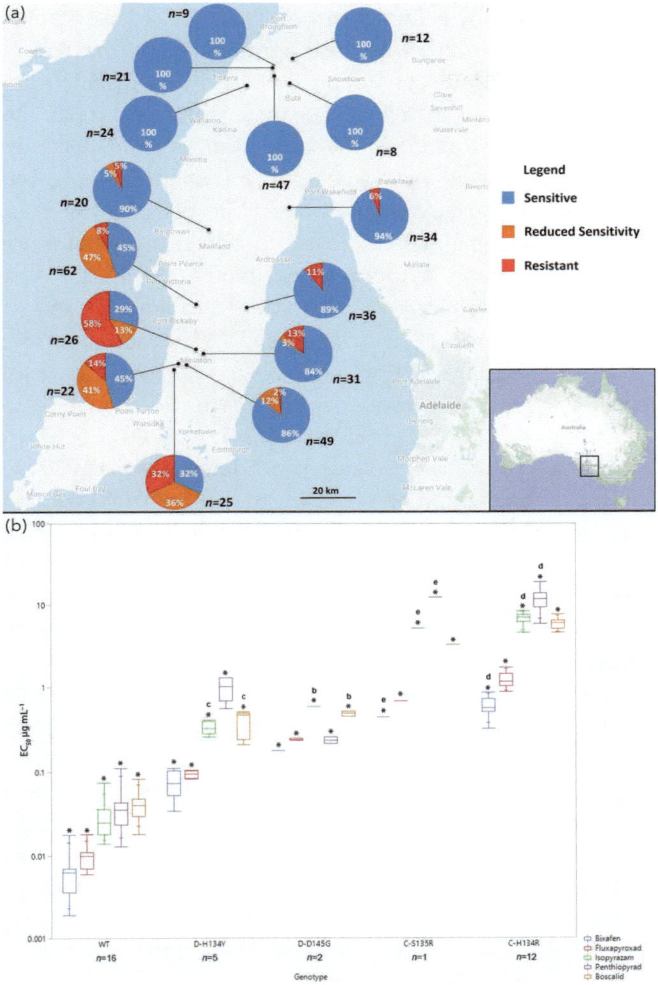

Figure 5 (a) Distribution and frequency of resistance and reduced sensitivity to SDHI fungicides in *Pyrenophora teres* f. *teres* from the Yorke Peninsula, South Australia (SA) in 2019. A field survey of *Pyrenophora teres* cultures (*n* = 427) derived from 15 sites (black points) on the Yorke Peninsula, SA in 2019. Pie charts: proportion of cultures at each site with resistance (red, growth at 10 µg mL^{-1}), reduced sensitivity (orange, growth at 2 µg mL^{-1}), or full sensitivity (blue), to fluxapyroxad. The overall frequency of resistance was 10%, with a further 31% showing reduced sensitivity to fluxapyroxad and 59% sensitive. SDHI resistance increased in frequency towards the Southern end of the Yorke Peninsula. (b) Five genotypes of the Sdh complex genes detected in *Pyrenophora teres* f. *teres* isolates from SA, with associated *in vitro* sensitivity to five SDHI fungicides. Quantile box plots of 50% effective concentration (EC_{50}) values (µg mL^{-1}) to bixafen (blue), fluxapyroxad (red), isopyrazam (green), penthiopyrad (purple), and boscalid (orange). *The mean difference between genotypes within treatments is significant at the 0.05 level (Dunnett's T3). Means with same letter within treatments are not significantly different at the 0.05 level: [a]WT, [b]D-H134Y, [c]D-D145G, [d]C-S135R, [e]C-H134R.

in 2018 (BAYER, 2018). The first cases of fluxapyroxad resistance in *Ptt* in Australia were reported in 2019 by local agronomists on the Yorke Peninsula of South Australia. A coordinated effort involving two research centres and local agronomists resulted in a field survey encompassing 15 sites on the Yorke Peninsula in 2019. To analyse the large number of field survey samples, a high-throughput workflow was developed for phenotyping of cultures on discriminatory doses of fluxapyroxad, as described by Knight et al. (2023); this methodology was subsequently employed to detect the emergence of SDHI resistance in Spot-form and Net-form of net blotch in Western Australia. The analysis found the overall frequency of resistance to fluxapyroxad (defined as growth at 10 μg mL^{-1}) to be 10%, with a further 31% showing reduced sensitivity to fluxapyroxad (defined as growth at 2 μg mL^{-1}) ($n = 427$) (Fig. 5a). Resistance and reduced sensitivity increased in frequency towards the Southern end of the Yorke Peninsula. Mutations *C-H134R*, *C-S135R*, *D-H134Y*, or *D-D145G* were identified in resistant or reduced-sensitivity isolates from South Australia in 2019, and were correlated with a range of reduced sensitivity to different SDHI compounds (Fig. 5b). The mutation *C-H134R* was associated with the highest levels of resistance to fluxapyroxad and bixafen, and was also the most frequently observed mutation. These results indicate a very rapid emergence and build-up of resistance, occurring within three years of the introduction of SDHIs for NFNB control. Based on these findings, hydrolysis probes and primers were developed to detect the presence of known Sdh resistance mutations directly from leaf samples using quantitative PCR and digital PCR; this now forms the basis of high-throughput genotypic surveillance for SDHI resistance in Australia. These findings have also informed the development of specific extension contents aimed at improving the management and avoidance of SDHI resistance in barley-growing regions of Australia, including dedicated webinars, factsheets, podcasts, and the Fungicide Resistance Management in Australian Grain Crops guide (Ireland et al., 2021), via the Australian Fungicide Resistance Extension Network (https://afren.com.au) and press releases and industry presentations via the Grains Research and Development Corporation (https://grdc.com.au).

6 Acknowledgements

Homology model and predicted docking for *Pyrenophora teres* Sdh courtesy of Dr Joel Haywood, Centre for Crop and Disease Management, Curtin University.

7 References

Abiko, K., Kishi, K. and Yoshioka, A. 1977. Occurrence of oxycarboxin-tolerant isolates of Puccinia horiana P. Hennings in Japan. *Japanese Journal of Phytopathology* 43, 145–150.

Ackrell, B. A. C. 2000. Progress in understanding structure-function relationships in respiratory chain complex II. *FEBS Letters* 466(1), 1-5.

Adaskaveg, J., Luo, Y. and Förster, H. 2019. Characterization of resistance to five SDHI sub-groups in Alternaria species causing leaf spot of almond in California. In: Deising, H., Fraaije, B., Mehl, A., Oerke, E., Sierotzki, H. and Stammler, G. (Eds). *Modern Fungicides and Antifungal Compounds IX*, Friedrichroda, Germany. DPG Spectrum Phytomedizin, 173–180.

Avenot, H. F., Sellam, A., Karaoglanidis, G. and Michailides, T. J. 2008. Characterization of mutations in the iron-sulphur subunit of succinate dehydrogenase correlating with Boscalid resistance in Alternaria alternata from California pistachio. *Phytopathology* 98(6), 736–742.

Avenot, H., Sellam, A. and Michailides, T. 2009. Characterization of mutations in the membrane-anchored subunits AaSDHC and AaSDHD of succinate dehydrogenase from Alternaria alternata isolates conferring field resistance to the fungicide Boscalid. *Plant Pathology* 58(6), 1134–1143.

Avenot, H. F., Thomas, A., Gitaitis, R. D., Langston, D. B. and Stevenson, K. L. 2012. Molecular characterization of Boscalid- and penthiopyrad-resistant isolates of Didymella bryoniae and assessment of their sensitivity to fluopyram. *Pest Management Science* 68(4), 645–651.

Avenot, H. F. and Michailides, T. J. 2007. Resistance to Boscalid fungicide in Alternaria alternata isolates from pistachio in California. *Plant Disease* 91(10), 1345–1350.

Avenot, H. F., Van Den Biggelaar, H., Morgan, D. P., Moral, J., Joosten, M. and Michailides, T. J. 2014. Sensitivities of baseline isolates and Boscalid-resistant mutants of Alternaria alternata from pistachio to fluopyram, penthiopyrad, and fluxapyroxad. *Plant Disease* 98(2), 197–205.

Baek, M., Dimaio, F., Anishchenko, I., Dauparas, J., Ovchinnikov, S., Lee, G. R., Wang, J., Cong, Q., Kinch, L. N., Schaeffer, R. D., Millán, C., Park, H., Adams, C., Glassman, C. R., Degiovanni, A., Pereira, J. H., Rodrigues, A. V., Van Dijk, A. A., Ebrecht, A. C., Opperman, D. J., Sagmeister, T., Buhlheller, C., Pavkov-Keller, T., Rathinaswamy, M. K., Dalwadi, U., Yip, C. K., Burke, J. E., Garcia, K. C., Grishin, N. V., Adams, P. D., Read, R. J. and Baker, D. 2021. Accurate prediction of protein structures and interactions using a three-track neural network. *Science* 373(6557), 871–876.

Bauske, M. J. and Gudmestad, N. C. 2018. Parasitic fitness of fungicide-resistant and -Sensitive isolates of Alternaria solani. *Plant Disease* 102(3), 666–673.

BAYER 2018. Global fungicide resistance experts tour Australia [Online]. Available at: https://www.crop.bayer.com.au/news-and-insights/news/news-container/2018/07/09/03/08/global-fungicide-resistance-experts-tour-australia. Accessed 19/09/2022. Bayer Australia, 2022.

Bi, L., Xu, J., Wang, Q., Li, T., Wang, J., Zhou, M. and Duan, Y. 2022. Baseline sensitivity and resistance risk assessment of Stemphylium solani to fluxapyroxad. *Crop Protection* 156, 105944.

Broomfield, P. L. E. and Hargreaves, J. A. 1992. A single amino-acid change in the iron-sulphur protein subunit of succinate dehydrogenase confers resistance to carboxin in Ustilago maydis. *Current Genetics* 22(2), 117–121.

Chen, C.-H. 2012. Phase II enzymes. *Activation and Detoxification Enzymes: Functions and Implications*. New York, NY: Springer New York.

Chen, F., Liu, X., Chen, S., Schnabel, E. and Schnabel, G. 2013. Characterization of Monilinia fructicola Strains resistant to both propiconazole and Boscalid. *Plant Disease* 97(5), 645–651.

Chen, W., Wei, L., Li, X., MA, ma, H., Lou, T., Zhang, P., Zheng, H., Zhu, X., Zhang, Y., Liu, F., Chen, C. and Yang, G. 2021a. Point mutations in FgSdhC2 or in the 5′ untranslated region of FgSdhC1 confer resistance to a novel succinate dehydrogenase inhibitor Flubeneteram in Fusarium graminearum. *Journal of Agricultural and Food Chemistry* 69(44), 13006–13019.

Chen, W., Wei, L., Zhao, W., Wang, B., Zheng, H., Zhang, P., Lou, T., Duan, Y., Hou, Y., Zhou, M. and Chen, C. 2021b. Resistance risk assessment for a novel succinate dehydrogenase inhibitor pydiflumetofen in Fusarium asiaticum. *Pest Management Science* 77(1), 538–547.

Cherrad, S., Hernandez, C., Vacher, S. and Steva, H. 2017. First detection of Boscalid-resistant strains of Erysiphe necator in French vineyards: biological and molecular characterization. In: Deising, H., Fraaije, B., Mehl, A., Oerke, E., Sierotzki, H. and Stammler, G. (Eds). *Modern Fungicides and Antifungal Compounds* (vol. VIII). Braunschweig: Deutsche Phytomedizinische Gesellschaft, 211–216.

Claus, A., Simões, K. and De Mio, L. L. M. 2022. SdhC-I86F mutation in Phakopsora pachyrhizi is stable and can be related to fitness penalties. *Phytopathology* 112(7), 1413–1421.

Cook, N. M., Chng, S., Woodman, T. L., Warren, R., Oliver, R. P. and Saunders, D. G. 2021. High frequency of fungicide resistance-associated mutations in the wheat yellow rust pathogen Puccinia striiformis f. sp. tritici. *Pest Management Science* 77(7), 3358–3371.

Costa, C., Dias, P. J., Sá-Correia, I. and Teixeira, M. C. 2014. MFS multidrug transporters in pathogenic fungi: Do they have real clinical impact? *Frontiers in Physiology* 5, 197.

De Miccolis Angelini, R. M., Habib, W., Rotolo, C., Pollastro, S. and Faretra, F. 2010. Selection, characterization and genetic analysis of laboratory mutants of Botryotinia fuckeliana (Botrytis cinerea) resistant to the fungicide Boscalid. *European Journal of Plant Pathology* 128(2), 185–199.

De Miccolis Angelini, R. M., Pollastro, S. and Faretra, F. 2015. Genetics of fungicide resistance. In: Ishii, H. and Hollomon, D. W. (Eds). *Fungicide Resistance in Plant Pathogens: Principles and a Guide to Practical Management*. Tokyo: Springer Japan.

De Miccolis Angelini, R. M., Rotolo, C., Masiello, M., Gerin, D., Pollastro, S. and Faretra, F. 2014. Occurrence of fungicide resistance in populations of Botryotinia fuckeliana (Botrytis cinerea) on table grape and strawberry in southern Italy. *Pest Management Science* 70(12), 1785–1796.

Del Sorbo, G., Schoonbeek, H.-J. and De Waard, M. A. 2000. Fungal transporters involved in efflux of natural toxic compounds and fungicides. *Fungal Genetics and Biology: FG and B* 30(1), 1–15.

Dooley, H., Shaw, M. W., Mehenni-Ciz, J., Spink, J. and Kildea, S. 2016. Detection of Zymoseptoria tritici SDHI-insensitive field isolates carrying the SdhC-H152R and SdhD-R47W substitutions. *Pest Management Science* 72(12), 2203–2207.

Einspanier, S., Susanto, T., Metz, N., Wolters, P. J., Vleeshouwers, V. G. A. A., Lankinen, Å., Liljeroth, E., Landschoot, S., Ivanović, Ž., Hückelhoven, R., Hausladen, H. and Stam, R. 2022. Whole-genome sequencing elucidates the species-wide diversity and evolution of fungicide resistance in the early blight pathogen Alternaria solani. *Evolutionary Applications* 15(10), 1605–1620.

Elena, S. F. and Lenski, R. E. 2003. Evolution experiments with microorganisms: The dynamics and genetic bases of adaptation. *Nature Reviews. Genetics* 4(6), 457–469.

Fan, Z., Yang, J. H., Fan, F., Luo, C. X. and Schnabel, G. 2015. Fitness and competitive ability of Alternaria alternata Field isolates with resistance to SDHI, QoI, and MBC fungicides. *Plant Disease* 99(12), 1744–1750.

Fraaije, B., Atkins, S., Hanley, S., Macdonald, A. and Lucas, J. 2020. The multi-fungicide resistance status of Aspergillus fumigatus populations in arable soils and the wider European environment. *Frontiers in Microbiology* 11, 599233.

Fraaije, B. A., Bayon, C., Atkins, S., Cools, H. J., Lucas, J. A. and Fraaije, M. W. 2012. Risk assessment studies on succinate dehydrogenase inhibitors, the new weapons in the battle to control Septoria leaf blotch in wheat. *Molecular Plant Pathology* 13(3), 263–275.

FRAC 2015. *List of Species Resistant to SDHIs April 2015* [Online]. Available at: https://www.frac.info/ [Accessed]. Fungicide Resistance Action Committee (FRAC).

FRAC 2016. [Accessed]. Succinate Dehydrogenase Inhibitor (SDHI) Working Group 9th Meeting on December 2, 2015 [Online]. Available at: http://www.frac.info/. Fungicide Resistance Action Committee (FRAC).

FRAC 2017. [Accessed]. Succinate Dehydrogenase Inhibitor (SDHI) Working Group 10th Meeting on December 14, 2016 [Online]. Available at: http://www.frac.info/. Fungicide Resistance Action Committee (FRAC).

FRAC 2018. [Accessed]. Succinate Dehydrogenase Inhibitor (SDHI) Working Group 10th Meeting on December 13, 2017 [Online]. Available at: http://www.frac.info/. Fungicide Resistance Action Committee (FRAC).

FRAC 2020. [Accessed]. Minutes of the 2020 SDHI Meeting, 21–22th of January 2020, with Recommendations for 2020 [Online]. Available at: http://www.frac.info/. Fungicide Resistance Action Committee (FRAC).

FRAC 2022. FRAC code list 2022 [Online]. Available at: https://www.frac.info/docs/default-source/publications/frac-code-list/frac-code-list-2022--final.pdf?sfvrsn=b6024e9a_2 [Accessed]. Fungicide Resistance Action Committee.

Glättli, A., Grote, T. and Stammler, G. 2011. SDH-inhibitors: history, biological performance and molecular mode of action. In: Dehne, H., Deising, H., Gisi, U., Kuck, K., Russell, P. and Lyr, H. (Eds). *Modern Fungicides and Antifungal Compounds* (vol. VI). Braunschweig: Deutsche Phytomedizinische Gesellschaft, 159–169.

Graf, S. 2017. *Characterisation of Metrafenone and Succinate Dehydrogenase Inhibitor Resistant Isolates of the Grapevine Powdery Mildew Erysiphe necator*.

Graf, S., Zito, R., Gold, R. and Stammler, G. 2017. Status of in vivo and molecular diagnosis of fungicide resistance in powdery mildews. In: Deising, H., Fraaije, B., Mehl, A., Oerke, E., Sierotzki, H. and Stammler, G. (Eds). *Modern Fungicides and Antifungal Compounds* (vol. VIII). Braunschweig: Deutsche Phytomedizinische Gesellschaft, 243–248.

GRDC 2022. *South Australian Crop Sowing Guide* (vol. 2023). Adelaide, Australia: Grains Research and Development Corporation.

Guo, J. and Lemire, B. D. 2003. The ubiquinone-binding site of the Saccharomyces cerevisiae succinate-ubiquinone oxidoreductase is a source of superoxide. *Journal of Biological Chemistry* 278(48), 47629–47635.

Guo, M., Zhu, X., Li, H., Tan, L. and Pan, Y. 2016. Development of a novel strategy for fungal transformation based on a mutant locus conferring carboxin-resistance in Magnaporthe oryzae. *AMB Express* 6(1), 57.

Gutiérrez-Alonso, O., Hawkins, N. J., Cools, H. J., Shaw, M. W. and Fraaije, B. A. 2017. Dose-dependent selection drives lineage replacement during the experimental evolution of SDHI fungicide resistance in Zymoseptoria tritici. *Evolutionary Applications* 10(10), 1055–1066.

Hägerhäll, C. 1997. Succinate: Quinone oxidoreductases: variations on a conserved theme. *Biochimica et Biophysica Acta* 1320(2), 107–141.

Horsefield, R., Yankovskaya, V., Sexton, G., Whittingham, W., Shiomi, K., Ōmura, S., Byrne, B., Cecchini, G. and Iwata, S. 2006. Structural and computational analysis of the quinone-binding site of Complex II (succinate-ubiquinone oxidoreductase): A mechanism of electron transfer and proton conduction during ubiquinone reduction. *Journal of Biological Chemistry* 281(11), 7309–7316.

Hou, Y. P., Chen, Y. L., Wu, L. Y., Wang, J. X., Chen, C. J. and Zhou, M. G. 2018a. Baseline sensitivity of Bipolaris maydis to the novel succinate dehydrogenase inhibitor benzovindiflupyr and its efficacy. *Pesticide Biochemistry and Physiology* 149, 81–88.

Hou, Y. P., Mao, X. W., Lin, S. P., Song, X. S., Duan, Y. B., Wang, J. X. and Zhou, M. G. 2018b. Activity of a novel succinate dehydrogenase inhibitor fungicide pyraziflumid against Sclerotinia sclerotiorum. *Pesticide Biochemistry and Physiology* 145, 22–28.

Hu, S., Zhang, J., Zhang, Y., He, S. and Zhu, F. 2018. Baseline sensitivity and toxic actions of Boscalid against Sclerotinia sclerotiorum. *Crop Protection* 110, 83–90.

Huang, X. P., Luo, J., Li, B. X., Song, Y. F., Mu, W. and Liu, F. 2019. Bioactivity, physiological characteristics and efficacy of the SDHI fungicide pydiflumetofen against Sclerotinia sclerotiorum. *Pesticide Biochemistry and Physiology* 160, 70–78.

Huf, A., Rehfus, A., Teichmann, M., Randall Gold, R. and Stammler, G. 2017. Sensitivity of Venturia inaequalis to fungicides. In: Deising, H., Fraaije, B., Mehl, A., Oerke, E., Sierotzki, H. and Stammler, G. (Eds). *Modern Fungicides and Antifungal Compounds* (vol. VIII). Braunschweig: Deutsche Phytomedizinische Gesellschaft, 197–198.

Ireland, K. B., Beard, C., Cameron, J., Chang, S., Davidson, J., Dodhia, K. N., Garrard, T. A., Hills, A. L., Hollaway, G., Jayasena, K. W., Kiss, L., Mair, W., Marcroft, S. J., Mclean, M. S., Milgate, A., Poole, N., Simpfendorfer, S., Snyman, L., Thomas, G. J., Wallwork, H., Van De Wouw, A. P., Zulak, K. G. and Lopez-Ruiz, F. J. 2021. *Fungicide Resistance Management in Australian Grain Crops*. Australia: Grains Research and Development Corporation.

Iverson, T. M., Maklashina, E. and Cecchini, G. 2012. Structural basis for malfunction in Complex II. *Journal of Biological Chemistry* 287(42), 35430–35438.

Iyanagi, T. 2007. Molecular mechanism of phase I and Phase II drug-metabolizing enzymes: implications for detoxification. *International Review of Cytology* 260, 35–112.

Kim, Y. K. and Xiao, C. L. 2011. Stability and fitness of pyraclostrobin- and Boscalid-resistant phenotypes in field isolates of Botrytis cinerea from apple. *Phytopathology* 101(11), 1385–1391.

Kimchi-Sarfaty, C., Oh, J. M., Kim, I. W., Sauna, Z. E., Calcagno, A. M., Ambudkar, S. V. and Gottesman, M. M. 2007. A "silent" polymorphism in the *MDR*1 gene changes substrate specificity. *Science* 315(5811), 525–528.

Kirikyali, N., Diez, P., Luo, J., Hawkins, N. and Fraaije, B. 2017. Azole and SDHI sensitivity status of Zymoseptoria tritici field populations sampled in France, Germany and the UK during 2015. In: Deising, H., Fraaije, B., Mehl, A., Oerke, E., Sierotzki, H. and Stammler, G. (Eds). *Modern Fungicides and Antifungal Compounds* (vol. VIII). Braunschweig: Deutsche Phytomedizinische Gesellschaft, 153–158.

Kleemann, J. and Mehl, A. 2019. SDHI sensitivity status of Zymoseptoria tritici and Botrytis cinerea. In: Deising, H., Fraaije, B., Mehl, A., Oerke, E., Sierotzki, H. and Stammler, G. (Eds). *Modern Fungicides and Antifungal Compounds IX*. Friedrichroda, Germany. DPG Spectrum Phytomedizin, 165–170.

Knight, N. L., Adhikari, K. C., Dodhia, K., Mair, W. J. and Lopez-Ruiz, F. J. 2023. Workflows for detecting fungicide resistance in net form and spot form net blotch pathogens. *bioRxiv*. doi:10.1101/2023.04.27.538624.

Krause, S., Willighagen, E. and Steinbeck, C. 2000. JChemPaint - using the collaborative forces of the Internet to develop a free editor for 2D chemical structures. *Molecules* 5(12), 93–98.

Kretschmer, M., Leroch, M., Mosbach, A., Walker, A. S., Fillinger, S., Mernke, D., Schoonbeek, H. J., Pradier, J. M., Leroux, P., De Waard, M. A. and Hahn, M. 2009. Fungicide-driven evolution and molecular basis of multidrug resistance in field populations of the grey mould fungus Botrytis cinerea. *PLoS Pathogens* 5(12), e1000696.

Krishnan, P., Meile, L., Plissonneau, C., MA, ma, X., Hartmann, F. E., Croll, D., Mcdonald, B. A. and Sánchez-Vallet, A. 2018. Transposable element insertions shape gene regulation and melanin production in a fungal pathogen of wheat. *BMC Biology* 16(1), 78.

Lalève, A., Fillinger, S. and Walker, A. S. 2014a. Fitness measurement reveals contrasting costs in homologous recombinant mutants of Botrytis cinerea resistant to succinate dehydrogenase inhibitors. *Fungal Genetics and Biology: FG and B* 67, 24–36.

Lalève, A., Gamet, S., Walker, A. S., Debieu, D., Toquin, V. and Fillinger, S. 2014b. Site-directed mutagenesis of the P225, N230 and H272 residues of succinate dehydrogenase subunit B from Botrytis cinerea highlights different roles in enzyme activity and inhibitor binding. *Environmental Microbiology* 16(7), 2253–2266.

Landschoot, S., Carrette, J., Vandecasteele, M., De Baets, B., Höfte, M., Audenaert, K. and Haesaert, G. 2017. Boscalid-resistance in Alternaria alternata and Alternaria solani populations: an emerging problem in Europe. *Crop Protection* 92, 49–59.

Lee, J., Elliott, M. R., Kim, M., Yamada, T. and Jung, G. 2021. A rapid molecular detection system for SdhB and SdhC point mutations conferring differential succinate dehydrogenase inhibitor resistance in Clarireedia populations. *Plant Disease* 105(3), 660–666.

Lemire, B. D. and Oyedotun, K. S. 2002. The Saccharomyces cerevisiae mitochondrial succinate:ubiquinone oxidoreductase. *Biochimica et Biophysica Acta* 1553(1–2), 102–116.

Leroux, P., Gredt, M., Leroch, M. and Walker, A. S. 2010. Exploring Mechanisms of Resistance to Respiratory Inhibitors in Field Strains of *Botrytis cinerea*, the Causal Agent of Gray Mold. *Applied and Environmental Microbiology* 76(19), 6615–6630.

Leroux, P. and Walker, A.-S. 2013. Activity of fungicides and modulators of membrane drug transporters in field strains of Botrytis cinerea displaying multidrug resistance. *European Journal of Plant Pathology* 135, 683–693.

Li, H. X., Nuckols, T. A., Harris, D., Stevenson, K. L. and Brewer, M. T. 2019. Differences in fungicide resistance profiles and multiple resistance to a quinone-outside inhibitor (QoI), two succinate dehydrogenase inhibitors (SDHI), and a demethylation inhibitor (DMI) for two Stagonosporopsis species causing gummy stem blight of cucurbits. *Pest Management Science* 75(11), 3093–3101.

Li, X., Gao, X., Hu, S., Hao, X., Li, G., Chen, Y., Liu, Z., Li, Y., Miao, J., Gu, B. and Liu, X. 2022. Resistance to pydiflumetofen in Botrytis cinerea: Risk assessment and detection of point mutations in sdh genes that confer resistance. *Pest Management Science* 78(4), 1448–1456.

Liang, X., Zou, L., Lian, W., Wang, M., Yang, Y. and Zhang, Y. 2022. Comparative transcriptome analyses reveal conserved and distinct mechanisms of the SDHI fungicide benzovindiflupyr inhibiting Colletotrichum. *Phytopathology* 112(6), 1255–1263.

Lichtemberg, P., Carraro, T., Doussoulin, H., Velasco, J., Puckett, R. and Michailides, T. 2019. With stable sensitivity and unclear disadvantages, the management of SDHI resistant mutant of Alternaria alternata will rely on an effective pistachio spray program. In: Deising, H., Fraaije, B., Mehl, A., Oerke, E., Sierotzki, H. and Stammler, G. (Eds). *Modern Fungicides and Antifungal Compounds IX*. Friedrichroda, Germany. DPG Spectrum Phytomedizin, 85–90.

Liu, K., Wen, Z., Ma, Z. and Shao, W. 2022a. Biological and molecular characterizations of fluxapyroxad-resistant isolates of Botrytis cinerea. *Phytopathology Research* 4(1), 2.

Liu, S., Ma, J., Jiang, B., Yang, G. and Guo, M. 2022b. Functional characterization of MoSdhB in conferring resistance to pydiflumetofen in blast fungus Magnaporthe oryzae. *Pest Management Science* 78(10), 4018–4027.

Mair, W. J., Lopez-Ruiz, F., Stammler, G., Clark, W., Burnett, F., Hollomon, D., Ishii, H., Thind, T. S., Brown, J. K., Fraaije, B., Cools, H., Shaw, M., Fillinger, S., Walker, A. S., Mellado, E., Schnabel, G., Mehl, A. and Oliver, R. P. 2016. Proposal for a unified nomenclature for target-site mutations associated with resistance to fungicides. *Pest Management Science* 72(8), 1449–1459.

Maklashina, E., Rajagukguk, S., Mcintire, W. S. and Cecchini, G. 2010. Mutation of the heme axial ligand of Escherichia coli succinate–quinone reductase: implications for heme ligation in mitochondrial complex II from yeast. *Biochimica et Biophysica Acta* 1797(6–7), 747–754.

Maklashina, E., Rothery, R. A., Weiner, J. H. and Cecchini, G. 2001. Retention of heme in Axial Ligand Mutants of succinate-ubiquinone Oxidoreductase (Complex II) from *Escherichia coli**. *Journal of Biological Chemistry* 276(22), 18968–18976.

Malandrakis, A. A., Apostolidou, Z. A., Louka, D., Markoglou, A. and Flouri, F. 2018. Biological and molecular characterization of field isolates of Alternaria alternata with single or double resistance to respiratory complex II and III inhibitors. *European Journal of Plant Pathology* 152(1), 199–211.

Malandrakis, A. A., Krasagakis, N., Kavroulakis, N., Ilias, A., Tsagkarakou, A., Vontas, J. and Markakis, E. 2022. Fungicide resistance frequencies of Botrytis cinerea greenhouse isolates and molecular detection of a novel SDHI resistance mutation. *Pesticide Biochemistry and Physiology* 183, 105058.

Malandrakis, A. A., Vattis, K. N., Markoglou, A. N. and Karaoglanidis, G. S. 2017. Characterization of Boscalid-resistance conferring mutations in the SdhB subunit of respiratory complex II and impact on fitness and mycotoxin production in Penicillium expansum laboratory strains. *Pesticide Biochemistry and Physiology* 138, 97–103.

Mallik, I., Arabiat, S., Pasche, J. S., Bolton, M. D., Patel, J. S. and Gudmestad, N. C. 2014. Molecular characterization and detection of mutations associated with resistance to succinate dehydrogenase-inhibiting fungicides in Alternaria solani. *Phytopathology* 104(1), 40–49.

Mao, X., Wu, Z., Bi, C., Wang, J., Zhao, F., Gao, J., Hou, Y. and Zhou, M. 2020. Molecular and biochemical characterization of pydiflumetofen-resistant mutants of Didymella bryoniae. *Journal of Agricultural and Food Chemistry* 68(34), 9120–9130.

Masiello, M., Somma, S., Haidukowski, M., Logrieco, A. F. and Moretti, A. 2020. Genetic polymorphisms associated to SDHI fungicides resistance in selected Aspergillus flavus strains and relation with aflatoxin production. *International Journal of Food Microbiology* 334, 108799.

McNutt, A. T., Francoeur, P., Aggarwal, R., Masuda, T., Meli, R., Ragoza, M., Sunseri, J. and Koes, D. R. 2021. GNINA 1.0: Molecular docking with deep learning. *Journal of Cheminformatics* 13(1), 43.

Metz, N., Adolf, B., Chaluppa, N., Hückelhoven, R. and Hausladen, H. 2019. Occurrence of sdh Mutations in German Alternaria solani Isolates and Potential Impact on Boscalid Sensitivity in vitro, in the Greenhouse, and in the Field. *Plant Disease* 103(12), 3065–3071.

Miao, J., Mu, W., Bi, Y., Zhang, Y., Zhang, S., Song, J. and Liu, X. 2021. Heterokaryotic state of a point mutation (H249Y) in SDHB protein drives the evolution of thifluzamide resistance in Rhizoctonia solani. *Pest Management Science* 77(3), 1392–1400.

Miyamoto, T., Hayashi, K., Okada, R., Wari, D. and Ogawara, T. 2020. Resistance to succinate dehydrogenase inhibitors in field isolates of Podosphaera xanthii on cucumber: monitoring, cross-resistance patterns and molecular characterization. *Pesticide Biochemistry and Physiology* 169, 104646.

Miyamoto, T., Ishii, H., Stammler, G., Koch, A., Ogawara, T., Tomita, Y., Fountaine, J. M., Ushio, S., Seko, T. and Kobori, S. 2010. Distribution and molecular characterization of Corynespora cassiicola isolates resistant to Boscalid. *Plant Pathology* 59(5), 873–881.

Mu, W., Wang, Z., Bi, Y., Ni, X., Hou, Y., Zhang, S. and Liu, X. 2017. Sensitivity determination and resistance risk assessment of Rhizoctonia solani to SDHI fungicide thifluzamide. *Annals of Applied Biology* 170(2), 240–250.

Murray, G. M. and Brennan, J. P. 2010. Estimating disease losses to the Australian barley industry. *Australasian Plant Pathology* 39(1), 85–96.

Nosanchuk, J. D. and Casadevall, A. 2003. The contribution of melanin to microbial pathogenesis. *Cellular Microbiology* 5(4), 203–223.

Omrane, S., Audéon, C., Ignace, A., Duplaix, C., Aouini, L., Kema, G., Walker, A. S. and Fillinger, S. 2017. Plasticity of the *MFS1* Promoter leads to multidrug resistance in the Wheat Pathogen *Zymoseptoria tritici*. *mSphere* 2(5), e00393-17.

Omrane, S., Sghyer, H., Audéon, C., Lanen, C., Duplaix, C., Walker, A. S. and Fillinger, S. 2015. Fungicide efflux and the MgMFS1 transporter contribute to the multidrug resistance phenotype in Zymoseptoria tritici field isolates. *Environmental Microbiology* 17(8), 2805–2823.

Orr, H. A. 2009. Fitness and its role in evolutionary genetics. *Nature Reviews. Genetics* 10(8), 531–539.

Outwater, C. A., Proffer, T. J., Rothwell, N. L., Peng, J. and Sundin, G. W. 2019. Boscalid Resistance in Blumeriella jaapii: distribution, Effect on Field Efficacy, and Molecular Characterization. *Plant Disease* 103(6), 1112–1118.

Oyedotun, K. S., Sit, C. S. and Lemire, B. D. 2007. The Saccharomyces cerevisiae succinate dehydrogenase does not require heme for ubiquinone reduction. *Biochimica et Biophysica Acta* 1767(12), 1436–1445.

Pao, S. S., Paulsen, I. T. and Saier, M. H. 1998. Major facilitator superfamily. *Microbiology and Molecular Biology Reviews: MMBR* 62(1), 1–34.

Pearce, T. L., Wilson, C. R., Gent, D. H. and Scott, J. B. 2019. Multiple mutations across the succinate dehydrogenase gene complex are associated with Boscalid resistance in Didymella tanaceti in pyrethrum. *PLoS ONE* 14(6), e0218569.

Piotrowska, M. J., Fountaine, J. M., Ennos, R. A., Kaczmarek, M. and Burnett, F. J. 2017. Characterisation of Ramularia collo-cygni laboratory mutants resistant to succinate dehydrogenase inhibitors. *Pest Management Science* 73(6), 1187–1196.

Platz, G., Snyman, L. and Fowler, R. 2017. Systiva performance in northern trials [Online]. Available at: https://grdc.com.au/resources-and-publications/grdc-update-papers/tab-content/grdc-update-papers/2017/02/systiva-performance-in-northern-trials#:~:text=Systiva%C2%AE%20gained%20APVMA%20approval,use%20on%20cereals%20in%20Australia. Grains Research and Development Corporation [Accessed 19/09/2022].

Popko, J. T., Sang, H., Lee, J., Yamada, T., Hoshino, Y. and Jung, G. 2018. Resistance of Sclerotinia homoeocarpa Field Isolates to succinate dehydrogenase Inhibitor Fungicides. *Plant Disease* 102(12), 2625–2631.

Pringle, A. and Taylor, J. W. 2002. The fitness of filamentous fungi. *Trends in Microbiology* 10(10), 474–481.

Rehfus, A. 2018. *Analysis of the Emerging Situation of Resistance to Succinate Dehydrogenase Inhibitors in Pyrenophora teres and Zymoseptoria tritici in Europe*. Universität Hohenheim.

Rehfus, A., Katamadze, A., Strobel, D., Bryson, R. and Stammler, G. 2019a. Evolution of SDHI adaptation in cereal pathogens in Europe. In: Deising, H., Fraaije, B., Mehl, A., Oerke, E., Sierotzki, H. and Stammler, G. (Eds). *Modern Fungicides and Antifungal Compounds IX*. Friedrichroda, Germany. DPG Spectrum Phytomedizin, 99–104.

Rehfus, A., Matusinsky, P., Strobel, D., Bryson, R. and Stammler, G. 2019b. Mutations in target genes of succinate dehydrogenase inhibitors and demethylation inhibitors in Ramularia collo-cygni in Europe. *Journal of Plant Diseases and Protection* 126(5), 447–459.

Rehfus, A., Miessner, S., Achenbach, J., Strobel, D., Bryson, R. and Stammler, G. 2016. Emergence of succinate dehydrogenase inhibitor resistance of Pyrenophora teres in Europe. *Pest Management Science* 72(10), 1977–1988.

Rehfus, A., Strobel, D., Bryson, R. and Stammler, G. 2018. Mutations in sdh genes in field isolates of Zymoseptoria tritici and impact on the sensitivity to various succinate dehydrogenase inhibitors. *Plant Pathology* 67(1), 175–180.

Sang, H., Chang, H. X., Choi, S., Son, D., Lee, G. and Chilvers, M. I. 2022. Genome-wide transcriptional response of the causal soybean sudden death syndrome pathogen Fusarium virguliforme to a succinate dehydrogenase inhibitor fluopyram. *Pest Management Science* 78(2), 530–540.

Sang, H., Hulvey, J., Popko, J. T., Lopes, J., Swaminathan, A., Chang, T. and Jung, G. 2015. A pleiotropic drug resistance transporter is involved in reduced sensitivity to multiple fungicide classes in Sclerotinia homoeocarpa (F. T. Bennett). *Molecular Plant Pathology* 16(3), 251–261.

Sang, H., Hulvey, J. P., Green, R., Xu, H., Im, J., Chang, T. and Jung, G. 2018. A xenobiotic detoxification pathway through transcriptional regulation in filamentous fungi. *mBio* 9(4), e00457-18.

Scalliet, G., Boehler, M., Bowler, J., Green, P. S., Kilby, P. M. and Fonné-Pfister, R. 2011. SDHIs and the fungal succinate dehydrogenase. In: Dehne, H., Deising, H., Gisi, U., Kuck, K., Russell, P. and Lyr, H. (Eds). *Modern Fungicides and Antifungal Compounds* (vol. VI). Braunschweig: Deutsche Phytomedizinische Gesellschaft, 171–178.

Scalliet, G., Bowler, J., Luksch, T., Kirchhofer-Allan, L., Steinhauer, D., Ward, K., Niklaus, M., Verras, A., Csukai, M., Daina, A. and Fonné-Pfister, R. 2012. Mutagenesis and functional studies with succinate dehydrogenase inhibitors in the wheat pathogen Mycosphaerella graminicola. *PLoS ONE* 7(4), e35429.

Shao, W., Wang, J., Wang, H., Wen, Z., Liu, C., Zhang, Y., Zhao, Y. and Ma, Z. 2022. Fusarium graminearum FgSdhC1 point mutation A78V confers resistance to the succinate dehydrogenase inhibitor pydiflumetofen. *Pest Management Science* 78(5), 1780–1788.

Shi, Y., Sun, B., Xie, X., Chai, A., Li, L. and Li, B. 2021a. Site-directed mutagenesis of the succinate dehydrogenase subunits B and D from Corynespora cassiicola reveals different fitness costs and sensitivities to succinate dehydrogenase inhibitors. *Environmental Microbiology* 23(10), 5769–5783.

Shi, Y., Zhu, F., Sun, B., Xie, X., Chai, A. and Li, B. 2021b. Two adjacent mutations in the conserved domain of SdhB confer various resistance phenotypes to fluopyram in Corynespora cassiicola. *Pest Management Science* 77(9), 3980–3989.

Shima, Y., Ito, Y., Kaneko, S., Hatabayashi, H., Watanabe, Y., Adachi, Y. and Yabe, K. 2009. Identification of three mutant loci conferring carboxin-resistance and development of a novel transformation system in Aspergillus oryzae. *Fungal Genetics and Biology: FG and B* 46(1), 67–76.

Sierotzki, H., Frey, R., Morchoisne, M., Olaya, G., Mösch, M. and Scalliet, G. 2011. Sensitivity of fungal pathogens to SDHI fungicides. In: Dehne, H., Deising, H., Gisi, U., Kuck, K., Russell, P. and Lyr, H. (Eds). *Modern Fungicides and Antifungal Compounds* (vol. VI). Braunschweig: Deutsche Phytomedizinische Gesellschaft, 179–186.

Sierotzki, H. and Scalliet, G. 2013. A review of current knowledge of resistance aspects for the next-generation succinate dehydrogenase inhibitor fungicides. *Phytopathology* 103(9), 880–887.

Simões, K., Hawlik, A., Rehfus, A., Gava, F. and Stammler, G. 2018. First detection of a SDH variant with reduced SDHI sensitivity in Phakopsora pachyrhizi. *Journal of Plant Diseases and Protection* 125(1), 21–26.

Skinner, W., Bailey, A., Renwick, A., Keon, J., Gurr, S. and Hargreaves, J. 1998. A single amino-acid substitution in the iron-sulphur protein subunit of succinate dehydrogenase determines resistance to carboxin in Mycosphaerella graminicola. *Current Genetics* 34(5), 393–398.

Stammler, G., Brix, H. D., Nave, B., Gold, R. and Schoefl, U. 2008. Studies on the biological performance of Boscalid and its mode of action. In: Dehne, H. W., Deising, H. B., Gisi, U., Russell, P. E. and Lyr, H. (Eds). *Modern Fungicides and Antifungal Compounds* (vol., 2008). Braunschweig: Germany Deutsche Phytomedizinische Gesellschaft, 45–51.

Stammler, G., Glättli, A., Koch, A. and Schlehuber, S. 2011. Mutations in the target protein conferring resistance to SDHI fungicides. In: Dehne, H., Deising, H., Gisi, U., Kuck, K., Russell, P. and Lyr, H. (Eds). *Modern Fungicides and Antifungal Compounds* (vol. VI). Braunschweig: Deutsche Phytomedizinische Gesellschaft, 195–198.

Stammler, G., Wolf, A., Glaettli, A. and Klappach, K. 2015. Respiration inhibitors: Complex II. In: Ishii, H. and Hollomon, D. W. (Eds). *Fungicide Resistance in Plant Pathogens: Principles and a Guide to Practical Management*. Tokyo: Springer Japan.

Steinhauer, D., Salat, M., Frey, R., Mosbach, A., Luksch, T., Balmer, D., Hansen, R., Widdison, S., Logan, G., Dietrich, R. A., Kema, G. H. J., Bieri, S., Sierotzki, H., Torriani, S. F. F. and Scalliet, G. 2019. A dispensable paralog of succinate dehydrogenase subunit C mediates standing resistance towards a subclass of SDHI fungicides in Zymoseptoria tritici. *PLoS Pathogens* 15(12), e1007780.

Stergiopoulos, I., Aoun, N., Van Huynh, Q., Neill, T., Lowder, S. R., Newbold, C., Cooper, M. L., Ding, S., Moyer, M. M., Miles, T. D., Oliver, C. L., Úrbez-Torres, J. R. and Mahaffee, W. F. 2022. Identification of putative SDHI target site mutations in the SDHB, SDHC, and SDHD subunits of the grape powdery mildew pathogen Erysiphe necator. *Plant Disease* 106(9), 2310–2320.

Sun, B., Zhu, G., Xie, X., Chai, A., Li, L., Shi, Y. and Li, B. 2022. Double mutations in succinate dehydrogenase are involved in SDHI resistance in Corynespora cassiicola. *Microorganisms* 10(1), 132.

Sun, H. Y., Cui, J. H., Tian, B. H., Cao, S. L., Zhang, X. X. and Chen, H. G. 2020. Resistance risk assessment for Fusarium graminearum to pydiflumetofen, a new succinate dehydrogenase inhibitor. *Pest Management Science* 76(4), 1549–1559.

Sun, H. Y., Lu, C. Q., Li, W., Deng, Y. Y. and Chen, H. G. 2017. Homozygous and heterozygous point mutations in succinate dehydrogenase subunits b, c and d of Rhizoctonia cerealis conferring resistance to thifluzamide. *Pest Management Science* 73(5), 896–903.

Szeto, S. S. W., Reinke, S. N., Sykes, B. D. and Lemire, B. D. 2007. Ubiquinone-binding site mutations in the Saccharomyces cerevisiae succinate dehydrogenase generate superoxide and lead to the accumulation of succinate. *Journal of Biological Chemistry* 282(37), 27518–27526.

Thomas, C. and Tampé, R. 2020. Structural and mechanistic principles of ABC transporters. *Annual Review of Biochemistry* 89, 605–636.

Torriani, S., Frey, R., Buitrago, C., Wullschleger, J., Scalliet, G. and Sierotzki, H. 2015. The threat of succinate-dehydrogenase inhibitors (SDHIs) resistance evolution in cereal pathogens. In: Feldmann, F. and Short Heinrichs, E. A. (Eds). *International Plant Protection Congress XVIII*. Berlin: DPG Spectrum Phytomedizin, 194.

Veloukas, T., Kalogeropoulou, P. and Karaoglanidis, G. S. 2014. Fitness and competitive ability of Botrytis cinerea field-isolates with dual resistance to SDHI and QoI fungicides, associated with several sdhB and the cytb G143A mutations. In: Dehne, H. W., Deising, H. B., Fraaije, B., Gisi, U., Hermann, D., Mehl, A., Oerke, E. C., Russell, P. E., Stammler, G., Kuck, K. H., et al. (Eds). *Modern Fungicides and Antifungal Compounds* (vol. VII). Braunschweig: Deutsche Phytomedizinische Gesellschaft, 207–212.

Walker, A. S., Gredt, M. and Leroux, P. 2011. Resistance to QoIs and SDHIs in populations of Botrytis cinerea. In: Dehne, H., Deising, H., Gisi, U., Kuck, K., Russell, P. and Lyr, H. (Eds). *Modern Fungicides and Antifungal Compounds* (vol. VI). Braunschweig: Deutsche Phytomedizinische Gesellschaft, 187–194.

Wang, Q., Mao, Y., Li, S., Li, T., Wang, J., Zhou, M. and Duan, Y. 2022. Molecular mechanism of Sclerotinia sclerotiorum resistance to succinate dehydrogenase inhibitor fungicides. *Journal of Agricultural and Food Chemistry* 70(23), 7039–7048.

Westhouse, R. A. and Car, B. D.. Chapter 13. Pharmacokinetics and safety assessment. In: Prendergast, G. C. and Jaffee, E. M. (Eds). *Cancer Immunotherapy* (2nd edn.) 2013. San Diego: Academic Press.

Xu, Q., Luo, C., Fu, Y. and Zhu, F. 2022. Risk and molecular mechanisms for Boscalid resistance in Penicillium digitatum. *Pesticide Biochemistry and Physiology* 184, 105130.

Yamashita, M. and Fraaije, B. 2018. Non-target site SDHI resistance is present as standing genetic variation in field populations of Zymoseptoria tritici. *Pest Management Science* 74(3), 672–681.

Yan, N. 2013. Structural advances for the major facilitator superfamily (MFS) transporters. *Trends in Biochemical Sciences* 38(3), 151–159.

Yan, N. 2015. Structural biology of the major facilitator superfamily transporters. *Annual Review of Biophysics* 44, 257–283.

Yang, J. H., Brannen, P. M. and Schnabel, G. 2015. Resistance in Alternaria alternata to SDHI fungicides causes rare disease outbreak in peach orchards. *Plant Disease* 99(1), 65–70.

Yankovskaya, V., Horsefield, R., Törnroth, S., Luna-Chavez, C., Miyoshi, H., Léger, C., Byrne, B., Cecchini, G. and Iwata, S. 2003. Architecture of succinate dehydrogenase and reactive oxygen species generation. *Science* 299(5607), 700–704.

Zhang, C. Q., Yuan, S. K., Sun, H. Y., Qi, Z. Q., Zhou, M. G. and Zhu, G. N. 2007. Sensitivity of Botrytis cinerea from vegetable greenhouses to Boscalid. *Plant Pathology* 56(4), 646–653.

Zhang, Y. J., Li, J., Zhao, W. and Zhou, M. G. 2010. A single amino acid substitution in the SdhB protein of succinate dehydrogenase determines resistance to amicarthiazol in Xanthomonas oryzae pv. oryzae. *Pest Management Science* 66(6), 627–633.

Zhang, Y., Lu, J., Wang, J., Zhou, M. and Chen, C. 2015. Baseline sensitivity and resistance risk assessment of Rhizoctonia cerealis to thifluzamide, a succinate dehydrogenase inhibitor. *Pesticide Biochemistry and Physiology* 124, 97–102.

Zhao, C., Li, Y., Liang, Z., Gao, L., Han, C. and Wu, X. 2022. Molecular mechanisms associated with the resistance of Rhizoctonia solani AG-4 isolates to the succinate dehydrogenase inhibitor thifluzamide. *Phytopathology* 112(3), 567–578.

Zhao, C., Zhang, X., Hua, H., Han, C. and Wu, X. 2019. Sensitivity of Rhizoctonia spp. to flutolanil and characterization of the point mutation in succinate dehydrogenase conferring fungicide resistance. *European Journal of Plant Pathology* 155(1), 13–23.

Zhu, F., Shi, Y., Xie, X., Chai, A. and Li, B. 2019. Occurrence, distribution, and characteristics of Boscalid-resistant Corynespora cassiicola in China. *Plant Disease* 103(1), 69–76.

Zhu, J., Li, J., MA, ma, D., Gao, Y., Cheng, J., Mu, W., Li, B. and Liu, F. 2022. SDH mutations confer complex cross-resistance patterns to SDHIs in Corynespora cassiicola. *Pesticide Biochemistry and Physiology* 186, 105157.

Zhu, X. L., Xiong, L., Li, H., Song, X. Y., Liu, J. J. and Yang, G. F. 2014. Computational and experimental insight into the molecular mechanism of carboxamide inhibitors of succinate-Ubquinone oxidoreductase. *ChemMedChem* 9(7), 1512–1521.

Chapter 13

Understanding resistance to Anilinopyrimidine fungicides[1]

Seiya Saito and Chang-Lin Xiao[2], USDA-Agricultural Research Service, San Joaquin Valley Agricultural Sciences Center, USA

1 Introduction

Anilinopyrimidines (AP) fungicides (AP fungicides) were introduced in the 1990s to control fungal strains that have developed multiple resistance to benzimidazoles, *N*-phenyl carbamates, and dicarboximides fungicides. AP fungicides belong to FRAC Code 9 and show no cross-resistance with other fungicide groups (FRAC, 2021). Three compounds of the AP fungicides currently available on the market are cyprodinil (Syngenta Crop Protection AG), mepanipyrim (Kumiai Chemical Industry Co. Ltd. and Ihara Chemical Industry Co. Ltd.), and pyrimethanil (AgrEvo GmbH) (Hayashi et al., 1997; Heye et al., 1994; Neumann et al., 1992; Waechter et al., 2001). The chemical structures of these three compounds differ only in their substituent at position 4 of the pyrimidine ring; cyclopropyl group for cyprodinil, a 1-propynyl group for mepanipyrim, and a methyl group for pyrimethanil (Fig. 1).

[1] Disclaimer: Mention of trade names or commercial products in this article is solely for the purpose of providing specific information and does not imply recommendations or endorsement by the U.S. Department of Agriculture. USDA is an equal opportunity provider and employer.

[2] United States Department of Agriculture - Agricultural Research Service, San Joaquin Valley Agricultural Sciences Center, Parlier, California, USA

http://dx.doi.org/10.19103/AS.2022.0116.13
Published by Burleigh Dodds Science Publishing Limited, 2023.

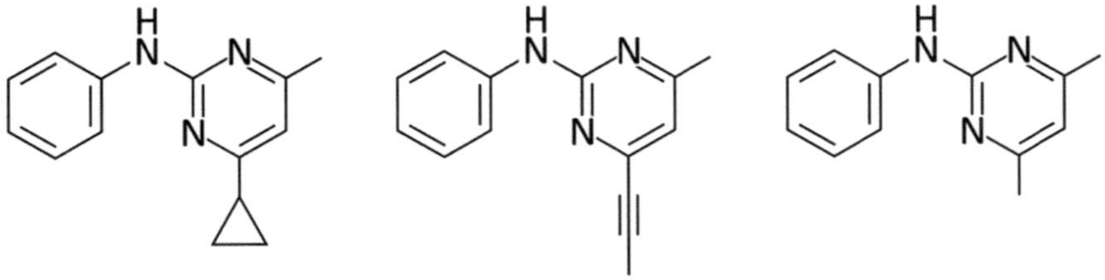

Figure 1 Chemical structure of cyprodinil, mepanipyrim, and pyrimethanil. *Source*: Modified from FRAC (2021).

AP fungicides are highly effective against a wide range of plant pathogens such as *Alternaria* spp., *Botrytis* spp., *Cladosporium carpophilum*, *Diplocarpon rosae*, *Erysiphe* spp., *Monilinia* spp., *Podosphaerea* spp., *Pseudocercosporella herpotrichoides*, *Pyrenophora teres*, *Rhynchosporium secalis*, *Septoria nodorum*, *Sphaerotheca* spp., and *Venturia* spp. (Hayashi et al., 1997; Waechter et al., 2010). Due to their broad spectrum, AP fungicides are currently used as a foliar fungicide in cereal grains, tree fruits, small fruits, stone fruits, vegetables, field crops, and ornamentals, and as a seed dressing on barley. Formulations for foliar applications are water-dispersible molecules (WG) or emulsifiable concentrates (EC); those for seed dressing are flowable concentrates for seed treatment (FS). Acibenzolar-S-methyl (FRAC Code P01), fludioxonil (FRAC Code 12), fenpropidin (FRAC Code 5), and triazole (FRAC Code 3) are common partners for the premixtures with AP fungicides.

2 Mode of action

The mode of action of AP fungicides has not been fully understood since the introduction of this class of fungicides. From early studies, two possible modes of action of AP fungicides were proposed: inhibition of enzyme secretion and inhibition of methionine biosynthesis. The latter mode of action is currently adopted by FRAC and thus AP fungicides are classified as potential methionine biosynthesis inhibitors (FRAC, 2021). However, these early mode of action studies on AP fungicides have never led to the discovery of the target site nor have they been supported by molecular characterization studies of fungal pathogens. Although a recent study has proposed an alternative mode of action for AP fungicide, which was then validated using reverse genetics, most studies assessing the mode of action have yielded interesting insight into the functions of the AP fungicides. We will therefore cover the two proposed modes of action from early studies as well as a new mode of action proposed in the most recent study.

Published by Burleigh Dodds Science Publishing Limited, 2023.

2.1 Enzyme secretion

Miura et al. (1994) analyzed enzyme activities in extracellular and intercellular crude enzyme solution prepared from a liquid medium cultured with *B. cinerea* in the presence of mepanipyrim. Results showed that mepanipyrim did not inhibit the activity of cell wall degrading enzymes such as cutinase, pectinases, and cellulase, but it prevented the secretion of these enzymes, leading to their intracellular accumulation. Because cell wall degrading enzymes play an important role in the early stage of a plant infection, secretory repression of such enzymes could inhibit or retard penetration of plant tissue by fungal pathogens (Miura et al., 1994). Mepanipyrim also prevented the secretion of hydrolytic enzymes such as lipases, protease, and invertase of *B. cinerea*, and electrophoretic analysis of accumulated invertase suggested that mepanipyrim could affect the late stage of the protein secretion (Miura et al. 1994).

Similar results were obtained by another study performed on *B. cinerea* using pyrimethanil. Milling and Richardson (1995) cultured an isolate of *B. cinerea* in a liquid medium or in plant tissues in the presence of pyrimethanil; examined the activities of enzymes such as polygalacturonase, cellulase, proteinase, and laccase; and found that pyrimethanil inhibited the secretion of extracellular enzymes. Since pyrimethanil did not inhibit enzyme activities, nor did it inhibit the synthesis of those enzymes, it was proposed that the fungicide inhibits protein secretion at a post-translational stage in the secretory pathway (Milling and Richardson, 1995).

The enzyme secretion inhibitors as a mode of action for AP fungicides was further supported by an observation of the pathogenesis of *B. fabae* on broad bean leaves under a scanning electron microscopy (Daniels and Lucas, 1995). At the beginning of the infection by *B. fabae*, cell wall degrading enzymes are secreted locally within the epidermal cell wall, resulting in a halo of softened cell wall around the penetration site, which is accompanied by the appearance of pin-point necrotic flecks on the leaf surface (Daniels and Lucas, 1995). Lesions on pyrimethanil-treated leaves did not develop beyond the pin-point necrotic fleck stage. This pyrimethanil-induced inhibition of *B. fabae* pathogenesis was even more evident when visible disease symptoms on detached leaves were assessed 5 days after inoculation (Daniels and Lucas, 1995). Based on those observations, the lytic function of infection hyphae in host epidermal cell walls appeared to be suppressed, which was inconsistent with the conclusions from the previous studies (Milling and Richardson, 1995; Miura et al., 1994).

2.2 Methionine biosynthesis

This mode of action was first suggested from early studies on *B. cinerea* (Leroux, 1994; Masner et al., 1994). Masner et al. (1994) investigated the effect of cyprodinil on mycelial growth of *B. cinerea* in minimal culture media by adding various compositions of amino acids and observed partial reversal growth inhibition of the fungus when sulfur-containing amino acids, particularly, methionine, were added to the media (Masner et al., 1994). To specify the site of inhibition, the potential intermediates in the methionine biosynthesis were also tested. Reversal of growth inhibition was observed when methionine and its upstream metabolite homocysteine were added to the media, suggesting that cyprodinil interfered with cystathionine-β-lyase (Fig. 2) (Masner et al., 1994).

Leroux (1994) conducted a similar study using pyrimethanil. Results showed that the addition of methionine or its upstream metabolite reversed the toxicity of pyrimethanil, suggesting that pyrimethanil inhibited the methionine biosynthesis pathway at the level of cystathionase and/or acetyl homoserine sulfhydrylase.

Analysis of sulfur assimilation from pyrimethanil-treated *B. cinerea* that had been cultured in a liquid medium in the presence of ^{35}S-labeled sodium sulfate as a sulfur source showed a twofold increase of cystathionine and a threefold decrease of methionine (Fritz et al., 1997). The absence of a reversal effect by cystathionine, a metabolite one step before homocysteine, suggested that the mode of action could be inhibition of cystathionine β-lyase (Fig. 2) (Fritz et al., 1997; Masner et al., 1994).

However, enzymatic studies using *B. cinerea* crude protein extracts showed that the enzyme activity of cystathionine β-lyase was not affected by pyrimethanil and cyprodinil at concentrations below 0.1 mM (Fritz et al., 2003). Although

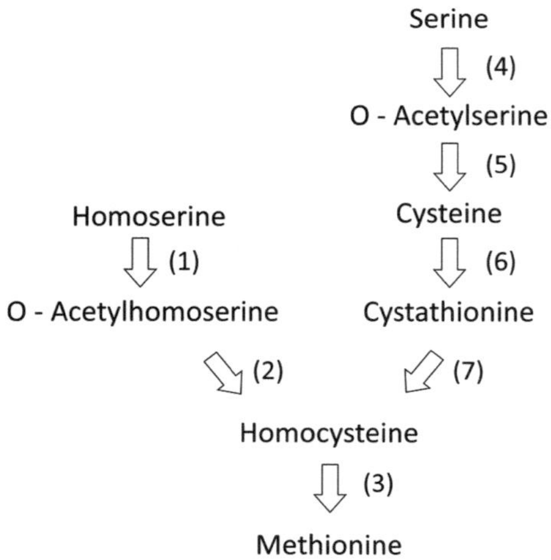

Figure 2 The pathway of methionine biosynthesis in *Aspergillus nidulans* (1) homoserine acetyltransferase; (2) O-acetyl homoserine sulfhydrylase; (3) homocysterine transmethylase; (4) serine acetyltransferase; (5) acetyl serine sulfhydrylase; (6) cystathionine γ-synthase; (7) cystathionin β-lyase. *Source*: Modified from Paszewski and Grabski (1974).

there was a little inhibitory effect of the fungicides at 0.1 mM, inhibition was too low to consider cystathionine β-lyase as the primary target site of AP fungicides (Fritz et al., 2003). Moreover, sequence analysis of the cystathionine β-lyase and cystathionine γ-synthase genes from sensitive and resistant isolates of *B. cinerea* to AP fungicides showed no correlation between amino acid substitutions found in the sequence and fungicide resistance phenotypes of the isolates tested (Fritz et al., 2003; Leroux et al., 2002). Similar findings were reported in *Aspergillus flavus*, *Penicillium digitatum*, and *Sclerotinia sclerotiorum* where there were no amino acid alterations in the sequences of those enzymes between AP fungicides-sensitive and resistant strains (Hou et al., 2018; Liu et al., 2014; Zhang et al., 2021). Moreover, Kanetis et al. (2008) examined the effects of amino acids and homocysteine on sensitivities of *B. cinerea* and *P. digitatum* to cyprodinil and pyrimethanil and found that the addition of certain amino acids significantly reduced the toxicity of the fungicides against *B. cinerea* but not *P. digitatum*, suggesting that the metabolite-chemical interaction observed in *B. cinerea* was due to a species-specific effect.

2.3 Mitochondrial function

In the 1990s, sensitivities to multiple fungicides in *B. cinerea* isolates from French vineyards were analyzed and three phenotypes with different levels of reduced sensitivity to AP fungicides were identified (Chapeland et al., 1999; Leroux et al., 1999). Those phenotypes of resistance to AP fungicides were denominated Ani R1, Ani R2, and Ani R3, among which isolates of Ani R1 exhibited the highest EC_{50} (effective concentration that inhibits 50% of fungal growth or germination) of AP fungicides, and only the isolates of this phenotype were found to be practical resistant in the field (Forster and Staub, 1996; Leroux et al., 1999). The isolates of Ani R2 and Ani R3 showed reduced sensitivity to AP fungicides and other unrelated fungicides (Leroux et al., 1999). The molecular mechanism responsible for the reduced sensitivities of Ani R2 and Ani R3 was later found to be due to the overexpression of multidrug transporters, which will be discussed in detail in Section 3.2 of this chapter. Despite the early description of Ani R1 phenotype in the mid-90s, the resistance mechanism has never been fully understood at the molecular level. Based on the analysis of progeny from sexual crosses between sensitive and resistant isolates, resistance to AP fungicides segregated in a Mendelian manner of 1:1 ratio, suggesting the presence of a resistance-conferring gene (Chapeland et al., 1999; Hilber and Hilber-Bodmer, 1998; Mosbach et al., 2017).

Because of its high resistance to AP fungicides, it has been suggested that the molecular mechanism of resistance in Ani R1 phenotype be identified in order to understand the mode of action of AP fungicides (Leroux et al., 2002). Recently, Mosbach et al. (2017) identified multiple point mutations associated with AP resistance in UV-generated laboratory mutants of *B. cinerea* resistant to AP fungicides using a whole genome sequencing approach. Eleven mutations in eight genes were found to be associated with resistance to AP fungicides, and mutations in four of them were found to confer resistance by reverse genetics (Mosbach et al., 2017). Additional mutations were also found in another gene from field isolates of Ani R1 phenotype (Mosbach et al., 2017). All nine genes were nuclearly encoding proteins and were homologous to yeast genes involved in various mitochondrial functions, indicating that the mitochondrion is the target of AP fungicides (Table 1) (Mosbach et al., 2017). Among AP fungicide-resistant *B. cinerea* isolates collected from grapes and strawberries in European countries, mitochondrial NADH kinase mutant *Bcpos5*^{L412F} was the most prevalent genotype accounting for 69.2% and 43.1% for grape and strawberry, respectively, followed by mitochondrial inner membrane ABC transporter *Bcmdl1*^{E407K} which accounted for 5.9% (grape) and

Table 1 Resistance-conferring mutations identified in UV-mutants and field isolates of *Botrytis cinerea* resistant to anilinopyrimidine fungicides

Gene name	Amino acid change	Identified in	
		UV mutants	Field isolates
Bcmix17	G79D[a]	✓	
	G83E[a]	✓	
Bcdnm1	E450G	✓	
Bcafg3	L305P	✓	
Bcphb2	L153S	✓	
Bcmcr1	A280 frameshift[a]	✓	
	G295 frameshift[a]	✓	
BcoliC	R33C[a]	✓	
Bcatm1	E414K[a]	✓	
Bcmdl1	E407K[a]	✓	✓
	G422R	✓	
	S466R[a]		✓
Bcpos5	V273I[a]		✓
	P293S[a]		✓
	P319A[a]		✓
	G408R[a]		✓
	G408V[a]		✓
	L412F[a]		✓
	L412S		✓
	L412V[a]		✓

Source: Modified from Mosbach et al. (2017).
[a]Resistance-confirming mutations were validated by reverse genetics.

14.4% (strawberry) (Mosbach et al., 2017). The authors proposed that *Bcpos5* would be the primary target of AP fungicides (Mosbach et al., 2017).

It is unique for a single class of fungicide to have various mutations within at least nine different genes that confer resistance. Other resistance-conferring mutations in other relevant genes would be likely yet to be discovered in the future. As the resistance to AP fungicides has also been reported in fungal species other than *Botrytis*, it is of great interest to determine whether the same target site is responsible for the resistance in other fungal genera and species.

3 Mechanism of resistance

The most common mechanism of fungicide resistance in phytopathogenic fungi is amino acid substitutions in the target protein, usually resulting in conferring a high level of resistance to a specific class of fungicides. Multidrug resistance (MDR) based on increased drug efflux mediated by membrane transporters has also been reported in a wide range of phytopathogenic fungi, such as *B. cinerea*, *Penicillium digitatum*, *P. expansum*, *Oculimacula yallundae*, *Sclerotinia homoeocarpa*, and *Zymoseptoria tritici* (Chapeland et al., 1999; Hulvey et al., 2012; Leroux et al., 2013: Nakaune et al., 1998; Omrane et al., 2015; Samaras et al., 2020). However, studies on MDR involving AP fungicides were conducted thus far exclusively on *B. cinerea* and other *Botrytis* spp.

Understanding the mechanism of resistance is of great importance in developing molecular-based diagnostic tools which facilitate monitoring fungicide resistance in fungal pathogen populations, thereby assisting in designing and implementing sound fungicide application programs. In this section, mechanisms of resistance in the target site as well as non-target site are reviewed.

3.1 Point mutations in the target site

As already covered in Section 2.3, the target sites of AP fungicides were identified in nine genes that are involved in mitochondrial functions. Among the nine genes, mutations in two genes, *pos5* and *mdl1*, were identified in field-resistant isolates of *B. cinerea* (Table 1, Mosbach et al., 2017). This was supported by a recent study conducted in Australian vineyards in which all isolates of *B. cinerea* with high levels of resistance to pyrimethanil showed either P319A or L412F amino acid substitutions in Pos5, and one isolate moderately resistant to pyrimethanil had T66A substitution in Mdl1 (Harper et al., 2021).

More resistance-conferring mutations were identified in UV mutants than in field-resistant *B. cinerea* isolates (Table 1, Mosbach et al., 2017). This is often the case for laboratory mutants, likely due to extreme selection pressure (Avila-Adame and Köller, 2003; Saito et al., 2009). For instance, more variations of amino acid substitutions associated with resistance to fenhexamid, a hydroxyanilide fungicide, were detected in laboratory mutants of *B. cinerea* than in field-resistant isolates (Albertini and Leroux, 2004; Fillinger et al., 2008; Saito et al., 2009). However, one of such amino acid substitutions identified in UV-induced mutants was later found in field-resistant isolates (Grabke et al., 2013). Further studies on point mutations in the AP target sites are needed to determine whether mutations identified in laboratory mutants are relevant at the field level and to find yet other novel mutations in other genes that could confer resistance to AP fungicides, as suggested by Mosbach et al. (2017).

Some isolates of *B. cinerea* moderately resistant to pyrimethanil did not have any mutations in either Pos5 or Mdl1 (Harper et al., 2021; Mosbach et al., 2017). It is reported that some moderate resistance to AP fungicides could be ascribed to alterations in the non-target site, the mechanism of which is discussed in Section 3.2.

3.2 Modifications in the non-target site

Drug efflux transporters are integrated membrane-bound proteins that facilitate the movement of macromolecules, proteins, ions, or small molecules across the membrane (Perlin et al., 2014). Two major groups of drug transporters in fungi are ATP-binding cassette (ABC) transporter and major facilitator superfamily (MFS) transporters.

Detailed studies of fungicide resistance in field isolates of *B. cinerea* identified three MDR phenotypes: MDR1 phenotype shows reduced resistance to fludioxonil and cyprodinil; MDR2 shows reduced resistance to fenhexamid, cyprodinil, and iprodione; and MDR3 is the combination of MDR1 and MDR2 (Kretschmer et al., 2009). MDR1 and MDR2 phenotypes were respectively Ani R2 and Ani R3, previously described fungicide resistance phenotypes of *B. cinerea* from French vineyards (Chapeland et al., 1999; Leroux et al., 1999). MDR1 was found to have activating mutations in the zinc cluster transcription factor, Mrr1, resulting in constitutive overexpression of the ABC transporter, AtrB (Kretschmer et al., 2009). MDR2 was associated with overexpression of the MFS gene, *mfsM2*, caused by an insertion of a retroelement-derived sequence (Mernke et al., 2011; Kretschmer et al., 2009). Two types of retroelement-induced promoter rearrangements were identified in the MDR2 strains, referred to as type A and type B. These stains were indistinguishable in their fungicide resistance phenotypes and at the levels of *mfsM2* overexpression. Both rearrangements convert the weak *mfsM2* promoter into a very strong constitutive promoter (Mernke et al., 2011). Sexual recombination of *mrr1* and *mfsM2* alleles led to the third group of MDR phenotype, MDR3, which shows a combined resistance profile of MDR1 and MDR2 (Kretschmer et al., 2009).

In addition to MDR strains of *B. cinerea*, a stronger variant of MDR1, called MDR1h, was found to be widely distributed in German strawberry fields

(Leroch et al., 2013). In the MDR1h stains, overexpression of AtrB was caused by a 3-bp deletion in *mrr1*, and the expression level of AtrB in MDR1h is even higher than that in MDR1 (Leroch et al., 2013). Phylogenetic analysis showed that the MDR1h stains were not *B. cinerea* but belonged to a novel clade, named *Botrytis* group S, which is closely related to *B. cinerea* and *B. fabae* (Leroch et al., 2013). The presence of MDR1h strains was also reported in the USA (Chen et al., 2016; Cosseboom and Hu, 2021; Fernández-Ortuño et al., 2015).

4 Fitness cost

While fungicide resistance provides a selective advantage under fungicide selection pressure, resistance-conferring mutations may sometimes result in fitness costs to the resistant strains. Because the development and evolution of fungicide resistance in a fungal population depends on the dynamics of competition between fungicide-sensitive and resistant strains, fitness costs associated with resistance have important implications for resistance management strategies and resistance risk assessments (Hawkins and Fraaije, 2018; Parnell et al., 2005).

A comprehensive study on fitness costs associated with resistance to AP fungicides was conducted on *B. cinerea* (Bardas et al., 2008). According to the study, there were no significant differences between cyprodinil-resistant and sensitive isolates of *B. cinerea* in mycelial growth, spore production in vitro, osmotic sensitivity, virulence, spore production in vivo, and spore germination rate (Bardas et al., 2008). Moreover, resistance to cyprodinil was found to be stable after consecutive transfers in vitro and in vivo in the absence of the fungicide (Bardas et al., 2008). Similarly, other studies on fitness cost in *Alternaria alternata*, *B. cinerea*, and *Penicillium expansum* concluded that there were no significant differences between AP fungicides-resistant and -sensitive isolates in terms of mycelial growth, spore production, spore germination rate, osmotic sensitivity, and pathogenicity (Avenot and Michailides, 2015; Avenot et al., 2018; Fernández-Ortuño et al., 2013; Karaoglanidis et al., 2010; Li and Xiao, 2008; Malandrakis et al., 2015; Moyano et al., 2004; Sun et al., 2010). In some of those studies, the isolates that show reduced sensitivity or resistance to AP fungicides were tested (Avenot and Michailides, 2015; Bardas et al., 2008; Fernández-Ortuño et al., 2013; Sun et al., 2010), suggesting that there is no fitness cost associated with resistance due to the alterations either in the target site or non-target site. In a field study, MDR3 strain of *B. cinerea* was recovered for consecutive years at a high frequency from the experimental vineyards where mixtures of sensitive and MDR3 strains had been artificially introduced, suggesting that the mutations in *mrr1* and *mfsM2* do not impair the fitness of MDR strains (Kretschmer et al., 2009).

The results of fitness costs were, however, not consistent for the laboratory mutants. For instance, laboratory mutants of *Aspergillus parasiticus*, *P. digitatum*, and *Sclerotinia sclerotiorum* that are resistant to AP fungicides showed some fitness costs in mycelial growth, sclerotia production, or pathogenicity compared to their wild-type parent strains (Hou et al., 2018; Markoglou et al., 2011; Zhang et al., 2021). In contrast, laboratory mutants of *A. flavus* and *P. expansum* resistant to AP fungicides showed no fitness costs compared to their wild-type parent strains (Li and Xiao, 2008; Liu et al., 2014). It is unclear if those fitness costs were associated with resistance to AP fungicides or other factors as the mutagenesis could often induce mutations at different locations in the genome.

Overall, there seemed to be no to little fitness costs associated with resistance to AP fungicides among field isolates resistant to the fungicides. Currently, AP fungicides are classified as medium-risk fungicides for the development of resistance, according to FRAC (FRAC, 2021).

5 Monitoring Anilopyrimidine fungicides resistance

Monitoring AP fungicides resistance is challenging because as mentioned previously, resistance to AP fungicides involves target point mutations, different types of MDR, or a combination of both. Indeed, the sensitivity of 199 isolates of *B. cinerea* to pyrimethanil showed a continuous curve with their EC_{50} values ranging from 0.1 µg/mL to 30 µg/mL (Grabke and Stammler, 2015), indicating that there is no definitive separation between sensitive and resistant isolates. Another study in *B. cinerea* showed that inhibition curves from the isolates that were classified as highly sensitive, sensitive, moderately resistant, and resistant to cyprodinil overlapped one another (Weber and Hahn, 2011). Nonetheless, the detection of practical resistance was conventionally achieved by *in vitro* tests using discriminatory concentrations obtained based on the baseline sensitivity established from the isolates that were collected prior to the use of AP fungicides. Since practical resistance is defined as diminished levels of disease control caused by the resistance to the fungicide (Köller, 1991), discriminatory dose-based resistance detected by *in vitro* tests should be verified by examining disease development on the host plant treated with the label rate of the relevant AP fungicide (Fernández-Ortuño et al., 2014).

The frequency of resistance to AP fungicides among field isolates of fungal pathogens from various crops in different countries is summarized in Table 2. In this list, only studies where researchers monitored the sensitivities of field isolates to AP fungicides and provided information about the frequency of resistance are included. It is noteworthy that among all those published papers, many different methods were used to determine the sensitivities of fungal pathogens to AP fungicides.

Table 2 Frequency of resistance to anilinopyrimidine fungicides in various fungal pathogens

Pathogen	Year of collection[a]	Location	Host plant	Fungicide[b]	Number of isolates	Frequency of resistance[c]	Reference
Alternaria alternata	2009-2010	IA, USA	Potato	P	21	19.0%	Fairchild et al. (2013)
A. alternata	2010-2011	Greece	Tomato	C	42	5.0%	Malandrakis et al. (2015)
A. alternata	2011-2014	USA	Potato	P	109	0%	Fonseka and Gudmestad (2016)
A. alternata	2012	CA, USA	Pistachio	C	129	10.1%	Avenot and Michailides (2015)
A. alternata	2013	CA, USA	Blueberry	C	143	0%	Wang et al. (2021)
A. solani	2009-2010	IA, USA	Potato	P	9	11.1%	Fairchild et al. (2013)
A. solani	2012-2014	USA	Potato	P	245	2.4%	Fonseka and Gudmestad (2016)
Botrytis cinerea	1990-1994	France, Switzerland	Grape	C	650	0-50.0%	Forster and Staub (1996)
B. cinerea	1993-1994	France	Grape	C, P	n. s.	5.0%	Leroux and Gredt (1995)
B. cinerea	1993-1995	Switzerland	Grape	C	303	1.0%	Hilber and Hilber-Bodmer (1998)
B. cinerea	1995-2001	Switzerland	Grape	C	751	0-84%	Baroffio et al. (2003)

(*Continued*)

Published by Burleigh Dodds Science Publishing Limited, 2023.

Table 2 (*Continued*)

Pathogen	Year of collection[a]	Location	Host plant	Fungicide[b]	Number of isolates	Frequency of resistance[c]	Reference
B. cinerea	1997–2001	Australia	Grape	P	910	0–12.0%	Sergeeva et al. (2002)
B. cinerea	1998–1999	Italy	Grape	P	62	88.7%	Gullino et al. (2000)
B. cinerea	Mainly 1999	BC, Canada	Mainly apple	C	32	0%	Sholberg et al. (2003)
B. cinerea	2000	Spain	Vegetables	P	165	12.0%	Moyano et al. (2004)
B. cinerea	2000–2001[d]	Chile	Grape	C, M, P	26	38.5%	Lattore et al. (2002)
B. cinerea	2001	Lianoning, China	Tomato	P	70	20.0%	Ji et al. (2003)
B. cinerea	2001–2003	WA, USA	Apple and pear	P	123	0%	Zhao et al. (2010)
B. cinerea	2002–2006	Japan	Vegetables	M	977	0.3%	Suzuki et al. (2011)
B. cinerea	2003	Hungary, Egypt	Mainly grape	P	23	60.9%	El-Naggar et al. (2008)
B. cinerea	2005	Greece	Vegetables	C, P	55	49.1–57.4%	Myresiotis et al. (2007)
B. cinerea	2006–2009	Germany	Grape	C	629	5.4–21.9%	Leroch et al. (2011)
B. cinerea	2007–2008	Shandong, China	Tomato and cucumber	P	108	53.2–100%	Sun et al. (2010)
B. cinerea	2010	Germany	Small berries	C	353	14.7%	Weber (2011)
B. cinerea	2010, 2012	FL, USA	Strawberry	C, P	392	52.7–59.5%	Amiri et al. (2013)

B. cinerea	2011	NC, SC, USA	Strawberry	C	217	46.5%	Fernández-Ortuño et al. (2013)
B. cinerea	2011-2016	Hebei, China	Tomato	P	512	88.5%	Zhao et al. (2019)
B. cinerea	2012	CA, USA	Grape	C	212	23.1%	Saito et al. (2019)
B. cinerea	2012-2013	Hubei, China	Tomato	C	221	29.0%	Fan et al. (2016)
B. cinerea	2012-2013	7 States in USA	Strawberry	C	1810	17.0-27.0%	Fernández-Ortuño et al. (2014)
B. cinerea	2012-2013	Zhejiang, China	Strawberry	P	486	44.0-90.7%	Yin et al. (2015)
B. cinerea	2012-2013	Hubei, China	Strawberry	C	240	42.4%	Fan et al. (2017)
B. cinerea	2013-2014	WA, CA, USA	Blueberry	C	200	20.1-29.2%	Saito et al. (2016)
B. cinerea	2013-2014	Henan, China	Tomato	P	263	85.9%	Liu et al. (2016)
B. cinerea	2013-2015	Brazil	Strawberry	P	99	20.0%	Baggio et al. (2018)
B. cinerea	2013-2016	Australia	Grape	P	725	7.7%	Harper et al. (2021)
B. cinerea	2013-2016	mainly Germany	Strawberry	C	340	41.5%	Weber and Entrop, (2017)
B. cinerea	2014	China	Stone fruit	C	164	8.5%	Yin et al. (2018)
B. cinerea	2014, 2018	MI, USA	Grape	C	240	62.6-69.6%	Alzohairy et al. (2021)
B. cinerea	2014-2015	FL, USA	Blueberry	C	613	10.0-30.0%	Amiri et al. (2018)
B. cinerea	2015-2016	CA, USA	Mandarin	P	200	71.0-93.0%	Saito and Xiao (2018)

(*Continued*)

Table 2 (*Continued*)

Pathogen	Year of collection[a]	Location	Host plant	Fungicide[b]	Number of isolates	Frequency of resistance[c]	Reference
B. cinerea	n.s.[e]	South Africa	Pear	P	122	12.3%	Rochefort et al. (2015)
B. cinerea	n.s.	Japan	Various crops	M	420	0%	Takagaki et al. (2004)
Botrytis spp.	2008–2011	Germany	Strawberry	C	173	25.0–68.4%	Leroch et al. (2013)
Botrytis spp.	2014–2019	MD, PA, USA	Small berries, grape	C	249	86.0%	Cosseboom and Hu (2021)
Didymella bryoniae	1994–2008	USA	Cucurbits	C	146	0%	Keinath (2015)
Monilinia laxa	2012	CA, USA	Almond	C	28	32.1%	Schnabel et al. (2019)
Monilinia spp.	1992–2014	Italy	Prunus spp.	C, P	15	6.7%	Abate et al. (2018)
Oculimacula spp.	2004–2008	France	Wheat	C	n.s.	12.0–84.0%	Leroux et al. (2012)
Penicillium digitatum	2017	China	Citrus	P	127	0%	Zhang et al. (2021)
P. expansum	2004–2005	WA, USA	Apple	P	120	0%	Li and Xiao (2008)
P. expansum	2008–2009	Greece	Apple	C	236	6.0%	Karaoglanidis et al. (2011)
P. expansum	2009	WA, USA	Apple	P	186	0.5%	Xiao et al. (2010)
P. expansum	2010–2011	WA, USA	Apple	P	779	0–95.8%	Caiazzo et al. (2014)

Sclerotinia sclerotiorum	2016-2017	Jiangsu, China	Rapeseed	C	100	0%	Hou et al. (2018)
Stemphylium vesicarium	2016	NY, USA	Onion	C, P	46	23.9-32.6%	Hay et al. (2019)
Venturia inaequalis	1978–1992	Mainly France	Apple	P	39	28.2%	Schnabel and Parisi (1997)
V. inaequalis	2002–2003	NY, USA	Apple	C, P	469	0%	Köller et al. (2005)
V. inaequalis	2011–2012	New Zealand	Apple	C, P	19	52.6%	Larsen et al. (2013)

[a]Years during which fungal pathogens were collected.
[b]Anilinopyrimidine fungicides used in the report, C = cyprodinil; M = mepanipyrim; P = pyrimethanil.
[c]Percentage of isolates resistant to anilinopyrimidine fungicides. Ranges of percentages show the variation depending on sampling year, sampling location, host plant, or type of fungicide tested.
[d]Fungal pathogens were most likely collected between 2000 and 2001.
[e]n.s. = not stated in the report.

For instance, various types of media have been used to determine *in vitro* sensitivity to AP fungicides. Since the reversal effects of fungicide were observed when the sensitivity of pathogens was tested on a nutrient medium rich in amino acids (Masner et al., 1994; Myresiotis et al., 2007), generally the media that lack important amino acids were used, such as minimal medium, synthetic medium containing L-asparagine, 0.5% sucrose agar medium, or CzapeK agar medium (Fernández-Ortuño et al., 2014; Hilber and Schüepp, 1996; Myresiotis et al., 2007; Weber and Hahn, 2011).

Discriminatory concentrations to determine resistance to AP fungicides also vary greatly depending on fungal species, fungicides, the medium used, or the type of assay. Regardless of fungal species, generally higher discriminatory concentrations (0.4–8.0 µg/mL) were used in mycelial growth assays compared to germ tube elongation assays (0.03–1.0 µg/mL). Some researchers obtained their own discriminatory concentrations based on the frequency distribution of EC_{50} values or MIC (minimum inhibitory concentration) derived from their fungal populations tested, others used previously reported discriminatory concentrations for the convenience with or without modifications.

Some researchers differentiated moderate resistance and high resistance to AP fungicides among fungal strains based on their EC_{50} values or their responses to two different discriminatory concentrations. Such a distinction between these phenotypes, however, may not be of practical importance as both phenotypes show comparable disease levels on detached fruit treated with recommended field dose of AP fungicides for *B.cinerea* (Weber, 2011; Fernández-Ortuño et al., 2013, 2014; Saito et al., 2016) and for *P. expansum* (Caiazzo et al., 2014).

Because there are many methods to determine resistance to AP fungicides, a direct comparison of the frequency of resistant isolates might not be appropriate in some cases. However, the trends of resistance to AP fungicides in phytopathogenic fungi are evident from long-term resistance-monitoring studies conducted by the same researchers or laboratories. For example, in *B. cinerea*, the frequency of resistance to cyprodinil increased from 0% to 84% from 1995 to 2001 in Swiss vineyards where cyprodinil had been applied consecutively over the years (Baroffio et al., 2003). Similarly, in German vineyards, the frequency of resistance increased from 5.4% to 21.9% from 2006 to 2008 among *B. cinerea* strawberry isolates (Leroch et al., 2011). In *P. expansum* from apples, no resistance was detected among the isolates collected in 2004 and 2005 prior to the use of AP fungicides (Li and Xiao, 2008). However, a high frequency of resistance (84.7–95.8%) was detected in 2010 and 2011 at packinghouses where pyrimethanil had been used for 4–5 consecutive years, whereas the frequency of resistance remained below 2% at the packinghouses where there was no previous history of pyrimethanil use (Caiazzo et al., 2014). The frequency of resistance to cyprodinil increased from

12% to 84% among the isolates of *Oculimacula* spp. at the experimental field where cyprodinil had been applied for 5 consecutive years (Leroux et al., 2013). These results clearly showed that the frequency of resistance increases after repeated use of AP fungicides. High frequency of resistance also resulted in decreased efficacy of AP fungicide to control gray mold in vineyards (Baroffio et al., 2003; Hilber and Schüepp, 1996).

Another major trend observed during the last two to three decades in monitoring fungicide resistance is that multiple resistance to fungicides with different modes of action has become more prevalent in phytopathogenic fungi. Most of the published studies listed in Table 2 examined the sensitivities of fungal pathogens to multiple classes of fungicides and reported the occurrence of multiple fungicide resistance in many fungal species. Fungicide resistance monitoring studies showed that the number of classes of fungicide to which pathogens have become resistant have also been increasing over the years (Amiri et al., 2013; Avenot and Michailides, 2015; Fernández-Ortuño et al., 2013, 2014; Rupp et al., 2016; Weber and Hahn, 2019). This multiple fungicide resistance in fungal pathogens has an important implication for AP resistance management as discussed in the next section.

6 Resistance risk management

General fungicide resistance management recommendations include limiting the number of applications per season, rotation, and/or use of a mixture of different chemical classes of fungicides (Brent and Holloman, 2007; Hobbelen et al., 2011). This is also applied to AP fungicides resistance management for *Botrytis* spp. and *Venturia* spp. advocated by FRAC (www.frac.info). A long-term resistance monitoring study conducted in Swiss vineyards showed that limiting the number of applications during the growing season slowed the selection process for resistance to AP fungicides, while frequent applications of AP fungicides resulted in an increase of resistant isolates in the *B. cinerea* populations, resulting in the failure of disease control (Baroffio et al., 2003). Moreover, alternating use of three classes of fungicides, cyprodinil, fenhexamid, and fludioxonil, was more effective for disease control than the solo use of the same class of fungicide (Baroffio et al., 2003).

However, these common strategies that may delay resistance in situations where the frequency of resistance to AP fungicides remains relatively low in a population may not be effective in situations where the frequency of resistance is already high and/or multiple fungicide resistance is prevalent in the pathogen populations. For instance, in many crops, *B. cinerea* has already developed resistance to AP fungicides as well as other classes of fungicides including boscalid and fluopyram (SDHI, FRAC 7), azoxystrobin and pyraclostrobin (QoI fungicide, FRAC11), fenhexamid (SBI Class III fungicide, FRAC 17), carbendazim and thiabendazole (MBC, FRAC 1), and iprodione (Dicarboximides, FRAC 2). Dual resistance to boscalid and pyraclostrobin is one of the most prevalent fungicide resistance phenotypes in *B. cinerea* populations (Amiri et al., 2013; Fernández-Ortuño et al., 2013; Saito et al., 2016, 2019), and with such populations, the solo use of either FRAC 7 or 11 or the premixture of both fungicides may not provide satisfactory control against gray mold (Amiri et al., 2013; Saito et al., 2016; 2019).

Fludioxonil (phenylpyrrole fungicide, FRAC 12) is one of the fungicides that is often used in combination with AP fungicides as pre-mixture. These mixtures have been effective against *A. alternata*, *B. cinerea,* and *P. digitatum* (Avenot and Michailides, 2015; Chapeland et al., 1999; Hilber and Schüepp, 1996; Kanetis et al., 2008) as there seemed to be no resistance to fludioxonil (Avenot et al., 2018; Caiazzo et al. 2014; Fan et al., 2016, 2017; Fernández-Ortuño et al., 2013; Saito et al., 2016, 2019; Wang et al., 2021). However, in recent years, resistance and/or reduced sensitivities to fludioxonil have been reported in *B. cinerea* due to alterations in either target or non-target genes (Harper et al., 2021; Hu et al., 2019; Kretschmer et al., 2009; Leroch et al., 2013). Continuous use of such mixtures may impose high selection pressure for the development of reduced sensitivity or resistance to fludioxonil (Fernández-Ortuño et al., 2013; Kretschmer et al., 2009).

Although a higher dose of fungicide may be required to obtain effective disease control, empirical, and theoretical evidence suggested that a high fungicide dose will increase resistance risk in most circumstances (van den Bosch et al., 2011). Recently, the effect of reduced dose in fungicide mixtures to control *Zymoseptoria tritici* on wheat was evaluated (Ballu et al., 2021). In this study, 'efficient-dose mixtures' consisting of reduced dose of fungicides delivered a similar level of disease control compared to that obtained by the solo use of each fungicide. Results suggested that the efficient-dose mixtures could provide a satisfactory level of disease control and resistance risk management while lowering environmental and economic costs (Ballu et al., 2021). However, to develop such efficient mixtures, it is necessary to have a better understanding of the specificities of the target pathogens, their interactions with fungicides, and interactions between fungicides.

In conclusion, there is no simple or universal solution for the management of resistance risk to AP fungicides. In the view of recent fungicide resistance situation reported worldwide, to manage AP resistance risk, it is indispensable to take into account every aspect of fungicide resistance of the target pathogens, including frequency of resistance to AP fungicides, MDR strains, multiple fungicide resistance strains, and alternation/mixture with fungicide classes that have no or low risk of resistance development. Since fungicide resistance situation is unique for each fungal pathogen, commodity, and location, it is necessary to monitor resistance in fungal pathogens to understand their fungicide resistance profiles and their spatial and temporal distributions, and such information should be used to design 'tailored' resistance management strategies and maintain the efficacy of AP fungicides.

7 Where to look for further information

7.1 Major readings

- Fungicide Resistance, the Assessment of Risk by Brent, K. J. and Hollomon, D. W. (2007). (second, revised edition), published by Fungicide Resistance Action Committee, https://www.frac.info/knowledge-database/downloads.
- Sensitivity Baselines in Fungicide Resistance Research and Management by Russell, P. E. (2004). published by Fungicide Resistance Action Committee, https://www.frac.info/knowledge-database/downloads.

7.2 Major international research projects

Fungicide Resistance Action Committee (FRAC) (www.frac.info) is an international consortium. FRAC works to prolong the effectiveness of fungicides liable to encounter resistance problems and to limit crop losses should resistance occur. FRAC provides background information as well as annually updated Fungicide Resistance Management Recommendations for fungicides of the major modes of action, including anilinopyrimidines.

8 References

Abate, D., Pastore, C., Gerin, D., De Miccolis, R. M., Rotolo, C., Pollastro, S. and Faretra, F. (2018). Characterization of *Monilinia* spp. populations on stone fruit in south Italy, *Plant Dis.* 102(9): 1708–1717.

Albertini, C. and Leroux, P. (2004). A *Botrytis cinerea* putative 3-keto reductase gene (ERG27) that is homologous to the mammalian 17β-hydroxysteroid dehydrogenase type 7 gene (17β-HSD7), *Eur. J. Plant Pathol.* 110(7): 723–733.

Alzohairy, S. A., Gillett, J., Saito, S., Naegele, R. N., Xiao, C. L. and Miles, T. D. (2021). Fungicide resistance profiles of *Botrytis cinerea* isolates from Michigan vineyards and development of a TaqMan assay for detection of fenhexamid resistance, *Plant Dis.* 105(2): 285–294.

Amiri, A., Heath, S. M. and Peres, N. A. (2013). Phenotypic characterization of multifungicide resistance in *Botrytis cinerea* isolates from strawberry fields in Florida, *Plant Dis.* 97(3): 393–401.

Amiri, A., Zuniga, A. I. and Peres, N. A. (2018). Potential impact of populations drift on *Botrytis* occurrence and resistance to multi- and single-site fungicides in Florida southern highbush blueberry fields, *Plant Dis.* 102(11): 2142–2148.

Avenot, H. F. and Michailides, T. J. (2015). Detection of isolates of *Alternaria alternata* with multiple-resistance to fludioxonil, cyprodinil, Boscalid and pyraclostrobin in California pistachio orchards, *Crop Prot.* 78: 214–221.

Avenot, H. F., Quattrini, J., Puckett, R. and Michailides, T. J. (2018). Different levels of resistance to cyprodinil and iprodione and lack of fludioxonil resistance in *Botrytis cinerea* isolates collected from pistachio, grape, and pomegranate fields in California, *Crop Prot.* 112: 274–281.

Avila-Adame, C. and Köller, W. (2003). Characterization of spontaneous mutants of *Magnaporthe grisea* expressing stable resistance to the Qo-inhibiting fungicide azoxystrobin, *Curr. Genet.* 42(6): 332–338.

Baggio, J. S., Peres, N. A. and Amorim, L. (2018). Sensitivity of *Botrytis cinerea* isolates from conventional and organic strawberry fields in Brazil to azoxystrobin, iprodione,pyrimethanil, and thiophanate-methyl, *Plant Dis.* 102(9): 1803–1810.

Ballu, A., Dérédec, A., Walker, A.-S. and Carpentier, F. (2021). Why do fungicide mixtures delay the evolution of resistance?, *An Exp. Evol. Approach bioRxiv*, DOI: 10.1101/2021.09.15.460529.

Bardas, G. A., Myresiotis, C. K. and Karaoglanidis, G. S. (2008). Stability and fitness of anilinopyrimidine-resistant strains of *Botrytis cinerea*, *Phytopathology* 98(4): 443–450.

Baroffio, C. A., Siegfried, W. and Hilber, U. W. (2003). Long-term monitoring for resistance of *Botryiotinia fuckeliana* to anilinopyrimidine, phenylpyrrole, and hydroxyanilide fungicides in Switzerland, *Plant Dis.* 87(6): 662–666.

Brent, K. J. and Holloman, D. W. (2007). *Fungicide Resistance in Crop Pathogens: How Can It Be Managed? FRAC Monograph No. 1* (2nd revised ed.). Fungicide Resistance Action Committee, CropLife International, Brussels, Belgium.

Caiazzo, R., Kim, Y. K. and Xiao, C. L. (2014). Occurrence and phenotypes of pyrimethanil resistance in *Penicillium expansum* from apple in Washington State, *Plant Dis.* 98(7): 924–928.

Chapeland, F., Fritz, R., Lanen, C., Gredt, M. and Leroux, P. (1999). Inheritance and mechanisms of resistance to anilinopyrimidine fungicides in *Botrytis cinerea* (*Botryotinia fuckelinana*), *Pestic. Biochem. Physiol.* 64(2): 85–100.

Chen, S. N., Luo, C. X., Hu, M. J. and Schnabel, G. (2016). Fitness and competitive ability of *Botrytis cinerea* isolates with resistance to multiple chemical classes of fungicides, *Phytopathology* 106(9): 997–1005.

Cosseboom, S. D. and Hu, M. (2021). Identification and characterization of fungicide resistance in *Botrytis* populations from small fruit fields in the Mid-Atlantic United States, *Plant Dis.* 105(9): 2366–2373.

Daniels, A. and Lucas, J. A. (1995). Mode of action of the anilino-pyrimidine fungicide pyrimethanil. 1. In-vivo activity against Botrytis fabae on broad bean (Vicia faba) leaves, *Pestic. Sci.* 45(1): 33–41.

El-Naggar, M., Kövics, G. J., Váczy, K. Z. and Karaffa, E. (2008). Pyrimethanil tolerance of *Botrytis cinerea* isolates from Egypt and Hungary, *Tiszántúli Növényvédelmi Fórum*, October, pp. 98–107. Available at: http://real.mtak.hu/11260/1/1231441.pdf.

Fairchild, K. L., Miles, T. D. and Wharton, P. S. (2013). Assessing fungicide resistance in populations of *Alternaria* in Idaho potato fields, *Crop Prot.* 49: 31–39.

Fan, F., Hamada, M. S., Li, N., Li, G. Q. and Luo, C. X. (2017). Multiple fungicide resistance in *Botrytis cinerea* from greenhouse strawberries in Hubei Province, China, *Plant Dis.* 101(4): 601–606.

Fan, F., Li, N., Li, G. Q. and Luo, C. X. (2016). Occurrence of fungicide resistance in *Botrytis cinerea* from greenhouse tomato in Hubei Province, China, *Plant Dis.* 100(12): 2414–2421.

Fernández-Ortuño, D., Chen, F. and Schnabel, G. (2013). Resistance to cyprodinil and lack of fludioxonil resistance in *Botrytis cinerea* isolates from strawberry in North and South Carolina, *Plant Dis.* 97(1): 81–85.

Fernández-Ortuño, D., Grabke, A., Bryson, P. K., Amiri, A., Peres, N. A. and Schnabel, G. (2014). Fungicide resistance profiles in *Botrytis cinerea* from strawberry fields of seven Southern U. S. States, *Plant Dis.* 98(6): 825–833.

Fernández-Ortuño, D., Grabke, A., Li, X. and Schnabel, G. (2015). Independent emergence of resistance to seven chemical classes of fungicides in *Botrytis cinerea*, *Phytopathology* 105(4): 424–432.

Fillinger, S., Leroux, P., Auclair, C., Barreau, C., Al Hajj, C. and Debieu, D. (2008). Genetic analysis of fenhexamid-resistant field isolates of the phytopathogenic fungus *Botrytis cinerea*, *Antimicrob. Agents Chemother.* 52(11): 3933–3940.

Fonseka, D. L. and Gudmestad, N. C. (2016). Spatial and temporal sensitivity of *Alternaria* species associated with potato foliar diseases to demethylation inhibiting and anilino-pyrimidine fungicides, *Plant Dis.* 100(9): 1848–1857.

Forster, B. and Staub, T. (1996). Basis for use strategies of anilinopyrimidine and phenylpyrrole fungicides against *Botrytis cinerea*, *Crop Prot.* 15(6): 529–537.

Fritz, R., Lanen, C., Chapeland-Leclerc, F. and Leroux, P. (2003). Effect of the anilinopyrimidine fungicide pyrimethanil on the cystathionine β-lyase of *Botrytis cinerea*, *Pestic. Biochem. Physiol.* 77(2): 54–65.

Fritz, R., Lanen, C., Colas, V. and Leroux, P. (1997). Inhibition of methionine biosynthesis in *Botrytis cinerea* by the anilinopyrimidine fungicide pyrimethanil, *Pestic. Sci.* 49(1): 40–46.

Fungicide Resistance Action Committee (FRAC) (2021). *FRAC Code List 2021: Fungal Control Agents Sorted by Cross Resistance Pattern and Mode of Action*. Available at: www.frac.info.

Grabke, A., Fernández-Ortuño, D. and Schnabel, G. (2013). Fenhexamid resistance in *Botrytis cinerea* from strawberry fields in the Carolinas is associated with four target gene mutations, *Plant Dis.* 97(2): 271–276.

Grabke, A. and Stammler, G. (2015). A *Botrytis cinerea* population from a single strawberry field in Germany has a complex fungicide resistance pattern, *Plant Dis.* 99(8): 1078–1086.

Gullino, M. L., Bertetti, D., Monchiero, M. and Garibaldi, A. (2000). Sensitivity to anilinopyrimidines and phenypyrroles in *Botrytis cinerea* in north-Italian vineyards, *Phytopathol. Mediterr* 39: 433–446.

Harper, L. A., Paton, S., Hall, B., McKay, S., Oliver, R. P. and Lopez-Ruiz, F. J. (2021). Fungicide Resistance Characterised Across Seven Chemical Classes in a Botrytis cinerea Population Isolated From Australian Vineyards, *Pest Manag. Sci.* 78(4): 1326–1340.

Hawkins, N. J. and Fraaije, B. A. (2018). Fitness penalties in the evolution of fungicide resistance, *Annu. Rev. Phytopathol.* 56: 339–360.

Hay, F. S., Sharma, S., Hoepting, C., Strickland, D., Luong, K. and Pethybridge, S. J. (2019). Emergence of Stemphylium leaf blight of onion in New York associated with fungicide resistance, *Plant Dis.* 103(12): 3083–3092.

Hayashi, S., Maeno, S., Kimoto, T. and Nagata, T. (1997). Development of a new fungicide, mepanipyrim, *J. Pestic. Sci.* 22(2): 165–175.

Heye, U. J., Speich, J., Siegle, H., Steinemann, A., Forster, B., Knauf-Beiter, G., Herzog, J. and Hubele, A. (1994). CGA 219417: A novel broad-spectrum fungicide, *Crop Prot.* 13(7): 541–549.

Hilber, U. W. and Hilber-Bodmer, M. (1998). Genetic basis and monitoring of resistance of *Botryotinia fuckeliana* to anilinopyrimidines, *Plant Dis.* 82(5): 496–500.

Hilber, U. W. and Schüepp, H. (1996). A reliable method for testing the sensitivity of *Botryotinia fuckeliana* to anilinopyrimidines in vitro, *Pestic. Sci.* 47(3): 241–247.

Hobbelen, P. H. F., Paveley, N. D. and van den Bosch, F. (2011). Delaying selection for fungicide insensitivity by mixing fungicides at a low and high risk of resistance development: A modeling analysis, *Phytopathology* 101(10): 1224–1233.

Hou, Y. P., Mao, X. W., Qu, X. P., Wang, J. X., Chen, C. J. and Zhou, M. G. (2018). Molecular and biological characterization of *Sclerotinia sclerotiorum* resistant to the anilinopyrimidine fungicide cyprodinil, *Pestic. Biochem. Physiol.* 146: 80–89.

Hu, M. J., Cosseboom, S. and Schnabel, G. (2019). AtrB-associated fludioxonil resistance in Botrytis fragariae not linked to mutations in transcription factor mrr1, *Phytopathology* 109(5): 839–846.

Hulvey, J., Popko, J. T., Sang, H., Berg, A. and Jung, G. (2012). Overexpression of *ShCYP51B* and *ShatrD* in *Sclerotinia homoeocarpa* isolates exhibiting practical field resistance to a demethylation inhibitor fungicide, *Appl. Environ. Microbiol.* 78(18): 6674–6682.

Ji, M. S., Qi, Z. Q., Wang, Y. Z., Cheng, G. W. and Zu, M. (2003). Resistance of *Botrytis cinerea* to pyrimethanil in tomato, *Acta Phytophylacica Sin.* 30: 396–400.

Kanetis, L., Forster, H., Jones, C. A., Borkovich, K. A. and Adaskaveg, J. E. (2008). Characterization of genetic and biochemical mechanisms of fludioxonil and pyrimethanil resistance in field isolates of *Penicillium digitatum*, *Phytopathology* 98(2): 205–214.

Karaoglanidis, G. S., Markoglou, A. N., Bardas, G. A., Doukas, E. G., Konstantiou, S. and Kalampokis, J. F. (2011). Sensitivity of *Penicillium expansum* field isolates to tebuconazole, iprodione, fludioxonil and cyprodinil and characterization of fitness parameters and patulin production, *Int. J. Food Microbiol.* 145(1): 195–204.

Keinath, A. P. (2015). Baseline sensitivity of *Didymella bryoniae* to cyprodinil and fludioxonil and field efficacy of these fungicides against isolates resistant to pyraclostrobin and Boscalid, *Plant Dis.* 99(6): 815–822.

Köller, W. (1991). Fungicide resistance in plant pathogens. In: *CRC Handbook of Pest Management in Agriculture* (Vol. 2, 2nd edn, pp. 679–720). CRC Press, Boca Raton, FL.

Köller, W., Wilcox, W. F. and Parker, D. M. (2005). Sensitivity of *Venturia inaequalis* populations to anilinopyrimidine fungicides and their contribution to scab management in New York, *Plant Dis.* 89(4): 357–365.

Kretschmer, M., Leroch, M., Mosbach, A., Walker, A. S., Fillinger, S., Mernke, D., Schoonbeek, H. J., Pradier, J. M., Leroux, P., de Waard, M. A. and Hahn, M. (2009). Fungicide-driven evolution and molecular basis of multidrug resistance in field populations of the grey mould fungus *Botrytis cinerea*, *PLoS Pathog.* 5(12): e1000696.

Larsen, H. J., Beresford, R. M., Wood, P. N., Wright, P. J. and Fisher, B. M. (2013). A synthetis agar assay for determining sensitivity of *Venturia inaequalis* to anilinopyrimidine fungicides in New Zealand apple orchards, *N. Z. Plant Prot.* 66: 293–302.

Lattore, B. A., Spadaro, I. and Rioja, M. E. (2002). Occurrence of resistant isolates of *Botrytis cinerea* to anilinopyrimidine fungicides in table grapes in Chile, *Crop Prot.* 21(10): 957–961.

Leroch, M., Kretschmer, M. and Hahn, M. (2011). Fungicide resistance phenotypes of *Botrytis cinerea* isolates from commercial vineyards in South West Germany, *J. Phytopathol.* 159(1): 63–65.

Leroch, M., Plesken, C., Weber, R. W. S., Kauff, F., Scalliet, G. and Hahn, M. (2013). Gray mold populations in German strawberry fields are resistant to multiple fungicides and dominated by a novel clade closely related to *Botrytis cinerea*, *Appl. Environ. Microbiol.* 79(1): 159–167.

Leroux, P. (1994). Influence du pH, d'acides amines et de diverses substances organiques sur la fongitoxicité du pyrimethanil, du glufosinate, du captafol, du cymoxanil et du fenpiclonil vis-à-vis de certaines souches de *Botrytis cinerea*, *Agronomie* 14(8): 541–554.

Leroux, P., Chapeland, F., Desbrosses, D. and Gredt, M. (1999). Patterns of cross-resistance to fungicides in *Botryotinia fuckeliana* (*Botrytis cinerea*) isolates from French vineyards, *Crop Prot.* 18(10): 687–697.

Leroux, P., Fritz, R., Debieu, D., Albertini, C., Lanen, C., Bach, J., Gredt, M. and Chapeland, F. (2002). Mechanisms of resistance to fungicides in field strains of *Botrytis cinerea*, *Pest Manag. Sci.* 58(9): 876–888.

Leroux, P. and Gredt, M. (1995). Étude in vitro de la resistance de *Botrytis cinerea* aux fongicides anilinopyrimidines, *Agronomie* 15(6): 367–370.

Leroux, P., Gredt, M., Remuson, F., Micoud, A. and Walker, A. S. (2013). Fungicide resistance status in French populations of the wheat eyespot fungi *Oculimacula acuformis* and *Oculimacula yallundae*, *Pest Manag. Sci.* 69(1): 15–26.

Li, H. X. and Xiao, C. L. (2008). Characterization of fludioxonil-resistant and pyrimethanil-resistant phenotypes of *Penicillium expansum* from apple, *Phytopathology* 98(4): 427–435.

Liu, S., Che, Z. and Chen, G. (2016). Multiple-fungicide resistance to carbendazim, diethofencarb, procymidone, and pyrimethanil in field isolates of *Botrytis cinerea* from tomato in Henan Province, China, *Crop Prot.* 84: 56–61.

Liu, Y., Liu, Z., Hamada, M. S., Yin, Y. N. and Ma, Z. H. (2014). Characterization of laboratory pyrimethanil-resistant mutants of *Aspergillus flavus* from groundnut in China, *Crop Prot.* 60: 5–8.

Malandrakis, A. A., Apostolidou, Z. A., Markoglou, A. and Flouri, F. (2015). Fitness and cross-resistance of *Alternaria alternata* field isolates with specific or multiple resistance to single site inhibitors and mancozeb, *Eur. J. Plant Pathol.* 142(3): 489–499.

Markoglou, A. N., Markoglou, E. G. and Malandrakis, A. A (2011). Effect of anilinopyrimidine resistance on aflatoxin production and fitness parameters in Aspergillus parasiticus Speare, *Int. J. Food Microbiol.* 146: 130–136.

Masner, P., Muster, P. and Schmid, J. (1994). Possible methionine biosynthesis inhibition by pyrimidinamine fungicides, *Pestic. Sci.* 42(3): 163–166.

Mernke, D., Dahm, S., Walker, A. S., Lalève, A., Fillinger, S., Leroch, M. and Hahn, M. (2011). Two promoter rearrangements in a drug efflux transporter gene are responsible for the appearance and spread of multidrug resistance phenotype MDR2 in *Botrytis cinerea* isolates in French and German vineyards, *Phytopathology* 101(10): 1176–1183.

Milling, R. J. and Richardson, C. J. (1995). Mode of action of the anilinopyrimidine fungicide pyrimethanil. 2. Effects on enzyme secretion in *Botrytis cinerea*, *Pestic. Sci.* 45(1): 43–48.

Miura, I., Kamakura, T., Maeno, S., Hayashi, S. and Yamaguchi, I. (1994). Inhibition of enzyme secretion in plant pathogens by mepanipyrim, a novel fungicide, *Pestic. Biochem. Physiol.* 48(3): 222–228.

Mosbach, A., Edel, D., Farmer, A. D., Widdison, S., Barchietto, T., Dietrich, R. A., Corran, A. and Scalliet, G. (2017). Anilinopyrimidine resistance in *Botrytis cinerea* is linked to mitochondrial function, *Front. Microbiol.* 8: 2361.

Moyano, C., Gomez, V. and Melgarejo, P. (2004). Resistance to pyrimethanil and other fungicides in *Botrytis cinerea* populations collected on vegetable crops in Spain, *J. Phytopathol.* 152(8-9): 484–490.

Myresiotis, C. K., Karaoglanidia, G. S. and Tzavella-Klonari, K. (2007). Resistance of *Botrytis cinerea* isolates from vegetable crops to anilinopyrimidine, phenylpyrrole, hydroxyanilide, benzimidazole, and dicarboximide fungicides, *Plant Dis.* 91(4): 407–413.

Nakaune, R., Hamamoto, H., Imada, J., Akutsu, K. and Hibi, T. (1998). A novel ABC-binding cassette transporter involved in multidrug resistance in the phytopathogenic fungus *Penicillium digitatum*, *Appl. Environ. Microbiol.* 64: 3983–3988.

Neumann, G. L., Winter, E. H. and Pittis, J. E. (1992). Pyrimethanil: A new fungicide, *Proc. Br. Crop Protec. Conf.* 5-1: 395–402.

Omrane, S., Sghyer, H., Audéon, C., Lanen, C., Duplaix, C., Walker, A. S. and Fillinger, S. (2015). Fungicide efflux and the MgMFS 1 transporter contribute to the multidrug resistance phenotype in *Zymoseptoria tritici* field Isolates, *Environ. Microbiol.* 17(8): 2805–2823.

Parnell, S., Gilligan, C. A. and van den Bosch, F. (2005). Small-scale fungicide spray heterogeneity and the coexistence of resistant and sensitive pathogen strains, *Phytopathology* 95: 63–639.

Paszewski, A. and Grabski, J. (1974). Regulation of S-amino acids biosynthesis in *Aspergillus nidulans*. Role of cysteine and-or homocysteine as regulatory effectors, *Mol. Gen. Genet.* 132(4): 307–320.

Perlin, M. H., Andrews, J. and San Toh, S. S. (2014). Essential letters in the fungal alphabet: ABC and MFS transporters and their roles in survival and pathogenicity. *Adv. Genet.* 85: 201–253.

Rochefort, J., Theron, J. A. and Lennox, C. L. (2015). Mycelial sensitivity assay of postharvest grey mould from 'Forelle' pears to anilinopyrimidine fungicide pyrimethanil, *Acta Hortic.* (1094): 411–413.

Rupp, S., Weber, R. W. S., Rieger, D., Detzel, P. and Hahn, M. (2016). Spread of *Botrytis cinerea* strains with multiple fungicide resistance in German horticulture, *Front. Microbiol.* 7: 2075.

Saito, S., Furuya, S., Takayanagi, T. and Suzuki, S. (2009). Phenotypic analyses of fenhexamid resistant *Botrytis cinerea* mutants. In: Carisse, O. (Ed.), *Fungicides*. InTech Publisher, Rijeka, Croatia, pp. 247–260.

Saito, S., Michailides, T. J. and Xiao, C. L. (2016). Fungicide resistance profiling in *Botrytis cinerea* populations from blueberry in California and Washington and their impact on control of gray mold, *Plant Dis.* 100(10): 2087–2093.

Saito, S., Michailides, T. J. and Xiao, C. L. (2019). Fungicide-resistant phenotypes in *Botrytis cinerea* populations and their impact on control of gray mold on stored table grapes in California, *Eur. J. Plant Pathol.* 154(2): 203–213.

Saito, S. and Xiao, C. L. (2018). Fungicide resistance in *Botrytis cinerea* populations in California and its influence on control of gray mold on stored mandarin fruit, *Plant Dis.* 102(12): 2545–2549.

Samaras, A., Ntasiou, P., Myresiotis, C. and Karaglanidis, G. (2020). Multidrug resistance of *Penicillium expansum* to fungicides: Whole trancriptome analysis of MDR strains reveals overexpression of efflux transporter genes, *Int. J. Food Microbiol.* 335: 108896.

Schnabel, G., Brannen, P. M., Avenot, H., Michailides, T. J. and Adaskaveg, J. E. (2019). Fungicide resistance in fungal pathogens of stone fruits. In: McGrath, M. T. and Wyenandt, C. A. (Eds.), *Fungicide resistance in North America*. The American Phytopathological Society, St. Paul, MI, pp. 209–221.

Schnabel, G. and Parisi, L. (1997). Sensitivity of *Venturia inaequalis* to five DMI fungicides, including the new triazole fluquinconazole, and pyrimethanil, *J. Plant Dis. Prot.* 104: 36–46.

Sergeeva, V., Nair, N. G., Verdanega, J. R., Shen, C., Barchia, I. and Spooner-Hart, R. (2002). First report of anilinopyrimidine-resistant phenotypes in *Botrytis cinerea* on grapevines in Australia, *Australas. Plant Pathol.* 31(3): 299–300.

Sholberg, P. L., Bedford, K. E. and Stokes, S. (2003). Effect of preharvest application of cyprodinil on postharvest decay of apples caused by *Botrytis cinerea*, *Plant Dis.* 87(9): 1067–1071.

Sun, H. Y., Wang, H. C., Chen, Y., Li, H. X., Chen, C. J. and Zhou, M. G. (2010). Multiple resistance of *Botrytis cinerea* from vegetable crops to carbendazim, diethofencarb, procymidone, and pyrimethanil in China, *Plant Dis.* 94(5): 551–556.

Suzuki, H., Kuroda, K., Kida, K., Matsuzawa, A. and Takagaki, M. (2011). Occurrence of mepanipyrim-resistant isolates of *Botrytis cinerea*, *Jpn. J. Phytopathol.* 77(1): 1–6.

Takagaki, M., Miura, I. and Nagayama, K. (2004). A method for monitoring the sensitivity of *Botrytis cinerea* to mepanipyrim, *J. Pestic. Sci.* 29(4): 369–371.

van den Bosch, F., Paveley, N., Shaw, M., Hobbelen, P. and Oliver, R. (2011). The dose rate debate: does the risk of fungicide resistance increase or decrease with dose?, *Plant Pathol.* 60: 597–606.

Waechter, F., Weber, E., Hertner, T. and May-Hertl, U. (2010). Cyprodinil: A fungicide of the anilinopyrimidine class. In: Krieger, R. (Ed.), *Hayes Handbook of Pesticide Toxicology* (3rd edn.). Academic Press, San Diego, pp. 1903–1913.

Wang, F., Saito, S., Michailides, T. J. and Xiao, C. L. (2022). Fungicide resistance in *Alternaria alternata* from blueberry in California and its impact on control of Alternaria rot, *Plant Dis.* 106(5): 1446–1453.

Weber, R. W. S. (2011). Resistance of *Botrytis cinerea* to multiple fungicides in Northern German small-fruit production, *Plant Dis.* 95(10): 1263–1269.

Weber, R. W. S. and Entrop, A.-P. (2017). Recovery of fungicide-resistant *Botrytis* isolates from strawberry nursery plants, *Eur. J. Plant Pathol.* 149(3): 739–742.

Weber, R. W. S. and Hahn, M. (2011). A rapid and simple method for determining fungicide resistance in *Botrytis*, *J. Plant Dis. Prot.* 118(1): 17–25.

Weber, R. W. S. and Hahn, M. (2019). Grey mould disease of strawberry in northern Germany: Causal agents, fungicide resistance and management strategies, *Appl. Microbiol. Biotechnol.* 103(4): 1589–1597.

Xiao, C. L., Kim, Y. K. and Boal, R. J. (2010). First report of occurrence of pyrimethanil resistance in Penicillium expansum from stored apples in Washington State, *Plant Dis.* 95(1): 72.

Yin, D., Chen, X., Hamada, M. S., Yu, M., Yin, Y. and Ma, Z. (2015). Multiple resistance to QoIs and other classes of fungicides in *Botrytis cinerea* populations from strawberry in Zhejiang Province, China, *Eur. J. Plant Pathol.* 141(1): 169–177.

Yin, W. X., Adnan, M., Shang, Y., Lin, Y. and Luo, C. X. (2018). Sensitivity of *Botrytis cinerea* from nextarine/cherry in China to six fungicides and characterization of resistant isolates, *Plant Dis.* 102(12): 2578–2585.

Zhang, Y., Fu, Y., Luo, C. and Zhu, F. (2021). Pyrimethanil sensitivity and resistance mechanisms in *Penicillium digitatum*, *Plant Dis.* 105(6): 1758–1764.

Zhao, H., Kim, Y. K., Huang, L. and Xiao, C. L. (2010). Resistance to thiabendazole and baseline sensitivity to fludioxonil and pyrimethanil in *Botrytis cinerea* populations from apple and pear in Washington State, *Postharvest Biol. Technol.* 56(1): 12–18.

Zhao, J., Bi, Q., Wu, J., Lu, F., Han, X. and Wang, W. (2019). Occurrence and management of fungicide resistance in *Botrytis cinerea* on tomato from greenhouses in Hebei, China, *J. Phytopathol.* 167(7–8): 413–421.

Chapter 14

Understanding resistance to oxysterol binding protein inhibitor fungicides

Jean-Luc Genet, Corteva Agriscience, France

1 Introduction

Oomycetes include some of the most widespread and highly destructive plant pathogens. The *Phytophthora* genus alone contains 260 described species and is one of the most important groups of oomycetes (Kamoun et al., 2015; http://www.mycobank.org). *P. infestans* is probably the most notorious *Phytophthora* species, as the cause of the potato Irish famine in the mid-1880s, and currently costs more than US$5 billion per year to global potato crop production (Robertson, 1991; Birch and Whisson, 2001). Downy mildews belonging to the genera *Plasmopara*, *Peronospora*, *Pseudoperonospora*, *Peronosclerospora*, *Sclerophthora*, and *Sclerospora* are obligate parasites and regularly cause severe diseases on a variety of crops including grapes and cucurbits as well as many other vegetables (Strange and Scott 2005; Koike et al., 2007).

Plasmopara viticola is an important pathogen of grapes in areas with significant rainfall during vegetative growth of the vines which include the

http://dx.doi.org/10.19103/AS.2022.0116.15

maritime regions of Europe (Lafon and Clerjeau, 1988). The genus *Pythium* is a group of more than 100 species of unspecialized soil fungi which attack seedlings and roots and cause some storage rots. The genera *Aphanomyces* and *Albugo*, the white rust organisms, are also part of the oomycete group and can cause significant yield losses in vegetable crops (Koike et al., 2007). Another related group is the phylum *Plasmodiophoramycota* which includes the clubroot pathogen of crucifers (*Plasmodiophora brassicae*).

According to Hermann and Gisi (2012), the most economically important foliar oomycete pathogen species are *Phytophthora infestans* (potato and tomato late blight), *Plasmopara viticola* (grape downy mildew), *Pseudoperonospora cubensis* (cucumber downy mildew), *Bremia lactucae* (lettuce downy mildew), *Peronospora* spp. (downy mildews on different crops such as peas, brassicas and tobacco) and soilborne pathogens of the genus *Pythium* (many crops including vegetables, ornamentals and turf).

Preventive management of diseases caused by oomycetes includes crop rotation, soil drainage, removing the source of overwintering inoculum and pruning practices aimed at reducing the presence of free water on the foliage (Lafon and Smith, 1988). In some crops such as sunflower, plant breeding has been successful in delivering cultivars with some degree of resistance to oomycete diseases however their effectiveness is often short-lived (Gascuel et al., 2015). Control of oomycete diseases therefore remains largely dependent on the use of fungicides.

On a global basis, the cost of control of oomycete diseases with fungicide applications was estimated in 2013 at approximately US$2 billion annually (Gisi and Sierotzki, 2015). Fungicide use started at the end of the nineteenth century after the discovery of the Bordeaux mixtures to control grape downy mildew (Lafon and Smith, 1988). This was followed by multisite protectants such as mancozeb, folpet and chlorothalonil which were widely adopted in the 1940–70s for the management of many vegetable diseases where they largely replaced the copper salts (Leadbeater, 2015; Gullino et al., 2010).

Single-site inhibitors with oomycete activity were introduced in the following decades often in combination with protectants and included the phenylamides (e.g. metalaxyl and mefenoxam), cymoxanil, carbamates (e.g. propamocarb), phosphonates (e.g. fosetyl Al), fluazinam, quinone outside inhibitors (QoIs) such as famoxadone, fenamidone and the strobilurins (e.g. azoxystrobin), inhibitors of microtubule assembly (zoxamide and ethaboxam) and the carboxylic acid amides (e.g. dimethomorph and mandipropamid). New modes of action have become available more recently, including the quinone inside inhibitors (QiIs) cyazofamid and amisulbrom, ametoctradin and benzamides (e.g. fluopicolide) (Hermann and Gisi, 2012). It is worth noting that while the older multisite fungicides provide control of a broad spectrum of plant diseases, single-site inhibitors often control only oomycete pathogens.

Oxathiapiprolin, discovered by DuPont Crop Protection and now commercialized by Corteva Agriscience and fluoxapiprolin, from Bayer AG, are the most recent fungicides introduced for the control of oomycete diseases (Bruhn et al., 2013; Wu et al., 2019), and are sold exclusively in mixtures with other fungicides active against the target diseases (Fig. 1). Both molecules belong to FRAC group 49 (Fungicide Resistance Action Committee, more information on FRAC later in this chapter) and are described as oxysterol-binding protein inhibitors (or OSBPIs). In this chapter, the challenges posed by fungicide resistance in oomycete pathogens and the biological attributes of OSBPIs are reviewed. In addition, the resistance risk associated with this novel mode of action is discussed, as well as OSBPI use recommendations to maximize their effective life.

2 Current status of fungicide resistance in oomycetes

In recent years, the crop protection industry has been quite successful in introducing and maintaining a diverse number of modes of action to control oomycete diseases (Leadbeater, 2015). As shown in Table 1 however, all single-site modes of action currently used for the control of oomycete diseases have encountered some resistance issues with the notable exception of beta-tubulin inhibitors.

Resistance of potato late blight to metalaxyl, a phenylamide fungicide (FRAC group 4), was first reported in 1980 in Ireland and the Netherlands, only two years after its introduction (Cook, 1991). Subsequently, phenylamide resistance was also reported in *P. cubensis*, *P. tabacina* and *P. viticola* and in most cases was associated with a decline in disease control (Hermann and Gisi,

Table 1 Single-site fungicide resistance in oomycetes

Fungicide/example	Group/mode of action	FRAC code	Mode of Action group resistance status	
			P. infestans	Downy mildews
Mefenoxam, etc.	Phenylamide	4	Common	Common
Fluazinam	2,6-dinitro-aniline	29	Rare	Not known
Cymoxanil	Cyanoacetamide oxine	27	Not known	Occasional
Azoxystrobin etc.	Strobilurin (QoI)	11	Not known	Common
Fosetyl	Phosphonates	P07	Not known	Rare
Cyazofamid	QiI	21	Not known	Occasional
Mandipropamid etc.	carboxylic acid amide	40	Occasional	Common
Zoxamide, ethaboxam	β-tubulin inhibitors	22	Not known	Not known
Ametoctradin	QoSI	45	Not known	Occasional

Adapted and updated from Leadbeater (2015).

2012). Resistance of *P. viticola* to carboxylic acid amide fungicides (FRAC group 40) was detected in 1994 soon after the commercial introduction of this mode of action (Gisi, 2012; Pang et al., 2013). Resistance to QoI fungicides (FRAC group 11) was reported in many plant pathogen species, including *P. viticola* and *P. cubensis,* only a few years after the first strobilurins were introduced in the late 1990s (Sierotzki et al., 2005; Leadbeater, 2012).

Cymoxanil, fluazinam, propamocarb and fluopicolide have all encountered some degree of resistance in target pathogen populations although after a longer period of use (Genet and Vincent, 1999; Schepers et al., 2018; Pavelková et al., 2014; Thomas et al., 2018; Wu et al., 2020). For example, cymoxanil resistance was first reported in 1997 (Gullino et al., 1997), 15 years after its introduction in Europe. While fluazinam (FRAC group 29) was first introduced commercially for the control of potato late blight in 1992, resistance of *P. infestans* to this mode of action was only reported in 2018 in the Netherlands in association with a new clonal lineage (EU_33_A2) (https://www.wur.nl/en/newsarticle/Reduced-sensitivity-to-fluazinam-detected-in-the-rapidly-increasing-Phytophthora-infestans-clonal-lineage-EU-37-Dark-Green-37.htm).

The first report of resistance to propamocarb (FRAC group 28) was published in 2002, well after the first registration of this active ingredient for commercial use and was followed recently by several other cases (Pavelková et al., 2014; Thomas et al., 2018).

Fluopicolide (FRAC group 43) was first introduced in 2006 yet resistance to this mode of action has only been described recently in *P. cubensis* (Thomas et al., 2018) and *P. capsici* (Wu et al., 2020; Wang and Ji, 2021; Siegenthaler and Hansen, 2021).

Only a few cases of resistance to fosetyl Al and phosphonate fungicides (FRAC group P07) have been published since the first report in *P. citrophthora* in 1988 (Khilare et al., 2003; Brown et al., 2004; Hao et al., 2021).

Ethaboxam and zoxamide (FRAC group 22) share the same mode of action and affect beta-tubulin assembly during mitosis (www.FRAC.info). According to FRAC, there are no reports of field resistance to this mode of action; however, *P. sojae* mutants resistant to ethaboxam were isolated in adaptation experiments on ethaboxam-amended medium (Peng et al., 2019). Interestingly, some *Pythium* species appear to be naturally insensitive to ethaboxam (Noel et al., 2019).

The QiI fungicides cyazofamid and amisulbrom (FRAC group 21) were introduced in the early 2000s. The first report of resistance to this mode of action was from the United States in 2008 (Kousik and Keinath, 2008) in *P. capsici* but there have been more recent reports from Europe in *P. viticola* (Lebaudy et al., 2018; Fontaine et al., 2019).

Finally, resistance to ametoctradin (FRAC group 45, QoS inhibitor) has also been reported in *P. viticola* populations from France (Fontaine et al., 2019).

Interestingly, an alternative oxidase (AOX)-related resistance mechanism has been described that affects both QoI and QoSI fungicides (Fontaine et al., 2019) and seems to be increasing in frequency over time in French vineyards.

It is noteworthy that *P. infestans* has not developed resistance to any fungicides except for the phenylamides in the late 1980s and more recently to fluazinam and mandipropamid (FRAC group 40, Abuley et al., 2023). In fact, it is precisely this observation that has prompted FRAC to change the resistance risk associated with this pathogen from high to medium (Herman and Gisi, 2012).

While a diversity of modes of action still exists to combat *P. infestans*, some downy mildews such as *P. viticola* and *P. cubensis* have developed some degree of resistance to almost all single-site fungicides, emphasizing the need for new modes of action. The recent ban of chlorothalonil and mancozeb in the European Union and the risk of their withdrawal in other geographies will logically result in growers' stronger reliance on single-site active ingredients thereby increasing the risk of developing resistance to these fungicides. The commercial introduction of oxathiapiprolin in 2016 and the recent announcement of fluoxapiprolin (Wu et al., 2019) for disease control in potato, should help relieve some of the selection pressure on fungicides with other modes of action and possibly help in extending their effective life.

3 Mode of action of oxysterol-binding protein inhibitor fungicides

The absence of cross-resistance between oxathiapiprolin and other anti-oomycete fungicides has suggested a novel mode of action (Sweigard et al., 2014). Initial identification of the target protein was determined by a combination of two different techniques. A biochemical approach used an analog of oxathiapiprolin to isolate binding proteins by affinity chromatography. One of these was identified by mass spectrometry as an oxysterol-binding protein (OSBP). A number of lab-generated *P. capsici* mutants resistant to oxathiapiprolin were isolated and determined by sequencing to have a single nucleotide polymorphism in the *P. capsici* OSBP gene. Finally, it was demonstrated that *P. capsici* cells transformed with a plasmid encoding a mutated OSBP gene conferred resistance to oxathiapiprolin thereby providing definitive evidence that OSBP was a critical component in the mechanism of action (Pasteris et al., 2016). Confocal microscopy showed localization of the target protein in the nucleus, Golgi and endoplasmic reticulum (Pasteris et al., 2016).

Oxysterol-binding proteins are lipid binding proteins that are conserved from yeast to humans and appear to be implicated in many cellular processes including vesicular trafficking, lipid metabolism, and non-vesicular sterol transport, among others (Raychaudhuri and Prinz, 2010). Inhibiting OSBP may disrupt various functions in the fungal cell, such as signalling, maintenance of cell membranes and the formation of complex lipids that are essential for the cell to survive (www.frac.info). Based on his study of the effect of oxathiapiprolin on *P. cubensis* zoospores, Cohen (2015) has speculated that these fungicides affect cellulose micro-fibrils trafficking towards the tip membrane of the germinating zoospore, thus preventing germ-tube emergence.

Fluoxapiprolin is another anti-oomycete fungicide, discovered and developed by Bayer AG, and has a very similar molecular structure to that of oxathiapiprolin (Wu et al., 2019). The work published by Miao et al. (2021) suggests that there is at least partial cross-resistance between oxathiapiprolin and fluoxapiprolin. Accordingly, fluoxapiprolin has been classified by FRAC as an OSBPI.

4 Biological activity of oxysterol-binding protein inhibitors

As indicated earlier, oxathiapiprolin provides specific control of plant diseases caused by oomycetes. Interestingly, it exhibited strong inhibitory effects against *Pythium splendens* and *P. ultimum* but was inactive against 14 other *Pythium* species. Strong activity has also been reported on *Phytopythium* spp. (Miao et al., 2020).

In all oomycetes studied, oxathiapiprolin is biologically active at exceptionally low concentrations. Complete inhibition of mycelial growth occurs *in vitro* at concentrations ranging from <0.0001 mg/L to 0.01 mg L^{-1} in *Phytophtora citrophthora, Phytophthora syringae, Phytophthora nicotianae, Phytophthora hibernalis, Phytophthora infestans, Phytophthora sojae, Plasmopara viticola, Plasmopara halstedii and Pseudoperonospora cubensis* (Perotin et al., 2015; Cohen, 2015; Vargas et al., 2016; Bittner and Mila, 2016; Gray et al., 2018; Pasteris et al., 2019; Spring and Gomez-Zeledon, 2020).

Studies with *P. infestans* and *P. cubensis* have shown that oxathiapiprolin can prevent direct germination of sporangia and encysted zoospores, inhibit zoospore release, mycelial growth, lesion formation and spore formation, while causing zoospores to swim slower and encyst faster (Salas et al., 2015; Cohen, 2015).

Strong curative effects have been reported (Cohen, 2015; Salas et al., 2015; Salas et al., 2019) however preventive applications are strongly recommended to prevent or delay the development of resistance (www.frac.info; Mboup et al., 2022).

Oxathiapiprolin is capable of partial acropetal translocation within the plants and provides protection of new growth that was not present at the time of application (Perotin et al., 2015; Cohen 2015; Qu et al., 2016; Salas et al., 2019). In the case of foliar sprays, this attribute may allow growers to increase application intervals *vs* standard practices, potentially leading to fewer sprays per season. While oxathiapiprolin may be relatively less mobile in plants than mefenoxam, another systemic anti-oomycete fungicide, the extent of translocation is sufficient to provide long-lasting disease control in new growth due to the extremely high potency of the molecule against target pathogens.

Greenhouse and field experiments have demonstrated that a single application of oxathiapiprolin to the root of potato and basil plants provided prolonged systemic protection against late blight and downy mildew (Cohen, 2020; Cohen and Rubin, 2020) while seed treatments have demonstrated to protect soybean, sunflower and basil against *Phytophthora* sojae, *Plasmopara halstedii* and *Peronospora belbahrii,* respectively (Patel et al., 2015; Humann et al., 2019; Spring and Gomez-Zeledon, 2020; Hegstad et al., 2021). Cohen et al. (2019) have also determined that oxathiapiprolin performed synergistically when applied to sunflower seeds in mixture with mefenoxam or acibenzolar-S-methyl for protection against downy mildew. Control of *Phytophthora citrophthora* root rot from soil applications of oxathiapiprolin has also been reported in citrus crops. Finally, field studies have also shown that application of oxathiapiprolin through drip irrigation tubes or by soil drench was able to reduce *Phytophthora* blight caused by *P. capsici* (Ji and Csinos, 2015) while soil-directed applications were efficacious for the reduction of black shank of tobacco (Bittner and Mila, 2017). Little has been published so far about the biological activity of fluoxapiprolin.

Taken together, these results illustrate the outstanding properties of oxathiapiprolin and the potential for diverse methods of use to control diseases caused by oomycetes. Its remarkable versatility and level of efficacy however might lead to product misuse in practical conditions, thereby increasing the risk of developing resistance to that new mode of action.

5 Case study: oxysterol-binding protein inhibitor resistance risk assessment

5.1 Principles

The resistance risk of a new fungicide is generally assessed by combining risk values for the pathogen, fungicide and agronomic system using a risk matrix (Kuck and Russell, 2006). In some geographies such as the European Union and China, an assessment of the resistance risk is required as part of the regulatory process for the approval of a new pesticide (Kuck, 2005). In the EU, such assessments are typically conducted following the European and Mediterranean Plant Protection Organisation (EPPO) standard PP 1/213(3) (Anonymous, 2015) and aim at providing information regarding the need and extent of anti-resistance strategies required to prevent or at least delay the development of resistance. In the following sections, we summarize the resistance risk associated with oxathiapiprolin using the aforementioned risk matrix.

5.2 Pathogen risk

The pathogen risk reflects the capacity of target pathogens to become resistant to fungicide treatments. According to Brent and Hollomon (1998) the risk is dependent on several epidemiological factors such as (1) the life cycle of the pathogen, (2) abundance of sporulation, (3) isolation of pathogen populations and (4) occurrence of a sexual stage in the life cycle.

FRAC has developed a 'pathogen risk list' which provides information about the risk of pathogens to develop resistance to fungicides under specific agronomic conditions (https://www.frac.info/docs/default-source/publications/pathogen-risk/frac-pathogen-list-2019.pdf). It is interesting to note that two important targets of the OSBPI fungicides (*P. viticola* and *P. cubensis*) are listed as having a high risk of development of resistance while *P. infestans*, *P. capsici* and several *Peronospora* species have been assigned a moderate risk and all soil- and seed-borne *Phytophthora* species are assigned a low risk (Table 2). The high risk associated with the grape and cucurbit downy mildew pathogens is consistent with the history of fungicide resistance associated with these two organisms (Section 2).

Table 2 Pathogen resistance risk according to FRAC's pathogen risk list

		Pathogen-associated risk		
Disease	Pathogen	High	Medium	Low
Grape downy mildew	*Plasmopara viticola*	•		
Cucumber downy mildew	*Pseudoperonospora cubensis*	•		
Potato/tomato late blight	*Phytophthora infestans*		•	
Lettuce downy mildew	*Bremia lactuca*		•	
Leek downy mildew	*Phytophthora porri*		•	
Soybean downy mildew	*Peronospora manshurica*		•	
Hops downy mildew	*Pseudoperonospora humuli*		•	
Other downy mildews	*Peronospora* spp.		•	
Pepper downy mildew (foliar)	*Phytophthora capsici*		•	
Soilborne *Phytophthora*	*Phytophthora* spp.			•

5.3 Fungicide risk

OSBPI fungicides were discovered several years after the risk matrix was developed by Kuck and Russell and therefore are not listed in the matrix. The assessment conducted at DuPont Crop Protection for oxathiapiprolin included mutagenesis experiments, pressurization studies and pre-commercialization sensitivity baselines establishment. The following sections provide a high-level summary of the results.

5.3.1 Mutagenesis experiments

Mboup et al. (2022) were able to produce *P. capsici* mutants resistant to oxathiapiprolin in the laboratory following UV mutagenesis. Sequencing of the target site revealed that 21 nonsynonymous mutations could occur across nine positions at the *OSBP* locus. Among the most frequent mutations, three had resistance factors greater than 100 and could have a significant impact on product efficacy should they be selected in the field. Two mutations have previously been reported in *P. capsici* adaptation mutants and one was confirmed by traditional overexpression transformation to confer oxathiapiprolin resistance (Miao et al., 2016; Miao et al., 2018). Mutations in the OSBP gene conferring resistance to oxathiapiprolin have also been reported for *P. nicotianae* and *P. sojae*, sometimes associated with high resistance factors (Bittner et al., 2017; Miao et al., 2020). More recently, seven heterozygous point mutations in the target protein were found in *P. capsici* mutants resistant to fluoxapiprolin (Miao et al., 2021). Several reports have indicated a potential fitness penalty associated with some OSBPI mutants suggesting that their survival potential in the field might be reduced, however this was not the case for all mutations (Bittner et al., 2017; Mboup et al., 2022). According to Brent and Hollomon (1998), the ease of resistant mutant production in mutagenesis experiments in the laboratory can indicate a potential for practical resistance problems therefore the above findings suggest a high risk for this scenario to happen in the case of the OSBPI fungicides.

5.3.2 Pressurization studies

A targeted sensitivity monitoring programme was conducted at DuPont Crop Protection, which focused on grape downy mildew and potato late blight pre-commercialization small plot trials where oxathiapiprolin applied at its label dose and recommended spray interval delivered sub-optimal efficacy (Mboup et al., 2022). Oxathiapiprolin-based products had been used intensively at these sites during several consecutive seasons with a high cumulative number of applications. Sequencing of the *OSBP* gene from 'survivor' isolates collected in treated plots at the end of the season revealed the presence of three point mutations in *P. viticola* (L863W, N837I and G770V) and one point mutation in *P. infestans* (N837L). Interestingly, all the positions of these point mutations had been predicted by the mutagenesis work with *P. capsici* although one of the amino acid substitutions had not.

These results indicate that OSBPI-resistant grape downy mildew and potato late blight isolates can be selected in small scale field trials conducted at the same site after only 3 to 5 years of continuous exposure to the fungicide.

5.3.3 Sensitivity baselines

Sensitivity baselines are not *per se* an indicator of resistance risk and are only intended to be used as a reference for future sensitivity monitoring studies following product commercialization. In the case of oxathiapiprolin, European sensitivity baselines were generated for both *P. viticola* and *P. infestans* as per EPPO 213 requirements (Anonymous, 2015) before commercial introduction of the fungicide. The sensitivity distribution for *P. viticola* and *P. infestans* single-lesion isolates collected between 2011 and 2013 is shown in Figure 1.

It is interesting to note that the sensitivity baseline distributions reported for both *P. viticola* and *P. infestans* were skewed towards lower sensitivity values indicating that less sensitive isolates were present at a detectable level prior to product introduction (Mboup et al., 2022) Figure 2. According to Brent and Hollomon (1998) such distributions provide a warning that resistance is

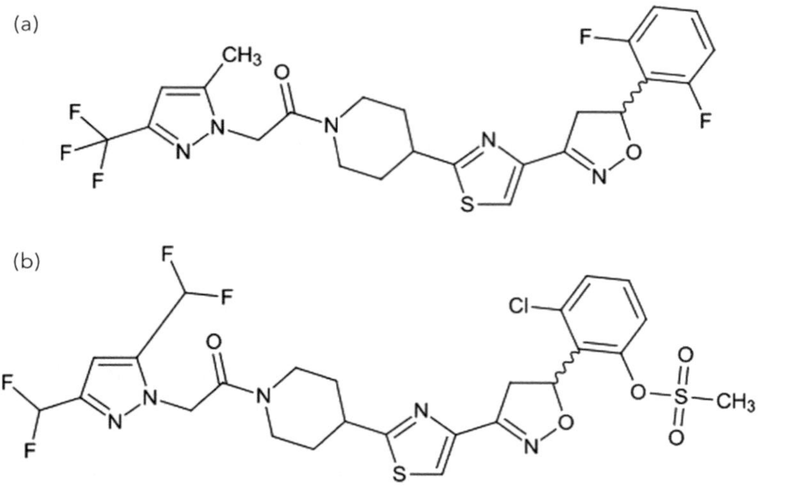

Figure 1 Molecular structure of (a) oxathiapiprolin and (b) fluoxapiprolin (https://pesticidecompendium.bcpc.org/).

possible if the outlying portion of the distribution can be associated with a true resistance response.

The non-normal distribution of sensitivity values observed for both pathogens prior to product commercial introduction (Fig. 2) is somewhat

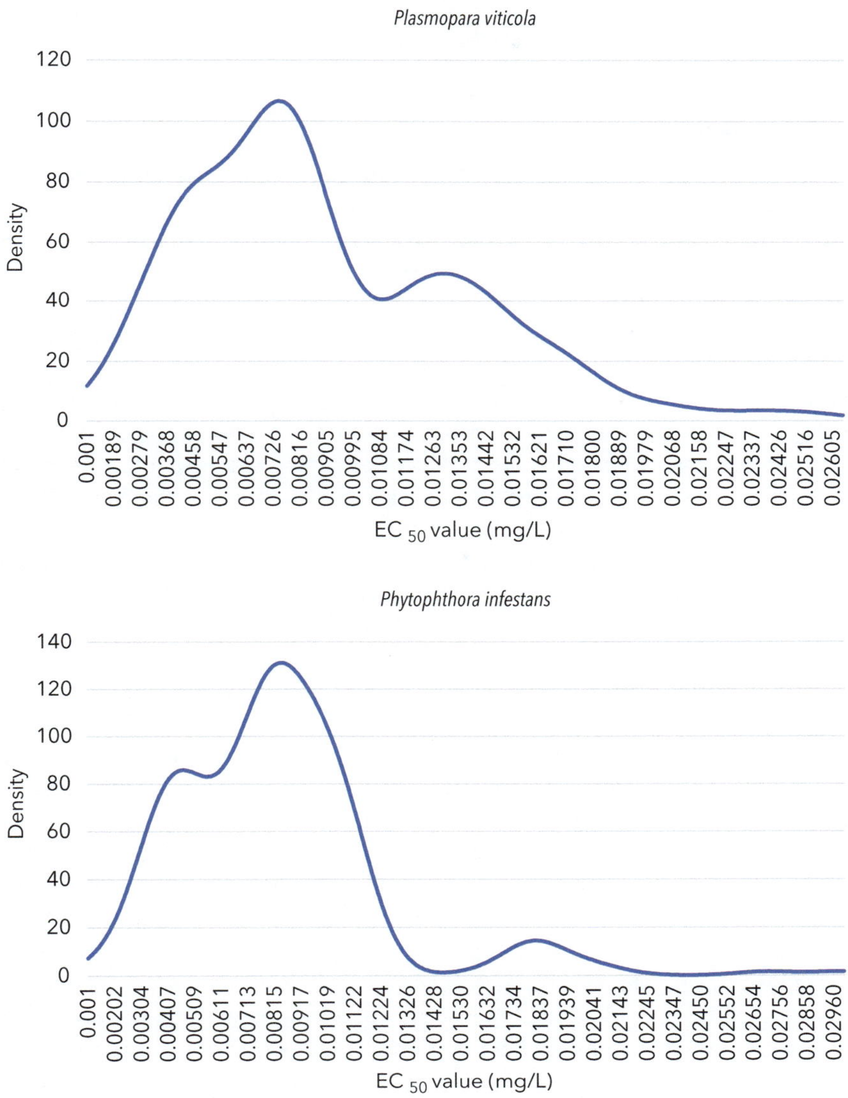

Figure 2 Baseline sensitivity to oxathiapiprolin in European *Plasmopara viticola* (top) and *Phytophthora infestans* (bottom) isolates (EC_{50} = dose needed to reduce fungal growth by 50%).

unexpected and the reason is unclear. The highest EC_{50} values were still low (below 0.03 mg/L) and were not found in any particular country or season. It is also worth noting that all isolates were fully controlled at a low dose of oxathiapiprolin (0.25 mg/L).

Sensitivity baselines have also been established for *Phytophthora* species causing brown rot and root rot of citrus (Gray et al., 2018) and *P. sojae* causing stem and root rot of soybean (McCoy et al., 2022).

5.4 Agronomic risk

According to Kuck and Russell (2006), the agronomic risk comprises two components: (1) an environment that is more or less conducive to the development and spread of disease; and (2) agronomic practices influenced by growers, including factors such as cropping intensity and choice of resistant or susceptible cultivars.

Resistance is more likely to develop first in areas of severe disease outbreaks where environmental conditions are conducive to the development and spread of the disease. In the case of oomycete diseases such as downy mildews and potato late blight, the presence of free water on the crop foliage or fruits is known to favour such disease epidemics and is a critical element of the agronomic risk. It is not surprising that resistance to phenylamides in *P. infestans* developed more quickly in the Netherlands, Ireland and Scotland (wet and moderate climate and generally high disease pressure) than in the regions with hot and dry summers (Hermann and Gisi, 2012). Similarly, the climatic conditions in the Veneto area of Italy are more favourable to grape downy mildew compared to southern Spain where the agronomic risk is probably lower. Even within a country such as France, the risk for *P. viticola* will vary greatly between, for example, the dry Provence area and the rainy Armagnac. The risk is also considered to be higher in areas with high genetic diversity of the primary inoculum and in the case of pathogen species having frequent sexual recombination, e.g. *P. infestans* in the Toluca Valley in Mexico or *P. viticola* in northern Italy (Grünwald et al., 2006; Gessler et al., 2011).

For *P. infestans*, a particularly high agronomic risk exists in tropical regions such as the Andean countries or Indonesia. Late blight is often endemic in these areas and, unlike in countries at higher latitudes which experience true winters, potatoes and tomatoes are grown year-round allowing continuous spread of the disease from field to field and from one crop planting to the next. This is further exacerbated by the fact that potato growers in these countries will often tolerate a low level of disease in their fields which could allow fungicide resistant 'escapes' to survive and propagate. Finally, there is the threat that climate change poses to conventional potato production, as global warming is making late blight more prevalent in potatoes in some parts of the world.

A range of farmer-determined agronomic factors can be used to mitigate the risk. They should aim at reducing pathogen populations to a minimum level thereby limiting reliance on fungicide use and the risk of resistance development. This includes removal of crop volunteers, alternative hosts and dumps; use of greenhouses or polythene tunnels which tend to isolate pathogen populations; use of drip versus sprinkler irrigation and adoption of appropriate crop rotations. For example, some crop rotations in greenhouses or in intensive crop cultivation areas can favour oomycete disease occurrence and severity possibly leading to more fungicide sprays. Traditional agronomic practices should not be neglected and the extent of use of hygienic practices can have a significant impact on the need for fungicide use and therefore on the risk of developing fungicide resistance.

Varietal resistance can in principle also be an effective tool when used in combination with other disease management practices as part of an Integrated Pest Management (IPM) programme, especially in potato (Bain et al., 2011). Resistance to potato late blight has been a priority in many breeding programmes and potato breeders have incorporated specific resistance genes in new varieties but in practice specific resistance has so far contributed little to controlling late blight (Dowley et al., 1991; Thurston and Schulz, 1986). Newer cultivars which have moderate levels of field resistance may require fewer fungicide sprays against late blight, but it remains difficult to grow potatoes without fungicide protection in regions with a humid environment during the growing season. The level of partial resistance of grapevines to downy mildew varies among *Vitis* species and cultivars. Breeding programmes based on the hybridization of American *Vitis* spp. with *V. vinifera* have recently delivered varieties which are partially resistant to downy mildew, enabling reduced use of fungicide (Bove and Rossi, 2020). It appears however that such resistance may be short-lived as new *P. viticola* isolates emerge that are able to overcome the resistance genes (Wingerter et al., 2021).

Finally, it has been shown that OSBPI fungicides provide outstanding disease control by foliar, soil or seed application (Qu et al., 2016; Hegstad et al., 2021; Humann et al., 2019; Cohen et al., 2019). This raises the possibility that a combination of these application methods could be used on the same crop which could result in prolonged exposure of the pathogen populations to these fungicides and accelerate the selection of resistant isolates. Such agronomic practices should therefore be avoided.

5.5 Overall risk

The overall resistance risk to a new fungicide derives from the combination of the risk values associated with the pathogen, the fungicide and the agronomic practices. Based on the above considerations, FRAC has assigned a moderate-to-high risk of resistance developing to OSBPI fungicides. The above conclusion was based primarily on studies conducted with oxathiapiprolin. It is reasonable to assume the same level of risk for fluoxapiprolin although no specific risk assessment has been published for this molecule. In practice, these two molecules should be considered together when managing resistance to OSBPI fungicides.

6 Sensitivity monitoring

The purpose of a fungicide resistance monitoring programme is to detect possible changes in fungal populations before they become a problem in the field. Monitoring of pathogen sensitivity to OSBPI fungicides started in 2015 mainly on *P. viticola*, *P. infestans*, *P. cubensis*, *Bremia lactucae*, *Peronospora destructor* and *P. halstedii* (www.frac.info). While these activities have focused primarily on Europe, a significant amount of sensitivity data has been generated in other regions including Asia (China, Korea, Japan, Indonesia, India), North America (United States), Latin America (Brazil, Mexico, Colombia, Ecuador) and Africa (South Africa). Bioassays have been established and validated by FRAC for most of the above pathogens and are made available to scientists around the world who may have interest in monitoring resistance development to this group of fungicides (https://www.frac.info/knowledge-database/monitoring-methods). Sensitivity monitoring can be performed using either single-spore (or single-lesion) isolates or bulk samples (micro-populations) consisting of multiple isolates collected throughout the field to be investigated. The use of bulk samples is generally preferred when the objective is the early detection of insensitive isolates.

Sensitivity monitoring data is reviewed by experts on the FRAC OSBPI working group on an annual basis and a summary of the findings is published on the FRAC webpage (see Section 9). While target pathogen populations have proved to be generally sensitive to OSBPI fungicides to date, there have been several reports of reduced sensitivity in *P. viticola* in recent years, mainly from trial sites in Europe where OSBPI fungicides had been used intensively during several seasons, often at doses higher than those specified on the label. However, the frequency of these insensitive isolates was low. In the case of *P. infestans*, single isolates with reduced sensitivity to OSBPIs were detected in four commercial areas of Central Java (Indonesia) and in the Antioquia region of Colombia (Mboup et al., 2022). A small number of *P. cubensis* isolates less sensitive to OSBPI fungicides have been found in 2019 in South Korea (protected cucumber) and the United States (Michigan, open field cucurbits) while in 2020 similar observations were made in China (cucumber crops grown under polytunnel). Finally, one case of localized resistance of *P. capsici* to OSBPI fungicides has recently been described in Tennessee, USA (Siegenthaler and

Hansen, 2021). Monitoring studies conducted in Europe with downy mildews of lettuce, onion and sunflower have so far shown fully sensitive populations.

As described earlier in this chapter, the repeated exposure of pathogen populations to oxathiapiprolin in small plot trials conducted at the same location during product development resulted in the selection of resistance to the fungicide prior to its commercial introduction in Europe (Mboup et al., 2021). The authors concluded that in trials needed for product registration, season-long applications of single-site fungicides should be restricted to the minimum required by regulatory authorities. Specifically, in the case of downy mildew of grapevine (a perennial crop), continuous applications of high-risk fungicides at the same location should be avoided or confined to areas where there is a low risk of spreading resistance to commercial vineyards.

7 Characterization of resistant mutants

The characteristics of resistant isolates are also an important aspect when evaluating the fungicide resistance risk. One of these aspects is the level of resistance conferred by the target site mutation(s) which is often referred to as the resistance factor. *P. capsici* mutants from either mutagenesis studies or fungicide adaptation experiments were found to be highly resistant to oxathiapiprolin (Miao et al., 2016; Mboup et al., 2022) or fluoxapiprolin (Miao et al., 2021). According to Mboup et al. (2022) some of the amino acid substitutions identified in *P. viticola* and *P. infestans* field mutants recovered from oxathiapiprolin pressurization studies produced resistance factors greater than 1000. Finally, several *P. sojae* transformants recovered using the CRISPR-Cas9 system have been characterized as having high resistance factors to oxathiapiprolin (Miao et al., 2020). While the presence at low levels of such mutants in field populations may have no practical impact, a reduction or even a total loss of disease control could become apparent as the frequency of highly resistant mutants increases. It will therefore be critical to develop methods capable of quantifying the frequency of highly resistant mutants in the field.

According to Brent and Hollomon (1998) the resistance risk is intensified if the resistant mutants have normal fitness with regard to growth, reproduction and pathogenicity. OSBPI-resistant *P. capsici* mutants generated by adaptation were reported to have significantly reduced fitness (Miao et al., 2016). Similar observations have been made for some *P. nicotianae* and *P. sojae* mutants resistant to oxathiapiprolin (Bittner et al., 2017; Miao et al., 2018). However, these examples involve laboratory-generated mutants and are not necessarily predictive of fitness for naturally occurring resistant mutants selected in the field.

As yet, there have been no reports of reduced field performance of OSBPI fungicides when used according to label recommendations. This suggests that the frequency of any isolates with reduced sensitivity to OSBPI fungicides in field populations has remained at a low level so far. The sensitivity monitoring data reported above suggest, however, that the situation is likely to evolve for some pathogens particularly in geographies with a high agronomic risk. It is therefore critical that the sensitivity of pathogen populations to OSBPI fungicides continues to be monitored on a regular basis especially in high-risk areas so that changes can be detected before product efficacy is affected. If there are indications of developing resistance, product use recommendations may need to be adjusted at the local level to slow down the selection process and limit spread to retain the utility of OSBPI-based fungicides as much as possible.

8 Molecular monitoring

In addition to traditional bioassays, molecular diagnostic tests have been developed to detect the presence of target site mutations at the OSBP locus in target pathogens (Miao et al., 2016). It is expected that additional molecular tools will be developed to quantify the frequency of resistant mutants in field populations to support large-scale resistance monitoring. The use of such tools however should be limited to mutations that are associated with significant resistance factors. More sophisticated molecular tools will also be needed as pathogen strains combining multiple target site mutations (resistant haplotypes) arise in the field.

9 Use recommendations

The FRAC is a specialist technical group of CropLife International and provides fungicide resistance management guidelines to prolong the effectiveness of 'at risk' fungicides and to limit crop losses should resistance occur (www .frac.info). The aim of FRAC is to (1) identify existing and potential fungicide resistance problems, (2) collate information and distribute it to those involved with fungicide research and product distribution, registration and use, (3) provide guidelines and advice on the use of fungicides to reduce the risk of resistance developing, and to manage it should it occur, (4) recommend procedures for use in fungicide resistance studies and (5) stimulate collaborations with universities, government agencies, advisors, extension services and product distributors.

A FRAC OSBPI working group was created in 2015 (mode of action group 49) and now oversees both oxathiapiprolin and fluoxapiprolin. Resistance monitoring data generated for relevant pathogens is shared annually by participating companies. The group also issues guidelines that provide information on how to prevent or delay the development of resistance to this group of fungicides (https://www.frac.info/frac-teams/working-groups/osbpi

-fungicides/recommendations-for-osbpi). The recommendations for the foliar use of OSBPI fungicides are summarized as follows:

- Apply OSBPIs at effective rates and intervals according to manufacturers' recommendations.
- Apply OSBPIs only preventatively.
- Apply OSBPIs only in mixtures with effective fungicides from different cross-resistance groups. The mixture partner should give effective control of the target disease(s) at the rate and interval selected.
- The number of foliar applications of OSBPI products within a total disease management programme must be limited: maximum two applications on grapes and four on other crops or maximum 33% of the total period of protection needed per crop, whichever is more restrictive.
- Apply no more than three consecutive applications of OSBPI fungicides.
- Applications of OSBPI products can be made in alternation with a fungicide with a different mode of action.

Additional recommendations can be found on the FRAC web page and include restrictions such as the use on nursery crops for transplantation to minimize the risk of disseminating resistance should it occur. The use of foliar applications of OSBPI following a soil- or a seed-treatment is also discouraged. Finally, the total number of applications per year targeting the same pathogen is restricted in case of multiple crops grown on the same acreage or greenhouse.

Effective disease management throughout the season is a critical component to delay the build-up and spread of resistant pathogen populations. FRAC recommendations should be integrated in an overall crop protection approach based on IPM including the use of disease tolerant varieties as well as traditional prophylactic measures which aim at reducing the amount of inoculum and limit pathogen spread. OSBPI fungicides (mixtures) should be integrated into a robust disease management programme in rotation with other fungicides having a different mode of action.

The above guidelines should be regarded as the minimum resistance management tactics required and more stringent practices may have to be recommended in specific cases such as areas with an extremely high agronomic risk. FRAC recommends growers to seek advice from their local crop advisor, extension agent, the product manufacturer or distributor to determine if a more restrictive approach is recommended. It must be noted that FRAC guidelines apply to all molecules within the mode of action group and that FRAC is not allowed to make recommendations for the use of individual products.

One of the key recommendations for the use of OSBPI fungicides is the use of mixtures with fungicides having a different mode of action and effectiveness against the target pathogens. This implies that the mixture partner remains effective and therefore monitoring of resistance to the companion molecule is required to fully understand the situation. From a resistance management point of view, multisite fungicides are preferred options for such mixtures (van den Bosch et al., 2014) since they are associated with a low resistance risk and have generally good efficacy against downy mildews. There is however strong regulatory pressure on these active ingredients and the list of multisite compounds to choose from has been recently reduced in geographies such as the European Union.

Another key component of the above recommendations is fungicide application timing and in particular the need to spray OSBPI products preventatively which, in practice, means before the appearance of any disease symptoms. In the case of diseases caused by oomycetes, infections often occur during or after a rain event and the best product performance will generally be observed when fungicide applications are made before rain. Both from a performance and a resistance management perspective, growers should not rely on the curative attribute of OSBPI products to protect their crop.

The strict observance by growers of product labels is critical to the long-term success of the resistance management strategy for OSBPI fungicides. Exceeding the use rate or the maximum number of applications beyond those specified on the label will likely lead to increased selection of resistant individuals in the fungal population (van den Bosch et al., 2011). On the contrary, much lower application doses and spray intervals longer than those advised on the label might result in imperfect disease control with the possible consequence that the following fungicide applications will be made on established disease thereby leading to a higher risk of selecting resistant individuals. All stakeholders including product distributors and advisers should play an active role in promoting the above guidelines, so that OSBPI fungicides retain their utility as long as possible.

10 Conclusion

In recent years, the crop protection industry has been rather successful in discovering, introducing and maintaining a diverse number of modes of action to control oomycete diseases, especially potato late blight and grape downy mildew. The OSBPI fungicide group (FRAC group 49) represents the most recent mode of action introduced for the control of oomycetes. Their intrinsic activity at low use rate and longer spray intervals combined with the emergence of more tolerant varieties has the potential to reduce the amount of chemicals used to protect crops against these diseases. Unfortunately, oomycetes are very adaptable plant pathogens and, as described in this chapter, there is a risk that resistance to this mode of action will eventually emerge and spread. While it is not always possible to mitigate environmental factors, sound agronomic practices can be used to reduce the resistance risk. The quick and full adoption of the above FRAC recommendations combined with relevant IPM practices will be essential to protect the effectiveness of these valuable new tools.

11 Where to look for further information

11.1 Further reading

FRAC Monograph 2. Fungicide resistance: The Assessment of Risk. Brent, K. J. and Hollomon, D. W. (1998). Published by GCPF (Brussels), September 1998.

Kuck, K. H. and Russell, P. E. (2006). FRAC: Combined Resistance Risk Assessment. *Asp Appl Biol* 78:3–10.

The book 'Fungicide Resistance in Crop Protection: Risk and Management' (ed. Thind, T. S.). *CAB International 2012* contains several chapters dealing with fungicide resistance in oomycetes.

Extensive information on fungicide resistance can be found on the FRAC website at: www.frac.info.

The EPPO Database on Resistance Cases aims to share information on documented resistance cases among the authorities responsible for registration of plant protection products across the EPPO region and between authorities and other stakeholders: https://resistance.eppo.int/.

12 References

Abuley, I. K., Lynott, J. S., Hansen, J. G., Cooke, D. E. L. and Lees, A. K. (2023). The EU43 genotype of Phytophthora infestans displays resistance to mandipropamid. *Plant Pathology*, doi: 10.1111/ppa.13737.

Anonymous. (2015). PP 1/213(4) resistance risk analysis. *EPPO Bull.* 45(3), 371–387.

Bain, R. A., Ritchie, F., Lees, A., Dyer, C. and Roberts, A. (2011). Development of foliar late blight (*P. infestans*) in relation to cultivar resistance and fungicide dose in potato. In: Schepers, H. T. A. M. (Ed.), *Proceedings of the 13th Workshop of a European Network for Development of an Integrated Control Strategy of Potato Late Blight*. PPO Special Report No. 15 139 (Vol. 144). Wageningen University, Lelystad, The Netherlands.

Bittner, R. J. and Mila, A. L. (2016). Effects of oxathiapiprolin on *Phytophthora* nicotianae, the causal agent of black shank of tobacco. *Crop Prot.* 81, 57–64.

Bittner, R. J. and Mila, A. L. (2017). Efficacy and timing of application of oxathiapiprolin against black shank of flue-cured tobacco. *Crop Prot.* 93, 9–18.

Bittner, R. J., Sweigard, J. A. and Mila, A. L. (2017). Assessing the resistance potential of *Phytophthora nicotianae*, the causal agent of black shank of tobacco to oxathiopropalin with laboratory mutants. *Crop Prot.* 102, 63–71.

Birch, P. R. J. and Whisson, S. C. (2001). *Phytophthora infestans* enters the genomic era. *Mol. Plant Pathol.* 2(5), 257–263.

Bove, F. and Rossi, V. (2020). Components of partial resistance to *Plasmopara viticola* enable complete phenotypic characterization of grapevine varieties. *Sci. Rep.* 10(1), 585. doi: 10.1038/s41598-020-57482-0.

Brent, K. J. and Hollomon, D. W. (1998). *Fungicide Resistance: The Assessment of Risk.* FRAC Monograph 2 Published by GCPF. Brussels, Belgium.

Brown, S., Koike, S. T., Ochoa, O. E., Laemmlen, F. and Michelmore, R. W. (2004). Insensitivity to the fungicide fosetyl-aluminum in California isolates of the lettuce downy mildew pathogen, Bremia lactucae. *Plant Dis.* 88(5), 502–508.

Bruhn, J. A., Carski, T. H., Genet, J. L., Hoffman, L. E., Perotin, B., Shepherd, C. P. and Sweigard, J. A. (2013). Oxathiapiprolin: The first member of a novel class of oomycete fungicides. In: *Admin. ICPP 10th International Congress of Plant Pathology*, Beijing (Vol. 2013).

Cohen, Y. (2015). The novel oomycide oxathiapiprolin inhibits all stages in the asexual life cycle of *Pseudoperonospora cubensis* – Causal agent of cucurbit downy mildew. *PLoS ONE* 10(10), e0140015. doi: 10.1371/journal.pone.0140015.

Cohen, Y. (2020). Root treatment with oxathiapiprolin, benthiavalicarb or their mixture provides prolonged systemic protection against oomycete foliar pathogens. *PLoS ONE* 15(1), e0227556. doi: 10.1371/journal.pone.0227556.

Cohen, Y. and Rubin, A. E. (2020). A new strategy for durable control of late blight in potato by a single soil application of an oxathiapiprolin mixture in early season. *PLoS ONE* 15(8), e0238148. https://doi.org/10.1371/journal.pone.0238148.

Cohen, Y., Rubin, A. E. and Galperin, M. (2019). Novel synergistic fungicidal mixtures of oxathiapiprolin protect sunflower seeds from downy mildew caused by *Plasmopara* halstedii. *PLoS ONE* 14(9), e0222827. doi: 10.1371/journal.pone.0222827.

Cook, L. R. (1991). Current problems in the control of late blight: The Northern Ireland experience. In: Lucas, J. A., Shattock, R. C., Shaw, D. S. and Cooke, L. R. (Eds.), *Phytophthora Symposium of the British Mycological Society, the British Society for Plant Pathology and the Society of Irish Plant Pathologists held at Trinity College, Dublin September 1989* (Vol. 1989). Cambridge University Press, Cambridge, pp. 337–348.

Dowley, L. J., O'Sullivan, E. O. and Kehoe, H. W. (1991). Development and evaluation of blight resistant cultivars. In: Lucas, J. A., Shattock, R. C., Shaw, D. S. and Cooke, L. R. (Eds.), *Phytophthora Symposium of the British Mycological Society, the British Society for Plant Pathology and the Society of Irish Plant Pathologists held at Trinity College, Dublin September 1989* (Vol. 1989). Cambridge University Press, Cambridge, pp. 373–382.

Fontaine, S., Remuson, F., Caddoux, L. and Barres, B. (2019). Investigation of the sensitivity of *Plasmopara viticola* to amisulbrom and ametoctradin in French vineyards using bioassays and molecular tools. *Pest Manag. Sci.* 75(8), 2115–2123.

Gascuel, Q., Martinez, Y., Boniface, M. C., Vear, F., Pichon, M. and Godiard, L. (2015). The sunflower downy mildew pathogen *Plasmopara* halstedii. *Mol. Plant Pathol.* 16(2), 109–122.

Genet, J. L. and Vincent, O. (1999). Sensitivity of European *Plasmopara viticola* populations to cymoxanil. *Pestic. Sci.* 55, 129–136.

Gessler, C., Pertot, I. and Perazzolli, M. (2011). *Plasmopara viticola*: A review of knowledge on downy mildew of grapevine and effective disease management. *Phytopathol. Mediterr.* 50, 3–44.

Gisi, U. (2012). Resistance to carboxylic acid amide (CAA) fungicides and anti-resistance strategies. In: Thind, T. S. (Ed.), *Fungicide Resistance in Crop Protection*. CAB International, pp. 96–103.

Gisi, U. and Sierotzki, H. (2015). Oomycete fungicides: Phenylamides, quinone outside inhibitors, and carboxylic acid amides. In: Ishii, H. and Hollomon, D. W. (Eds.), *Fungicide Resistance in Plant Pathogens*. Springer, Tokyo, Japan, pp. 145–174.

Gray, M. A., Hao, W., Förster, H. and Adaskaveg, J. E. (2018). Baseline sensitivities of new fungicides and their toxicity to selected life stages of *Phytophthora* species from citrus in California. *Plant Dis.* 102(4), 734–742.

Grünwald, N. J., Sturbaum, A. K., Montes, G. R., Serrano, E. G., Lozoya-Saldaña, H. and Fry, W. E. (2006). Selection for fungicide resistance within a growing season in field populations of *Phytophthora infestans* at the center of origin. *Phytopathology* 96(12), 1397–1403.

Gullino, M. L., Mescalchin, E. and Mezzalama, M. (1997). Sensitivity to cymoxanil in populations of *Plasmopara viticola* in northern Italy. *Plant Pathol.* 46(5), 729–736.

Gullino, M. L., Tinivella, F., Garibaldi, A., Kemmitt, G. M., Bacci, L. and Sheppard, B. (2010). Mancozeb past, present and future. *Plant Dis.* 94(9), 1076–1087.

Hao, W., Forster, H. and Adaskaveg, J. E. (2021). Resistance to potassium phosphite in *Phytophthora* species causing citrus Brown rot and integrated practices for management of resistant isolates. *Plant Dis.* 105(4), 972–977.

Hegstad, J. M., Gaspar, A. P., Feng, L. Z., Lackermann, K., Hudson, A. and Howieson, M. (2021). Agronomic and efficacy evaluations of oxathiapiprolin as a soybean seed treatment. *Agron. J.* 113(6), 4850–4864. doi: 10.1002/agj2.20866.

Hermann, D. and Gisi, U. (2012). Fungicide resistance in oomycetes with special reference to *Phytophthora infestans* and Phenylamides. In: Thind, T. S. (Ed.), *Fungicide Resistance in Crop Protection*. CAB International, pp. 133–140.

Humann, R. M., Johnson, K. D., Wunsch, M. J., Meyer, S. M., Jordahl, J. G., Bauske, E. C., Halvorson, J. M., Friskop, A. J., O'Bryan, K. A., Gulya, T. J. and Markell, S. G. (2019). Evaluation of oxathiapiprolin for the management of sunflower downy mildew. *Plant Dis.* 103(10), 2498–2504.

Ji, P. and Csinos, A. S. (2015). Effect of oxathiapiprolin on asexual life stages of *Phytophthora* capsici and disease development on vegetables. *Ann. Appl. Biol.* 166(2), 229–235.

Kamoun, S., Furzer, O., Jones, J. D., Judelson, H. S., Ali, G. S., Dalio, R. J., Roy, S. G., Schena, L., Zambounis, A., Panabières, F., Cahill, D., Ruocco, M., Figueiredo, A., Chen, X. R., Hulvey, J., Stam, R., Lamour, K., Gijzen, M., Tyler, B. M., Grünwald, N. J., Mukhtar, M. S., Tomé, D. F., Tör, M., Van Den Ackerveken, G., McDowell, J., Daayf, F., Fry, W. E., Lindqvist-Kreuze, H., Meijer, H. J., Petre, B., Ristaino, J., Yoshida, K., Birch, P. R. and Govers, F. (2015). The Top 10 oomycete pathogens in molecular plant pathology. *Mol. Plant Pathol.* 16(4), 413–434.

Khilare, V. C., Deokate, A. S. and Ganawane, L. V. (2003). Occurrence of aluminium phosethyle (Alliette) resistance in *Plasmopara viticola* causing downy mildew of grapevine in Maharashtra. *J. Phytol. Res.* 16, 239–241.

Koike, S. T., Gladders, P. and Paulus, A. O. (2007). *Vegetable Diseases – A Color Handbook*. Academic Press, Boston, San Diego.

Kousik, C. S. and Keinath, A. P. (2008). First report of insensitivity to cyazofamid among isolates of *Phytophthora capsici* from the southeastern United States. *Plant Dis.* 92(6), 979.

Kuck, K. H. (2005). Fungicide resistance management in a new regulatory environment. In: Dehne, H. W., Gisi, U., Kuck, K. H., Russell, P. E. and Lyr, H. (Eds.), *Modern Fungicides and Antifungal Compounds IV: 14th International Reinhardsbrunn Symposium, Friedrichroda*, Thuringia, Germany, April 25–29, pp. 35–43.

Kuck, K. H. and Russell, P. E. (2006). FRAC: Combined resistance risk assessment. *ASp Appl. Biol.* 78, 3–10.

Lafon, R. and Clerjeau, M. (1988). Downy mildew. In: Pearson, R. C. and Goheen, A. C. (Eds.), *Compendium of Grape Diseases*. APS Press, Saint Paul.

Lafon, R. and Smith, I. M. (1988). Oomycetes. In: Smith, I., Dunez, J., Lelliott, R. A., Philipps, D. H. and Archer, S. A. (Eds.), *European Handbook of Plant Diseases*. Blackwell Scientific Publications, pp. 226–228.

Leadbeater, A. (2012). Resistance risk to QoI fungicides and anti-resistance strategies. In: Thind, T. S. (Ed.), *Fungicide Resistance in Crop Protection*. CAB International, pp. 141–154.

Leadbeater, A. (2015). Recent developments and challenges in chemical disease control – A review. *Plant Prot. Sci.* 51(4), 163–169.

Lebaudy, A., Araki, S., Amorin, V., de Derumier, A., Archer, E. and Giraud, F. (2018). Identification of a mutation in the CYTB gene associated to resistance to cyazofamid in grape downy mildew. In: *12e Conférence Internationale sur les Maladies Des Plantes*, 11 et 12 Décembre 2018, Tours, France, pp. 460–469.

Mboup, M. K., Sweigard, J. W., Carroll, A., Jaworska, G. and Genet, J. L. (2022). Genetic mechanism, baseline sensitivity and risk of resistance to oxathiapiprolin in oomycetes. *Pest Manag. Sci.* 78(3), 905–913. doi: 10.1002/ps.6700.

McCoy, A. G., Noel, Z. A., Jacobs, J. L., Clouse, K. M. and Chilvers, M. I. (2022). *Phytophthora* sojae pathotype distribution and fungicide sensitivity in Michigan. *Plant Dis.* 106(2), 425–431. doi: 10.1094/PDIS-03-21-0443-RE.

Miao, J., Cai, M., Dong, X., Liu, L., Lin, D., Zhang, C., Pang, Z. and Liu, X. (2016). Resistance assessment for oxathiapiprolin in *Phytophthora capsici* and the detection of a point mutation (G769W) in PcORP1 that confers resistance. *Front. Microbiol.* 7, 615. doi: 10.3389/fmicb.2016.00615.

Miao, J., Chi, Y., Lin, D., Tyler, B. M. and Liu, X. (2018). Mutations in ORP1 conferring oxathiapiprolin resistance confirmed by genome editing using CRISPR/Cas9 in *Phytophthora capsici* and *P. sojae*. *Phytopathology* 108(12), 1412–1419.

Miao, J., Li, C., Liu, X., Zhang, X., Li, G., Xu, W., Zhang, C. and Liu, X. (2021). Activity and resistance-related point mutations in target protein PcORP1 of fluoxapiprolin in *Phytophthora capsici*. *J. Agric. Food Chem.* 69(13), 3827–3835.

Miao, J., Liu, X., Du, X., Li, G., Li, C., Zhao, D. Y. and Lui, X. (2020). Sensitivity of *Pythium* spp. and *Phytopythium* spp. and tolerance mechanism of *Pythium* spp. to oxathiapiprolin. *Pest Manag. Sci.* 76(12), 3975–3981.

Miao, J., Liu, X., Li, G., Du, X. and Liu, X. (2020). Multiple point mutations in PsORP1 gene conferring different resistance levels to oxathiapiprolin confirmed using CRISPR–Cas9 in *Phytophthora sojae*. *Pest Manag. Sci.* 76(7), 2434–2440.

Noel, Z. A., Sang, H., Roth, M. G. and Chilvers, M. I. (2019). Convergent evolution of C239S mutation in *Pythium spp.* beta-tubulin coincides with inherent insensitivity to ethaboxam and implications for other peronosporalean oomycetes. *Phytopathology* 109(12), 2087–2095.

Pang, Z., Shao, J., Chen, L., Lu, X., Hu, J., Qin, Z. and Liu, X. (2013). Resistance to the novel fungicide Pyrimorph in *Phytophthora capsici*: Risk assessment and detection of point mutations in CesA3 that confer resistance. *PLoS ONE* 8(2), e56513. doi: 10.1371/journal.pone.0056513.

Pasteris, R. J., Hanagan, M. A., Bisha, J. J., Finkelstein, B. L., Hoffman, L. E., Gregory, V., Andreassi, J. L., Sweigard, J. A., Klyashchitsky, B. A., Henry, Y. T. and Berger, R. A. (2016). Discovery of oxathiapiprolin, a new oomycete fungicide that targets an oxysterol binding protein. *Bioorg. Med. Chem.* 24(3), 354–361.

Pasteris, R. J., Hoffman, L. E., Sweigard, J. A., Andreassi, J. L., Ngugi, H. K., Perotin, B. and Shepherd, C. P. (2019). Oxysterol binding protein inhibitors: Oxathiapiprolin – A new oomycete fungicide that targets an oxysterol-binding protein. In: Jeschke, P., Witschel, M., Krämer, W. and Schirmer, U. (Eds.), *Modern Crop Protection Compounds* (Vols. 1–3), pp. 979–987; DPG Verlag, Braunscheig, Germany.

Patel, J. S., Costa de Novaes, M. I. and Zhang, S. (2015). Evaluation of the new compound oxathiapiprolin for control of downy mildew in basil. *Plant Health Prog.* 16(4), 165–172. doi: 10.1094/PHP-RS-15-0026.

Pavelková, J., Lebeda, A. and Sedláková, B. (2014). Efficacy of fosetyl-Al, propamocarb, dimethomorph, cymoxanil, metalaxyl and metalaxyl-m in Czech *Pseudoperonospora* cubensis populations during the years 2005 through 2010. *Crop Prot.* 60, 9–19.

Peng, Q., Wang, Z., Fang, Y., Wang, W., Cheng, X. and Liu, X. (2019). Point mutations in the beta-tubulin of *Phytophthora sojae* Confer resistance to ethaboxam. *Phytopathology* 109(12), 2096–2106.

Perotin, B., Pierre, S., Shepherd, C., Summers, B., Salas, S. and Majeau, G. (2015). Zorvec™ (oxathiapiprolin), a new fungicide from DuPont de Nemours against oomycetes'. In: *AFPP – 11th Conférence Internationale sur les Maladies des Plantes*, 11 et 12 Décembre 2018, Tours, France, pp. 463–472.

Qu, T., Grey, T. L., Csinos, A. S. and Ji, P. (2016). Translocation of oxathiapiprolin in bell pepper plants and systemic protection of plants against Phytophthora Blight. *Plant Dis.* 100(9), 1931–1936.

Raychaudhuri, S. and Prinz, W. A. (2010). The diverse functions of oxysterol-binding proteins. *Annu. Rev. Cell Dev. Biol.* 26, 157–177.

Robertson, N. F. (1991). The challenge of *Phytophthora infestans*. In: Ingram, D. S. and Williams, P. H. (Eds.), *Advances in Plant Pathology: Phytophthora infestans, the Cause of Late Blight of Potato* (Vol. 7). Academic Press.

Salas, S. E., Shepherd, C. P., Donofrio, N. M. and Powell, D. H. (2015). Use of scanning electron microscopy to visualize how oxathiapiprolin affects the initial infection process of potato leaves by *Phytophthora infestans*. *Phytopathology* 105, S4.122.

Salas, S. E., Shepherd, C. P., Ngugi, H. K. and Genet, J. L. (2019). Disease control attributes of oxathiapiprolin fungicides for management of cucurbit downy mildew. *Plant Dis.* 103(11), 2812–2820.

Schepers, H. T. A. M., Kessel, G. J. T., Lucca, F., Förch, M. G., van den Bosch, G. B. M., Topper, C. G. and Evenhuis, A. (2018). Reduced efficacy of fluazinam against *Phytophthora infestans* in the Netherlands. *Eur. J. Plant Pathol.* 151(4), 947–960.

Siegenthaler, T. B. and Hansen, Z. R. (2021). Sensitivity of *Phytophthora capsici* from Tennessee to mefenoxam, fluopicolide, oxathiapiprolin, dimethomorph, mandipropamid, and cyazofamid. *Plant Dis.* 105(10), 3000–3007. doi: 10.1094/PDIS-08-20-1805-RE.

Sierotzki, H., Kraus, N., Assemat, P., Stanger, C., Cleere, S., Windass, J. and Gisi, U. (2005). Evolution of resistance to QoI fungicides in *Plasmopara viticola* in Europe. In: Dehne, H. W., Gisi, U., Kuck, K. H., Russel, P. E. and Lyr, H. (Eds.), *Modern Fungicides and Antifungal Compouds IV, 14th International Reinhardsbrunn Symposium,* Friedrichroda, Thuringia, Germany, April 25–29, pp. 73–80.

Spring, O. and Gomez-Zeledon, J. (2020). Influence of oxathiapiprolin on preinfectional and early infection stages of *Plasmopara* halstedii, downy mildew of the sunflower. *Plant Prot. Sci.* 56(2), 83–91.

Strange, R. N. and Scott, P. R. (2005). Plant diseases: A threat to global food security. *Annu. Rev. Phytopathol.* 43, 83–116.

Sweigard, J. A., Andreassi, J., Pember, S., Gutteridge, S., Pasteris, R. J. and Hanagan, M. A. (2014). Discovery of the target site of oxathiapiprolin (DuPont Zorvec disease control). 13th IUPAC Congress of Pesticide Chemistry, San Francisco, abstract Agro 911.

Thomas, A., Neufeld, K. N., Seebold, K. W., Braun, C. A., Schwarz, M. R. and Ojiambo, P. S. (2018). Resistance to fluopicolide and propamocarb and baseline sensitivity to ethaboxam among isolates of *Pseudoperonospora* cubensis from the Eastern Unites States. *Plant Dis.* 102(8), 1619–1626.

Thurston, H. D. and Schulz, O. (1986). *Phytophthora infestans*. In: Howker, W. J. (Ed.), *Compendium of Potato Diseases*. American Phytopathological Society, Saint Paul, pp. 40–42.

van den Bosch, F., Paveley, N., Shaw, M., Hobbelen, P. and Oliver, R. (2011). The dose rate debate: Does the risk of fungicide resistance increase or decrease with dose? *Plant Pathol.* 60(4), 597–606.

van den Bosch, F., Paveley, N., van den Berg, F., Hobbelen, P. and Oliver, R. (2014). Mixtures as a fungicide resistance management tactic. *Phytopathology* 104(12), 1264–1273.

Vargas, A., Eyre, M. and Dorrance, A. E. (2016). Efficacy of oxathiapiprolin toward *Phytophthora sojae* and *Phytophthora sansomeana*. *Phytopathology* 106, S4.122.

Wang, L. and Ji, P. S. (2021). Fitness and competitive ability of field isolates of *Phytophthora capsici* resistant or sensitive to fluopicolide. *Plant Dis.* 105(4), 873–878.

Wingerter, C., Eisenmann, B., Weber, P., Dry, I. and Bogs, J. (2021). Grapevine Rpv3-, Rpv10- and Rpv12-mediated defense responses against *Plasmopara viticola* and the impact of their deployment on fungicide use in viticulture. *BMC Plant Biol.* 21(1), 470. doi: 10.1186/s12870-021-03228-7.

Wu, Q., Hao, Z. and Fan, Z. (2019). Research progress on oxysterol binding protein based novel piperidinylthiazole pesticide development. *Chin. J. Pestic. Sci.* 21(5–6), 595–608.

Wu, J., Xue, Z. L., Miao, J. Q., Zhang, F., Gao, X. and Liu, X. (2020). Sensitivity of different developmental stages and resistance risk assessment of *Phytophthora capsici* to fluopicolide in China. *Front. Microbiol.* 11, 185. doi: 10.3389/fmicb.2020.00185.

Index